Global Atmospheric Change and its Impact on Regional Air Quality

NATO Science Series

A Series presenting the results of scientific meetings supported under the NATO Science Programme.

The Series is published by IOS Press, Amsterdam, and Kluwer Academic Publishers in conjunction with the NATO Scientific Affairs Division

Sub-Series

I. Life and Behavioural Sciences	IOS Press
II. Mathematics, Physics and Chemistry	Kluwer Academic Publishers
III. Computer and Systems Science	IOS Press
IV. Earth and Environmental Sciences	Kluwer Academic Publishers
V. Science and Technology Policy	IOS Press

The NATO Science Series continues the series of books published formerly as the NATO ASI Series.

The NATO Science Programme offers support for collaboration in civil science between scientists of countries of the Euro-Atlantic Partnership Council. The types of scientific meeting generally supported are "Advanced Study Institutes" and "Advanced Research Workshops", although other types of meeting are supported from time to time. The NATO Science Series collects together the results of these meetings. The meetings are co-organized bij scientists from NATO countries and scientists from NATO's Partner countries – countries of the CIS and Central and Eastern Europe.

Advanced Study Institutes are high-level tutorial courses offering in-depth study of latest advances in a field.
Advanced Research Workshops are expert meetings aimed at critical assessment of a field, and identification of directions for future action.

As a consequence of the restructuring of the NATO Science Programme in 1999, the NATO Science Series has been re-organised and there are currently five sub-series as noted above. Please consult the following web sites for information on previous volumes published in the Series, as well as details of earlier sub-series.

http://www.nato.int/science
http://www.wkap.nl
http://www.iospress.nl
http://www.wtv-books.de/nato-pco.htm

Series IV: Earth and Environmental Sciences – Vol. 16

Global Atmospheric Change and its Impact on Regional Air Quality

edited by

I. Barnes

Bergische University Wuppertal,
Physical Chemistry Department,
Wuppertal, Germany

Kluwer Academic Publishers

Dordrecht / Boston / London

Published in cooperation with NATO Scientific Affairs Division

Proceedings of the NATO Advanced Research Workshop on
Global Atmospheric Change and its Impact on Regional Air Quality
Irkutsk, Russian Federation
August 21–27, 2001

A C.I.P. Catalogue record for this book is available from the Library of Congress.

ISBN 1-4020-0958-5 (HB)
ISBN 1-4020-0959-3 (PB)

Published by Kluwer Academic Publishers,
P.O. Box 17, 3300 AA Dordrecht, The Netherlands.

Sold and distributed in North, Central and South America
by Kluwer Academic Publishers,
101 Philip Drive, Norwell, MA 02061, U.S.A.

In all other countries, sold and distributed
by Kluwer Academic Publishers,
P.O. Box 322, 3300 AH Dordrecht, The Netherlands.

Printed on acid-free paper

Printed in the Netherlands.

Global Atmospheric Change and its Impact on Regional Air Quality

CONTENTS

Emissions

Arctic Regions

Kinetic and Mechanistic Studies in the Laboratory

Poster Contributions

Kinetic and Mechanistic Studies

Global and Regional Air Pollution Studies

Methods and Techniques

Impact on Ecosystems

Preface

The NATO ARW in Irkutsk was an excellent occasion for the coming together of Eastern and Western scientists who are involved in tropospheric science; the workshop has greatly contributed to the scientific and social understanding among the participants from the many different countries. Many new personal contacts were made which will help to strengthen future collaborations.

In particular, the Lake Baikal area and the Limnological Institute offer splendid opportunities for environmental research which, in part, is already on going. For most participants it was the first time to see the impressive nature of the Lake Baikal region. Hopefully, there will be a chance for a follow-up event in Siberia where researchers from the East and West can again meet and engage in fruitful scientific dialogue.

The book contains extended abstracts of the lectures and the poster presentations presented at the NATO ARW "Global Atmospheric Change and its Impact on Regional Air Quality" Irkutsk, Lake Baikal, Russian Federation, August 21-27, 2001. The ARW was composed of 22 oral presentations by key lecturers and 6 additional shorter oral presentations from participants. In a special poster session the 36 poster contributions were presented and discussed. Unfortunately not all contributors submitted extended abstracts, however, to compensate two contributions have been added from 2 participants who were originally invited but were unable to attend. The content and structure of the book is the same as the NATO ARW programme, and I will therefore dispense with a discussion of the individual contributions at the workshop. In order to impart a better impression of what the NATO ARW achieved we have included below the summary of the official Discussion and Conclusions of the workshop:

General Discussion and Conclusions

Global changes of the atmosphere, the topic of our workshop, create adverse effects, however, uncontrolled changes without knowing the reasons would be even more frightening. We all can agree with Valentin Rasputin, the famous Russian environmental campaigner and book author, that mankind is endangered somehow as soon as his environment gets damaged by pollution. The Lake Baikal region with all its beauty and essentials for flora and fauna is a worldwide outstanding example for such a potential danger.

In order to better understand our environment, at least the troposphere, where most of the ARW participants are involved with their work, the following needs for the next future which came out from the interesting presentations of the workshop have been summarised:

- The finer details of chemical feedback processes which control global changes in the atmosphere are not understood.
- For reliable modelling anthropogenic and natural emission data have to be evaluated and permanently tested worldwide. Even in Europe the biogenic VOC emissions, but also the anthropogenic VOC emissions from non-traffic sources require further evaluation. New field measurements indicate that also the natural NO_x emissions can have a significant impact on oxidation processes in rural areas.
- The finer details of atmospheric processes are strongly related to the radical budget which is responsible for the tropospheric oxidation capacity. It is this property which determines the turn over of degradation processes with its impact on the average concentration of many trace compounds which, in part, are involved in the radiation balance of the atmosphere. Many photochemical radical sources, e.g. by oxygenated VOCs, are presently not known. On the other side also surface reactions as sinks of radicals or their precursors have to be better investigated. Under conditions of higher urban pollution most processes are fairly clear at present, maybe also under very clean

conditions in remote maritime areas. However, what we heard about the chemistry of halogen radicals in the Arctic regions and the sulfur chemistry in marine air gives some doubts about the completeness of our knowledge.

- All models require more or less complete chemical mechanisms. To further develop, test and simplify such complex mechanisms under controlled conditions needs more work in the laboratory and the use of appropriate simulation chambers which are now available in Europe.
- The further improvement of field measurements by setting up appropriate quality assurance of the employed analytical instruments together with the operating persons who handle the measurements are urgently required for gaseous species in the sub ppb range and even more for particle measurements. The development of new techniques and methods is always a requirement, however, this topic was not touched upon in this workshop. The use of isotope separation techniques, as we heard at this workshop, can really improve the analysis of complex oxidation processes which involve numerous VOCs.
- The interaction between modellers and field experimentalists to define and select representative monitoring sites could improve the possibilities to further develop and test models.

In general, a better exchange of data in all respects between larger countries seems to be necessary. It would help the understanding to know much more what e.g. happens in Siberia with respect to biogenic emissions, surface reactions on snow and ice and photochemistry during summertime. The Japanese colleagues are using the Siberian air as background air to detect pollution from China during the long-range transport. Someone might wonder what influence could be observed for Europe either directly or by a flow of air masses over the Arctic or vice versa.

A very important issue is that young scientists should be more attracted to our research area, and also we should be able to offer them the possibility to visit places in all of the countries which are represented here at the ARW. These young persons should use well-equipped laboratories with the aim to realise their ideas for new experiments. An exchange programme is necessary, but in all directions, which can initiate the strengthening of research in our area, even without much financial support. This workshop was only the beginning of such activities. Most participants are expecting a follow-up workshop in another area of Russia or a former Soviet Union country, maybe after 2 to 3 years from now, hopefully supported again by NATO.

Karl H. Becker and Igor I. Morozov
Directors of the NATO ARW

Acknowledgements

The NATO Science Committee is gratefully acknowledged for financing the workshop. Other sponsors who gave additional financial support either by direct contributions or by extra support within their research programmes or their institutions are also gratefully acknowledged. Prof. Grachev and his coworkers are sincerely thanked for hosting the NATO ARW in the Limnological Institute, Irkutsk and making it such an enjoyable and successful event.

Financial support of the workshop was by the following institutions:

- NATO-Scientific and Environmental Affairs Division, Brussels
- Russian Foundation for Basic Research (RFBR), Moscow
- N. Semenov Institute of Chemical Physics, RAS, Moscow
- Limnological Institute of SB RAS, Irkutsk
- European Commission, Brussels
- Department of Education, Science and Research, State of Northrhine-Westphalia, Duesseldorf
- Department of Physical Chemistry of the University of Wuppertal
- Association of Friends of the University of Wuppertal
- BMW-Bavarian Motor Company, Munich
- DaimlerChrysler, Stuttgart
- Ford Motor Company through the Ford Research Center Aachen
- Savings Bank Wuppertal

NATO ARW - Main Author list with addresses: *version Monday, 08 July 2002*

Main Author	**Institute**
Angeletti, G.	European Commission, Directorate General for Research, Rue de la Loi, 200 B – 1049 Brussels, Belgium
Arsene, C.	Bergische Universität, Physikalische Chemie / FB 9, Gaußstraße 20, D-42097 Wuppertal, Germany
Butkovskaya, N. I.	Bergische Universität, Physikalische Chemie / FB 9, Gaußstraße 20, D-42097 Wuppertal, Germany
Cavalli, F.	Bergische Universität, Physikalische Chemie / FB 9, Gaußstraße 20, D-42097 Wuppertal, Germany
Dogeroglu, T.	Anadolu University Environmental Engineering Department Iki Eylul Campus 26470 Eskisehir-TURKEY
Egorov, V.I.	Institute of Global Climate and Ecology Glebovskaya 20-B, 107258 Moscow, Russia.
Gad, A	National Research Center Giza Egypt
Gershenzon, Yu.M	N.Semenov Institute of Chemical Physics RAS, Moscow 119991, Russian Federation
Gorshkov, A.	Limnological Institute, SB RAS, Ulan-Butorskaya st.3,

	Irkutsk, 664033 (Russia)
Herrmann, H.	Institut für Troposphärenforschung Permoserstr. 15 D-04318 Leipzig, Germany
Jancar-Webster, B.	State University of New York, 282 Atateka Drive, Chestertown, NY 12817 USA
Karimov, K.	Institute of Physics of National Academy of Sciences Chui Prosp., 265-A, Bishkek, 720071 Kyrgyz Republic
Kharytonov, M.	State Agrarian University, Voroshilov st.25, Dnipropetrovsk, 49600,Ukraine
Koutsenogii, K	Institute of Chemical Kinetics and Combustion SB RAS, Institutskaya ., 3 630090 Novosibirsk,
Koutsenogii, P.	Institute of Chemical Kinetics and Combustion SB RAS, Institutskaya, 3 630090 Novosibirsk,
Kuchmenko, E. V.	Limnological Institute of Siberian Branch of RAS Box 4199 Irkutsk 664033, Russia
Larin, I. K.	Institute of Energy Problems of Chemical Physics of RAS, Leninskii avenue 38, bdlg 2, 117829, Moscow, Russia

Laritchev, M. N.	INEP CP RAS, Moscow, Leninsky Prospect 38, 2, Moscow, Russia
Laritcheva, O. O.	Ministry of Health of Russian Federation Russian Research Center of Emergency Situations, 46,4 Shabolovka str., 117419 Moscow, Russia
Le Bras, G.	Laboratoire de Combustion et Systèmes Réactifs, LCSR-CNRS, 45071 Orléans Cédex 2 (France)
Lesclaux, R.	LPCM, Université Bordeaux I 33405 Talence Cedex (France)
Levanov, A. V.	Chemistry Department of Lomonosov Moscow State University Vorobjevi gori, 119899 Moscow, Russia
Lunin, V. V.	Moscow State University,Department of Chemistry, Vorobjevi gori, 119899 Moscow, Russia
Markova, T. A.	A.M. Oboukhov Institute of Atmospheric Physics, RAS Pyzhevsky 3, 109017, Moscow, Russia
Matvienko G.G.	Institute of Atmospheric Optics SB RAS, 1. Akademicheskii ave Tomsk, 634055, Russia
Mikhailova, T.	Siberian Institute for Plants Physiology and Biochemistry of SB RAS, Lermontov St. 132,664033, P.O.Box 1243, Irkutsk, Russia

Mocano, R.	Department of Analytical Chemistry, Faculty of Chemistry, "Al. I. Cuza" University, 6600 Iasi, Romania
Moortgat, G. K.	Max-Planck-Institut für Chemie, Atmospheric Chemistry Department, P.O.Box 3060, D-55020 Mainz, Germany
Morozov, I. I.	Laboratory of Cluster and Radical Processes, N. Semenov Institute of Chemical Physics RAS, Kosygin Str.4, 117334, Moscow, Russia
Morozova, O. S.	N. Semenov Institute of Chemical Physics of RussiaAcademy of Sciences, Kosygin st. 4, 117334 Moscow, Russia
Nielsen, C. J.	Department of Chemistry, University of Oslo, Blindern, N-0315 Oslo, Norway
Pasiuk-Bronikowska, W.	Institute of Physical Chemistry, Polish Academy of Science Kasprzaka 44/52, 01-224 Warsaw, Poland
Petrea, M.	Bergische Universität, Physikalische Chemie / FB 9, Gaußstraße 20, D-42097 Wuppertal, Germany
Pilling, M.	School of Chemistry, University of Leeds, Leeds, LS29JT, UK

att, U.	Institut für Umweltphysik, University of Heidelberg, INF 229, D-69120 Heidelberg
otemkin, V.	Limnological Institute of Siberian Branch of Russian Academy of Sciences Box 4199 Irkutsk 664033 Russia
urmal, A. P.	N. Semenov Institute of Chemical Physics of RAS,. Kosygina Str. 2, 117829 Moscow, Russia
aes, F.	Institute for Environment and Sustainability, Joint Research Centre European Commission 21020 Ispra (VA) Italy
osenbohm E.	Ford Research Centre Aachen, Environmental Science, Suesterfeldstr. 200, D-52072 Aachen, Germany
udolph, J.	Centre for Atmospheric Chemistry, York University, 4700 Keele Street, Toronto, Ontario, M3J 1P3 Canada
alter, L.	Cornwall College, Cornwall TR15 3RD, UK.
avilov S.V	Moscow State University,Department of Chemistry, Vorobjevi gori, 119899 Moscow, Russia

Semenov, M.	Limnological Institute of Siberian Branch of Russian Academy of Sciences Box 4199 Irkutsk 664033, Russia
Sidebottom, H.	Department of Chemistry, University College Dublin, Belfield, Dublin 4, Ireland
Simpson, D.	EMEP MSC-W, The Norwegian Meteorological Institute P.B. 43, N-0313 Oslo, Norway
Smirnova, T. Y.	Central Asian Research Hydrometeorological Institute Uzbekistan
Trubina, L.	Siberian State Geodetic Academy Plakhotnogo, 10 630108, Novosibirsk, Russian Federation
Wiesen, P.	Bergische Universität Wuppertal, Phys. Chemie / FB 9, Gauss Strasse 20 D-42097 Wuppertal, Germany
Zelenov, V. V.	Institute of Energy Problems of Chemical Physics (Branch), Russian Academy of Sciences, Chernogolovka 142432, Moscow District, Russian Federation

Atmospheric Chemistry: General Aspects

European Research on Atmospheric Global Change: EU Supported Projects and Future Planning

Giovanni Angeletti

European Commission, Directorate General for Research, Rue de la Loi, 200
B – 1049 Brussels, Belgium

Introduction

The European Commission (EC) at the time of writing supports atmospheric research mainly within the key action on Global Change, Climate and Biodiversity of the 5th Framework Programme (FP) 1999-2003. The content of the work programme on atmospheric composition change is based on a strategic document which was prepared through the EC Science Panels on Atmospheric Research and Stratospheric Ozone and was published in November 1997 (AIRES, Mégie et al.).

The atmospheric issues having a regional, global dimension are treated in integrated manner with other issues of the “Global Change” key action as climate, ecosystems, carbon cycle, global observing systems, while local, urban scale problems related to air quality in cities are dealt in the key action “City of tomorrow and cultural heritage”.

Concerning the issue of atmospheric composition, two calls for research proposals have been launched, the first in Summer 1999 and the second with deadline 15 February 2001.

Here, a description is given of the outcome of these two calls and an indication of the perspectives for the future in this field within the European Research Area (ERA).

Research on atmospheric composition within the key action on “Global Change” of FP5

Concerning the 1st call of 1999, the evaluation process permitted the selection of 12 projects in this area out of 70 research proposals, with an EC contribution of about 10 Meuros and an average of 8 partners per project, while the 2nd call resulted in the support of 10 projects out of 40 proposals, with an EC contribution of about 17 Meuros and the same number of partners.

These projects together with some other projects in related areas as “Global observing systems”, “Support for research infrastructure” and generic activities have been grouped into 5 clusters, whose objectives and activities are briefly described here after.

Rapporteurs selected among the members of the Science Panel ensure the exchange of information and dissemination of progress and results.

1. Ozone budget

The main objective of the 5 projects included in this cluster is to evaluate and predict the impact of different precursors emissions on the ozone budget in the troposphere. Investigations include the quantification of the contribution of biogenic emissions and the validation of the efficacity of past and ongoing EU air quality legislation with respect to ozone and its precursors.

I. Barnes (ed.), Global Atmospheric Change and its Impact on Regional Air Quality, 3–5.

The physical, chemical and meteorological processes responsible for the spatial and temporal variability of photochemical pollutants in the Mediterranean area and their effect on long-range transport characteristics are also considered, as well the development of methodologies for an integrated assessment of European air pollution control strategies in support of the Clean Air for Europe (CAFE) policy.

2. Sources of pollutants

Here different origins of pollutants are considered, involving various compounds. Investigations, mainly in the laboratory, and through modelling are made on:
- aromatic compounds and their oxidation effects;
- formation of secondary organic aerosols;
- alternative fluorinated alcohols and ethers;
- emissions of nitrogen oxides from forests;
- oxygenated species.

These investigations are complemented by simulation experiments making use of the EUPHORE facility in Valencia.

3. Oxidation processes

This cluster aims to better understand the oxidation of different atmospheric species and their impact on climate as far as regards dimethyl sulphide, on the oxidation capacity of the atmosphere concerning nitrous acid and on photochemical ozone regarding formaldehyde as a tracer.

Another project of this cluster investigates the heterogeneous processes on cirrus ice particles in the upper troposphere, having an impact on the concentration of ozone and thus on the radiation budget.

4. Aerosols

Projects in this cluster aim to characterize the physico/chemical properties and composition of aerosols, both organic and inorganic, e.g. the nucleation in the boundary layer, condensational growth and surface reactivity.

A project supported under the "City of tomorrow" key action studies the particulate matter characteristics, indoor/outdoor, in different cities, while another project deals with biogenic aerosols and air quality in the Mediterranean area.

5. Chemistry/climate interactions

This is a new cluster which includes three projects selected from the last call and aims to better understand the interactions and feedback mechanisms of atmospheric processes with climate.

One project is a European contribution to the Large Scale Biosphere – Atmosphere Experiment in Amazonia (LBA) and aims to study the link between the composition and abundance of biomass burning aerosol, the lowering of the cloud droplets size and the climatic consequences of the resulting perturbation of cloud physics.

The other two projects deal, one with the role of carbonaceous aerosols on climate over Europe through reconstruction of trends of relevance species from Alpine ice cores, the other with trends of greenhouse gases through a series of ice core extractions in both hemispheres, in Greenland and Antarctica.

Perspectives for the future

At present the EC is developing a strategy for a frontier – free research policy in Europe, the European Research Area (ERA). Improved cooperation among researchers in the Member States, integration of national and EU programmes, access to large research infrastructures, networking of centres of excellence, large integrated projects, stimulation of women and young people's career in science, as well new relationships between science and society are among the main principles for future European research.

Within the ERA, one of the thematic programmes related to the next proposed Framework Programme, FP6, 2002-2006 deals with "Sustainable development and global change".

It is expected that atmospheric research will find a suitable niche within the envisaged actions on global climate change and observing systems.

To this effect, it is expected also that the new strategic document on Atmospheric Interdisciplinary Research, AIRES in ERA recently published (Hov et al, 2001), which defines new research challenges, will play an important role in providing a contribution to ERA and FP6.

References

Mégie G., Amanatidis G.T., Angeletti G., Becker K.H., Cox R.A., Harris N., Hov Ø. and Versino B. (eds) (1997); A Global Strategy for European Atmospheric Interdisciplinary Research, AIRES. EC, DGXII, Report EUR 17645, Brussels, 45 p.

Hov Ø., Amanatidis G.T., Angeletti G., Brasseur G., Harris N., Mégie G., Schumann U. and Slanina S. (eds) (2001); A Global Strategy for Atmospheric Interdisciplinary Research in the European Research Area, AIRES in ERA. EC, DG Research, Report EUR 19436, Brussels, 58 p.

Probing Chemical Mechanisms in Clean and Polluted Environments through Field Measurement / Model Comparisons

Nicola Carslaw, Dwayne E. Heard, James D Lee, Alastair C. Lewis, Michael J. Pilling and Sandra M. Saunders

School of Chemistry, University of Leeds, Leeds, LS29JT, UK

Introduction

The oxidation of volatile organic compounds (VOCs) in the atmosphere is primarily initiated by reaction with OH. NO_X and sunlight are also needed. The process leads to ozone formation in the boundary layer and the free troposphere with impacts on both regional air quality and climate change. The oxidation of some VOCs, especially aromatics and terpenes, leads to formation of secondary organic aerosol.

For many VOCs, the OH reaction provides their major removal mechanism from the atmosphere. In consequence, the atmospheric lifetime, τ_X, of a VOC, X, is equal to $\{k_X[OH]\}^{-1}$ where k_X is the rate coefficient for OH + X. For a mean [OH] of 10^6 cm^{-3}, VOC lifetimes vary greatly e.g. 6y for CH_4, 10 days for benzene and 3h for isoprene. These values affect the global/regional distribution of the VOC and also influence the range over which regional ozone is formed. Conversely, [OH], which is a major contibutor to the oxidizing capacity of the atmosphere, is equal to $P_{OH}/(\Sigma k_X [X])$, where P_{OH} is the rate of production of OH and the sum is taken over all VOCs contributing to the removal of OH.

It is clear that the oxidation of VOCs plays a central role in the chemistry of the troposphere. This paper examines that chemistry in outline and the use of field measurements of OH to probe the mechanisms.

Mechanism of oxidation of VOCs in the troposphere

The mechanism has elements of a chain reaction and it is instructive to set up an outline mechanism in the form of a straight chain reaction. The major route for OH production

Radical generation

The major primary route to formation of radicals is the oxidation of ozone:

$$O_3 + h\nu \rightarrow O^1D + O_2$$
$$O^1D + H_2O \rightarrow 2OH$$

Propagation

The propagation steps involve abstraction of H by OH, or addition of OH to the VOC, RH to form a carbon centred radical which rapidly reacts with OH to form a peroxy radical:

$$OH + RH\ (+O_2) \rightarrow RO_2 + H_2O$$

The peroxy radical oxidises NO to NO_2, at the same time forming an alkoxy radical, RO:

$$RO_2 + NO \rightarrow NO_2 + RO$$

I. Barnes (ed.), Global Atmospheric Change and its Impact on Regional Air Quality, 7–12.

The alkoxy radical reacts, by a meachanism that depends on its identity, which includes isomerization, dissociation and reaction with O_2. The net effect, after one or more steps, is the formation of HO_2 and of a carbonyl compound.:

$$RO \rightarrow\rightarrow HO_2 + R'CHO$$

The propagation cycle is then completed by reaction of HO_2 with NO to regenerate OH:

$$HO_2 + NO \rightarrow OH + NO_2$$

Secondary radical generation

The carbonyl compounds are photolabile and can act as secondary sources of radicals:

$$R'CHO + h\nu\ (+O_2) \rightarrow HO_2 + R'O_2 + CO$$

Termination

The termination mechanism depends on the conditions. At high levels of pollution)High NO_x, termination involves formation of HNO_3, which is removed from the atmosphere, e.g. by rainout.

$$OH + NO_2 + M \rightarrow HNO_3 + M$$

At low NO_x, quadratic termination ccurs via peroxy-peroxy reactions:

$$HO_2 + HO_2 \rightarrow H_2O_2$$

$$RO_2 + H/RO_2 \rightarrow \text{Products}$$

The peroxides formed in these reactions can be photolysed, to regenerate radicals; the termination is completed by their deposition.

Ozone formation

The final stage in the process is the photolysis of NO_2, formed in the reactions of RO_2 and HO_2 with NO, into ozone, on of the principal components of photochemical smog:

$$NO2 + h\nu \rightarrow NO + O$$

$$O + O2 + M \rightarrow O3 + M$$

The details of the oxidation process depend on the VOC and these can affect the yield of ozone. Key questions, relating to this detailed chemistry, include the effects of $[NO_x]$ and [VOC], the chain length, which affects the efficiency of the oxidative process, and the influence of the secondary chemistry, involving the carbonyl compounds.

Master chemical mechanism (MCM)

The master chemical mechanism is a near explicit mechanism which describes the detailed oxidation chemistry of 124 principal VOCs. The majority of these compounds are anthropogenic in origin, but isoprene and α and β pinene are also included. The MCM is based on a development protocol (Jenkin *et al.* 1997), which incorporates initiation by photolysis and reaction with O_3 and NO_3, as well as with OH. It uses laboratory data for photolysis processes and for the kinetics and mechanisms of the component elementary reactions, together with estimates, where experimental data are not available, based on *ab initio* calculations, structure activity relations, analogy etc. The MCM is constructed for use with a FACSIMILE integrator and the mechanisms are set up for direct incorporation into the FACSIMILE code.

The latest version, MCM3, was recently completed and is mounted on the web (http://www.chem.leeds.ac.uk/Atmospheric/MCM/mcmproj.html). Its development has been funded by the UK Department of the Environment, Food and Rural Affairs, and is used by them in policy applications, especially in relation to the determination of the ozone forming potentials of different VOCs.

Comparisons of the MCM with results of field experiments

While, as far as possible, the MCM is based on experimental data, it is important that it is tested against field experiments. One of the difficulties is that models must recognise the interaction between atmospheric dynamics and chemistry, thus complicating the issue of tests of chemical mechanisms. This problem can be circumvented, at least to some extent, by comparing measurements and models of the hydroxyl radical.

OH is short lived ($\tau \sim 1s$) and atmospheric transport is not significant on this timescale. This allows us to test the chemistry using a box model, with longer lived species constrained to measured concentrations, using a comparison between measured and modelled OH as the basis of our test.

OH is measured using the FAGE (Fluorescence Assay by Gas Expansion) technique, which is based on resonant laser induced fluorescence (LIF). The Leeds instrument is described by Creasey *at al.* (1997). Ambient air is expanded through a nozzle (~1mm diameter) to a pressure of ~1 Torr and irradiated with 308 nm radiation from a high repetition (~7kHz) copper vapour laser pumped dye laser. This radiation excites OH to the v=0 vibrational level of the $A^2\Sigma^+$ state and fluorescence is collected at the same wavelength. The gas expansion increases the lifetime of the excited state, by reducing collisional quenching by air, and allows the laser scatter and fluorescence signals, detected by the photomultiplier, to be separated by gating. [OH] is very small ($\sim 10^6$ cm^{-3}) and the fluorescence signals are weak, hence the need for the high laser pulse repetition rate. Nevertheless, good signals can be recovered with ~ 2 min averaging.

The box model (Carslaw *at al.* (1999, 2002) is constructed from the MCM, by incorporating the mechanisms for those VOCs that significantly affect the chemistry of OH in the conditions applying under the field conditions. The model essentially describes the chemistry occurring in the immediate vicinity of the FAGE sampling point. The longer-lived species are constrained in the model to measured values, which are ideally determined in the same region. Table 1 shows the techniques used in field campaigns at Mace Head in Ireland, by a consortium of UK and Irish Universities. The species marked by asterisks are constrained in the box model. Those marked with a double asterisk are constrained where possible – the experimental dataset is sometimes not adequate for constraint. These are all secondary species and, if necessary, they are simulated in the model.

The field experiments are highly dependent on the prevailing atmospheric conditions. Figure 1 shows 5 day back trajectories for the air masses for the whole of a campaign at Mace Head in May 1997. The figure also shows measurements of acetylene, which is a marker of anthropogenic pollution. There are three main categories of air, corresponding to polar, UK and France air masses. The first category leads to generally clean air, as demonstrated by the low acetylene concentrations, while the levels of pollution are much higher for the UK and France categories

Table 1. Techniques employed in field campaigns

Species	Technique
OH, HO_2	Laser induced fluorescence. HO_2 by conversion to OH with NO
*VOCs	Gas Chromatography (GC)
*J(O^1D), J(NO_2),	Radiometry
*NO_X, NO_Y	Chemiluminescence
*O_3	Absorption spectroscopy
*Aerosol surface area	Epiphaniometer
**HCHO, aldehydes, ketones	GC/HPLC
**H_2O_2, CH_3OOH, ROOH	HPLC
**PAN	GC
*CH_4	GC
*CO	Infra red
NO_3	Long path differential absorption spectroscopy
Peroxy radicals	Peroxy radical chemical amplifier

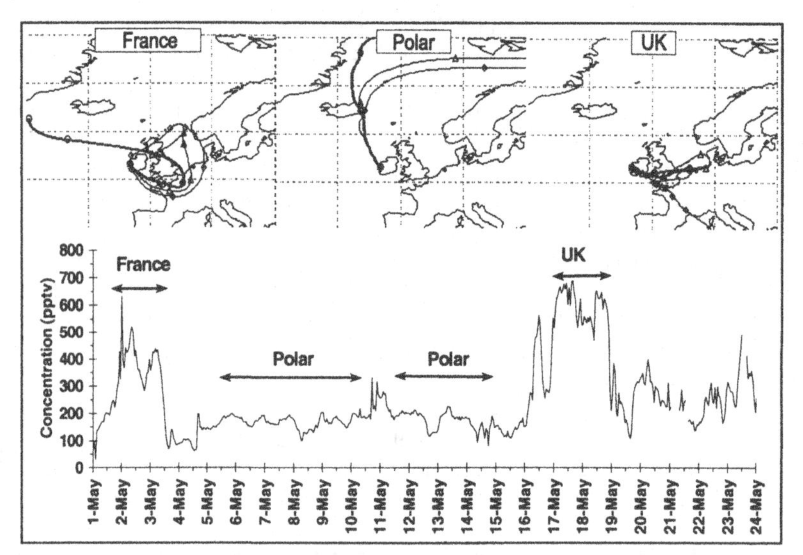

Figure 1. Air mass trajectories and acetylene concentrations, Mace Head, 1997

Carslaw *et al.*(2002) have carefully examined the effects of the conditions on the chemical mechanisms. For the polar air masses, where pollutant levels (VOCs and NO_x) are low; quadratic termination dominates and the chain length is short (~1) For the more polluted UK and France air masses, however, termination occurs via formation of HNO_3 and, because the rate determining propagation steps (RO_2/HO_2 + NO) are much faster because of the higher [NO], the chain length is longer (~5).

Carslaw *et al.* (2000) examined the effects of secondary chemistry on day of year 199 in 1966, again at Mace Head. Day 199 experienced anticyclonic conditions, with air arriving from the south east. The air was comparatively unpolluted, but had picked up significant concentrations of isoprene as it crossed Ireland. Isoprene has a short atmospheric lifetime (q.v.) and so is only detected comparatively close to its source.

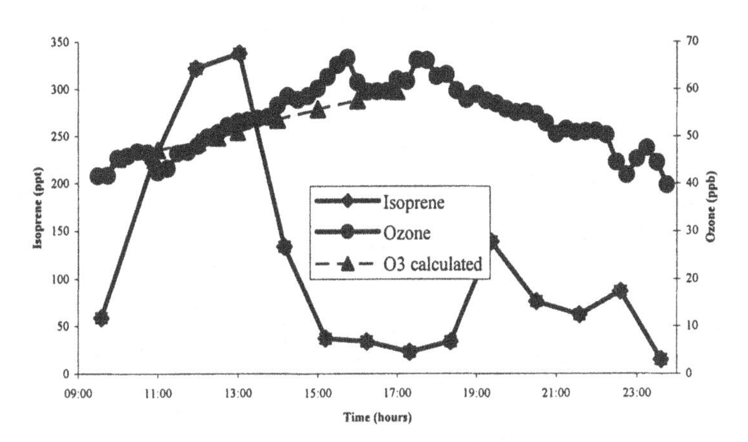

Figure 2. Isoprene and ozone concentrations, day of year 199, 1996, Mace Head

Figure 2 shows the isoprene and ozone concentrations on day 199, and includes ozone calculated from the model. Carslaw *et al.* showed that the rate of radical formation from the secondary carbonyls formed in the oxidation of isoprene (methacrolein, methyl vinyl ketone, methyl glyoxal, formaldehyde) became important as a source of radicals and, in the afternoon, exceeded ozone photolysis. The carbonyls also act as significant co-reactants for OH, as shown in Figure 3.

Comparisons between the measured and modelled [OH] show mixed success. For very clean ('baseline') conditions in Tasmania (Cape Grim), the agreement is excellent. Termination occurs exclusively via peroxy-peroxy reactions and the chain length is very short (~0.1). Even modest pollution, either at Cape Grim or Mace Head, leads to much poorer agreement, with model [OH] a factor of up to two greater than measured [OH]. The origin of this discrepancy is not clear. At Mace Head, complex air flow, coupled with measurement sites sampling different air masses and heterogeneous effects have been proposed. A campaign in 2002 will investigate these explanations. An alternative possibility, which becomes more applicable under polluted conditions, is failure to determine all of the VOCs. Lewis *et al.* (2000) demonstrated, using 2-dimensional (comprehensive) gas chromatography, that conventional GC techniques have problems determining the total atmospheric loading in the hydrocarbon range above ~ C_8.

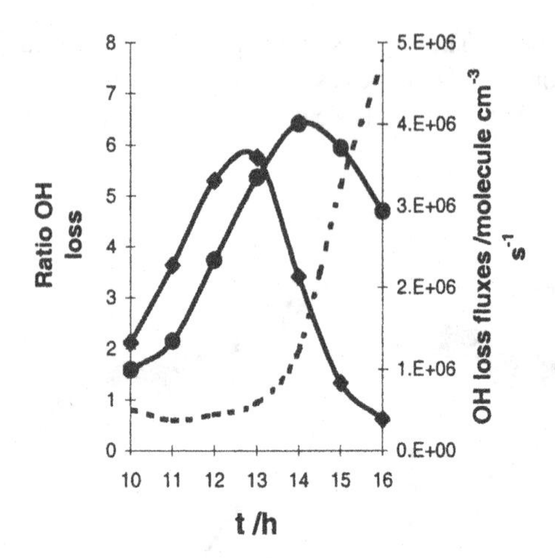

Figure 3. Rates of OH loss by reaction with isoprene (diamonds) and carbonyls (circles) formed in the oxidation of isoprene. The dashed line shows the ratio of the rates. Mace Head, day 199, 1996.

References

Carslaw, N., D.J. Creasey, D.E. Heard, A.C. Lewis, J.B. McQuaid, M.J. Pilling, P.S. Monks, B.J. Bandy, S.A. Penkett; Modelling OH, HO_2 and RO_2 radicals in the marine boundary layer: 1. Model construction and comparison with field measurements, *J Geophys Res*, **104,** (1999) 30241-30255

Carslaw N., N. Bell, A.C. Lewis, J.B. McQuaid and M.J. Pilling; A detailed study of isoprene chemistry during the EASE96 Mace Head campaign: July 17th 1996, a case study, *Atmos. Env.* **34**, (2000) 2827-2836

Carslaw N., D.J. Creasey, D.E. Heard, P.J. Jacobs, J.D. Lee, A.C. Lewis, J.B. McQuaid, M.J. Pilling, S. Bauguitte, S.A. Penkett, P.S. Monks, G. Salisbury.; The Eastern Atlantic Spring Experiment 1997 (EASE97): 2. Comparisons of model concentrations of OH, HO_2 and RO_2 with measurements, *J.Geophys. Res* (2002, submitted)

Creasey D.J., P.A.Halford-Maw, D.E. Heard, M.J. Pilling and B.J. Whitaker, Measurement of OH and HO_2 in the troposphere by laser-induced fluorescence, *J. Chem. Soc. Faraday Trans.* **93**, (1997) 2907.

Jenkin M.E., S.M. Saunders and M.J.Pilling; Tropospheric degradation of volatile organic compounds: a protocol for mechansim development, *Atmos. Environ.*, **31**, (1997) 81-104.

Lewis, A.C. Kinghorn, P Lewis A.C. N Carslaw, P J Marriott, R M Morrison, A L Lee, K D Bartle, and M J Pilling; Evidence of Underestimation in the Reactive Organic Carbon Budget of Urban Atmospheres., *Nature*, **405**, (2000) 779-781

Formation and Cycling of Aerosols in the Global Troposphere

Frank Raes
Institute for Environment and Sustainability, Joint Research Centre, European Commission, 21020 Ispra (VA) Italy

Abstract

In their role as cloud condensation nuclei (CCN), aerosols are linked to, and often control, the hydrologic cycle and therefore major fluxes of the Earth's radiation balance. Clouds, in turn, affect the levels and geographical distribution of aerosols by removing them in precipitation and by driving the general circulation. Aerosols are formed, evolve, and are eventually removed within the general circulation of the atmosphere. The characteristic time of many of the microphysical aerosol processes is days up to several weeks, hence longer than the residence time of the aerosol within a typical atmospheric compartment (e.g. the marine boundary layer, the free troposphere etc. ...).To understand aerosol properties, one cannot confine the discussion to such compartments, but one needs to view aerosol microphysical phenomena within the context of atmospheric dynamics that connects those compartments. This paper attempts to present an integrated microphysical and dynamical picture of the global tropospheric aerosol system. It does so by reviewing the microphysical processes and those elements of the general circulation that determine the size distribution and chemical composition of the aerosol

Microphysics of Aerosol Formation and Evolution

Traditionally, atmospheric aerosols have been divided into two size classes: coarse (D_p>1 μ m) and fine (D_p<1 μm), reflecting the two major formation mechanisms: primary and secondary. Both populations strongly overlap however in the 0.1-1 μm diameter range.

Primary particles that are derived from the break-up and suspension of bulk material by the wind, such as sea salt, soil dust, and biological material, have most of their mass associated with particles of diameters exceeding 1 μm, however their highest number concentrations occur in the 0.1-1 μm range. For such emission mechanisms, the particle number concentration increases nonlinearly with increasing wind speed (O'Dowd and Smith, 1993; Schultz *et al*, 1998). Because of their low concentrations and large sizes, primary particles generally do not coagulate with one another, but they can mix with other species through exchange of mass with the gas phase. A particular and important type of primary particles are soot particles formed in combustion. Initially they are formed at high concentrations within the combustion process as particles with a diameter of 5 to 20 nm. They coagulate however rapidly to form fractal-like aggregates, which, in turn, will collapse to more compact structures of several tens of nanometers due to capillary forces of condensing vapours.

Secondary aerosol mass is formed by transformation of gaseous compounds to the liquid or solid phase. This occurs when the concentration of the compound in the gas phase exceeds its equilibrium vapour pressure above the aerosol surface. In the atmosphere, several processes can lead to such a state of supersaturation:

1) gas-phase chemical reactions leading to an increase in the gas-phase concentration of compounds with low equilibrium vapour pressure. Examples are:

$SO_2 + OH \rightarrow$ (known reaction scheme) $\rightarrow H_2SO_4$

$NO_2 + OH \rightarrow HNO_3$

I. Barnes (ed.), Global Atmospheric Change and its Impact on Regional Air Quality, 13–18.

α- Pinene + O_3 → (unknown reaction scheme) → Pinic Acid

2) lowering the ambient temperature leading to a lowering of the equilibrium vapour pressure above the aerosol,

3) formation of multi-component aerosol, so that the equilibrium vapour pressure of the single compounds above the aerosol is lowered by the presence of other species in the aerosol (Raoult effect).

The equilibrium vapour pressure over a spherical particle increases with increasing curvature of the particle (Kelvin effect), hence the equilibrium vapour pressure above molecular clusters formed by random collisions is much larger than that above a film on a pre-existing particle or flat surface. Consequently, molecular clusters will generally evaporate. Their growth to a stable size, i.e. nucleation, will be favoured primarily by the absence of pre-existing aerosol surface, and by extreme realisations of the three processes described above. Classical nucleation theory predicts that nucleation is highly non-linearly dependent on the concentration of the nucleating species in the gas phase.

When nucleation does occur, the new particles grow by condensation and self-coagulation. As particles reach a diameter of the order of the mean free path length of the condensing molecule (typically about 60 nm), condensation becomes diffusion-limited and slows down. Also, self-coagulation, which is a second-order process, eventually quenches as number concentrations fall. Hence, under background tropospheric conditions, particles formed initially by nucleation require days to weeks to grow larger than about 0.1 μm solely by condensation and coagulation. Under polluted, urban type conditions, this growth can occur within a day (Raes *et al.*, 1995).

One straightforward way of accumulating secondary aerosol species in the 0.1-1 μm diameter range is by condensation on primary particles emitted in that range. Another more elusive but important growth process is by chemical processing in non-precipitating clouds (Mason, 1971, Friedlander, 1977, Hoppel *et al.,* 1986, 1994). This process begins with the activation of aerosols, which is the uncontrolled uptake of water once water vapour becomes supersaturated above a certain critical limit. According to traditional Köhler theory, the critical supersaturation for activation depends on the amount of soluble material in the particle and its hygroscopicity, i.e. tendency of the material, once dissolved, to lower the equilibrium water vapour pressure over the solution (Pruppacher and Klett, 1980). The critical supersaturation needed to activate all particles larger than a given dry size increases with decreasing particle size. When the supersaturation in an air parcel rises, cloud activation will therefore preferentially occur on larger particles. The rapid condensation of water quenches a further increase of the supersaturation (which usually does not exceed 2%), so that activation is limited to a subset of particles (cloud condensation nuclei, CCN). For example, for pure ammonium sulfate aerosol and a maximum supersaturation of 0.2%, typical for marine stratus clouds, only particles larger than about 80 nm in diameter will activate. Once a droplet is formed, gaseous species like SO_2 can dissolve and be oxidised in the aqueous phase. When the droplets evaporate, the residue particles are larger than the original CCN upon which the droplets formed as a result of the additional oxidised material, e.g. sulfate from the following aqueous phase reactions:

S(IV) + O_3 → S(VI) + O_2

S(IV) + H_2O_2 → S(VI) + H_2O

Reactions occurring in clouds might also occur in non-activated aerosol solution droplets, however with different efficiencies because of the larger ionic strength in such droplets.

Moreover, some gases might also react on the aerosol surface producing products that might either remain on the particle or return into the gas phase. Examples are the heterogeneous conversion of NO_x to HONO on fresh soot aerosol (Ammann *et al.*, 1998) and halogen release from sea salt (Vogt *et al.*, 1996).

Aerosols are removed from the atmosphere by dry and wet processes. Small particles ($D_p < 0.1\ \mu m$) diffuse to the earth's surface, a process which becomes less efficient as the particle size increases. Large particles ($D_p > 1\ \mu m$) settle gravitationally, a process which becomes less efficient as the particle size decreases. In the range $0.1 < D_p < 1\ \mu m$, dry removal is very slow, and the formation and growth processes discussed above tend to accumulate the aerosol in this size range. These particles, when they have the right hygroscopic properties, will be removed mainly by activation in clouds and subsequent precipitation.

It has become generally accepted to name particles with a diameter in the range $0.1 < D_p < 1\ \mu m$ *Accumulation mode particles*, particles in the range $0.01 < D_p < 0.1\ \mu m$ *Aitken mode particles*, and particles with $D_p < 0.01\ \mu m$ *Nucleation Mode particles*. Particles with $D_p > 1\ \mu m$ are called *Coarse Mode particles*. The idea to represent the aerosol size distribution with a number of log-normally distributed modes is supported by measurements (see Section 3) and was first ventured by Whitby (1978)

Each of the processes described above has a characteristic time. Loss processes (coagulation, deposition) which occur on a time scale that is small compared to the compartment residence time, indicate an unstable, i.e. decaying, property. Time scales that are long compared to compartment residence times lead to stable properties within the compartment and inter-compartment exchange of those properties. For instance, the extremely short residence time of nucleation mode particles in the boundary layer, due to coagulation with larger particles, implies that they can only be observed in the immediate vicinity of their sources. Hence, as these ultrafine particles are frequently observed in the polluted continental boundary layer, the urban and con-urban continental compartment must contain sources for nucleation mode aerosol. The absence of a persistent nucleation mode in the MBL indicates that in this case no strong in-situ sources for this mode are available. However, occasional nucleation bursts could temporally occur leading to an unstable and rapidly decaying nucleation mode. Because of their short life time, nucleation mode particles are generally not exchanged between the atmospheric compartments, unless they happen at the boundaries of such compartments. Accumulation mode particles on the other hand decay much slower and as such they can travel from one compartment to the other, mixing their properties with the ones of newly formed aerosol within the next compartment.

Global Atmospheric Circulation and the Life Cycle of the Tropospheric Aerosol

Tropospheric general circulation is characterized by rapid, localized upward motion due to convection (in the tropics) or slantwise ascent along frontal surfaces (in the mid-latitudes), which is compensated by relatively slow and large-scale subsidence in the sub-tropical and polar regions. Horizontal transport in the lower and upper troposphere connects areas of upward and downward transport, in what are supposed to be toroidal circulation patterns. Long-term averages of both the meridional and zonal wind fields in the tropics/sub-tropics reveal the existence of these patterns, which are called the Walker and Hadley circulations, respectively. In a snapshot of the global wind fields, these toroidal circulations are less evident (Newell *et al.*, 1996, Wang *et al.*, 1998).

Subsidence over the sub-tropical oceans leads to the existence of a temperature inversion and the creation of a marine boundary layer, which is topped by vast stratiform clouds. Thus in the subtropics there is a clear separation between the marine boundary layer and the free troposphere aloft, whereas in convective regions this separation is less clear.

The general circulation is described by global observations of fields of winds and other meteorological parameters (Oort, 1983), or it can be reproduced by General Circulation Models (GCM's) from basic physical principles. Using the observed climatologies, or off-line versions of the GCM's, Chemical Transport models (CTM's) have been build in which the descriptions of emissions, transport, transformations and removal of chemical species have also been considered (Zimmerman, 1984, Heimann *et al.*, 1990).

Major progress has been achieved in simulating the global distribution of tropospheric aerosol mass using global CTM's. The first simulation of the global distribution of biogenic and anthropogenic sulfur (Langner and Rodhe, 1991) lead to the recognition that anthropogenic sulfate aerosols may have a significant impact on the global radiation balance (Charlson *et al.*, 1991). This spurred a large interest, and simulations of the global mass distributions for the aerosols types listed in Table 1 followed. Despite the simplification of considering each aerosol type independently, these studies were important to relate emissions to global distributions, to construct global and regional budgets, to estimate the contribution of anthropogenic sources to the burden of aerosol species that are also produced naturally, and to draw attention to elements of the general circulation that are important in aerosol transport, in particular deep convection (Feichter and Crutzen, 1990).

The role of deep convection

The high updraft velocities in convective clouds and the corresponding supersaturations up to 2 % lead to activation of soluble aerosol particles with diameters as small as 0.01 μm; partly soluble particles activate at somewhat larger diameters. Soluble trace gases in these updrafts will be taken up by the cloud droplets. Precipitation, which, on a global scale, is produced mainly in convective clouds, eventually removes the activated particles and dissolved gases. During convective cloud transport a separation therefore occurs between soluble species that are rained out and insoluble species that are pumped into the free troposphere (Rodhe, 1983)

Organic aerosol has been observed to be ubiquitous in the upper troposphere (Novakov *et al.*, 1997; Murphy *et al.*, 1998; Putaud *et al.*, 2000. See Section 3.2). Although the accuracy of organic matter determination in aerosols is in question, these studies do claim that there is relatively more organic matter in the upper troposphere than sulfate. This would be consistent with a separation between soluble and insoluble species during vertical transport.

Deep convection leads to an increase of the photo-oxidizing capacity of the upper troposphere by pumping up nitrogen oxides produced by anthropogenic sources or biomass burning at the surface or by lightning within the convective cloud (Lelieveld and Crutzen, 1994, Jacob *et al.*, 1996). Given the link between photochemistry and aerosol formation , as well as the increase in gas-phase concentrations of H_2SO_4 discussed in the previous section, a link between convective clouds and aerosol nucleation might be expected.

Recent observations have shown that the free troposphere is chemically not homogeneous. Quasi-horizontal layers are frequently observed which are characterised by various combinations of ozone and water vapour (Newell *et al.*, 1999), and other chemical species (Wu *et al.*, 1997). Layers with lower O_3 and higher H_2O than the background are tentatively interpreted as due to convection from the boundary layer. Layers with higher O_3 and lower H_2O, which are the most abundant are interpreted as originating in the stratosphere. Vertical profiles of aerosols also indicate layered structures in e.g. aerosol number concentration (e.g. Clarke *et al.*, 1996) but a detailed study of their relationship with other gas-phase species has not been made yet. Convective transport of (insoluble) pollution aerosols or nucleation near clouds might be two ways of producing such layers in the middle and upper troposphere.

Summary and outlook

During the past decade enormous progress has been made in the understanding of the life cycle of aerosols in the global atmosphere. In the previous sections we argued that even a basic understanding of aerosols at a global scale requires the understanding and integration of both microphysical and large scale dynamics processes. This is primarily because the time scales of aerosol evolution are in many cases longer than the residence time in particular atmospheric compartments. Furthermore, important phenomena such as nucleation and particle wet removal are occurring at the boundary of such compartments.

At present, however, the picture of the aerosol life cycle remains fragmentary. Observational data sets are incomplete and models need to take simple approaches, favoring one aspect of the aerosol (e.g. calculation of aerosol mass) at the expense of others (e.g. calculation of aerosol number). A full version of this paper has been published in Atmospheric Environment 34 (2000) 4215-4240.

References

Ammann M., Kalberer, M., Jost D.T., Tobler L., Rossler E., Piguet D., Gaggeler, H.W., and Baltensperger U., (1998) Heterogeneous production of nitrous acid on soot in polluted air masses, Nature, 395, 157-160.

Charlson R.J., Langner, J., Rodhe, H., Leovy, C.B., and Warren S.G., (1991) Perturbation of the nothern hemisphere radiative balance by backscattering from anthropogenic sulfate aerosols, Tellus, 43AB, 152-163.

Clarke, A.D., Porter, J.N., Valero, F.P.J., and P. Pilewskie, P., (1996) Vertical profiles, aerosol microphysics, and optical closure during ASTEX: measured and modeled column optical properties, Journal of Geophysical Research, 101, 4443-4453.

Clarke, A.D., Uehara, T., and Porter J.N., (1996) Lagrangian evolution of an aerosol column during the Atlantic Stratocumulus Transition Experiment, Journal of Geophysical Research, 101, 4351-4362.

Feichter J., and Crutzen, P.J., (1990) Parameterization of vertical tracer transport due to deep cumulus convection in a global transport model and evaluation with radon measurements, Tellus 42 B, 100-117.

Friedlander, S.K. (1977) Smoke, Dust and Haze: fundamentals of aerosol behaviour, John Wiley & Sons, New York.

Heimann, M., Monfray P., and Polian, G., (1990) Modeling the long-range transport of Rn-222 to subantarctic and antarctic areas, Tellus, 42B, 83-99, 1990.

Hoppel W.A., Frick, F.M., and Larson, R.E., (1986) Effect of nonprecipitating clouds on the aerosol size distribution in the marine boundary layer, Geophysical Research Letters, 13, 125-128.

Hoppel W.A., Frick G.M., Fitzgerald J.W., and Wattle B.J. (1994) A cloud chamber study of the effect that nonprecipitating water clouds have on the aerosol size distribution, Aerosol Science and Technol., 20, 1-30,

Jacob D.J., Heikes, B.G., Fan, S.-M., Logan, J.A., Mauzerall, D.L., Bradshaw, J.D., Singh, H.B., Gregory, G.L., Talbot, R.W., Blake, D.E., and Sachse, G.W., (1996) Origin of ozone and NOx in the tropical troposphere: A photochemical analysis of aircraft observations over the South Atlantic basin, Journal of Geophysical Research, 101, 24235-24250.

Langner, J., and Rodhe, H., (1991) A global three-dimensional model of the tropospheric sulfur cycle, Journal of Atmospheric Chemistry, 13, 225-263.

Lelieveld, J., and Crutzen, P.J., (1994) Role of deep cloud convection in the ozone budget of the troposphere, Science, 264, 1759-1761.

Mason B. J. (1971) The physics of clouds, Clarendon Press, Oxford.

Murphy D.M., Thomson, D.S., and Mahoney M.J., (1998) In situ measurements of organics, meteoritic material, mercury, and other elements in aerosols at 5 to 19 kilometers, Science, 282, 1664-1669.

Newell R.E., Zhu, Y., Browell, E.V., Read, W.G., and Waters, J.W., (1996) Walker circulation and tropical upper tropospheric water vapor, Journal of Geophysical Research, 101, 1961-1974.

Newell R.E., V. Thouret, V., Cho, J.Y.N., Stoller, P., Marenco, A., and Smit, H.G., (1999) Ubiquity of quasi-horizontal layers in the tropophere, Nature, 398, 316-319.

Novakov T., Hegg, D.A., and Hobbs, P.V., (1997) Airborne measurements of carbonaceous aerosols on the East Coast of the United States, Journal of Geophysical Research, 102, 30023-30030.

O'Dowd, C.D. and Smith, M.H., (1993) Physicochemical properties of aerosols over the Northeast Atlantic: Evidence for wind-speed-related submicron sea-salt aerosol production, Journal of Geophysical Research,98, 1137-1149.

Pruppacher, H.R. and Klett, J.D., (1980) Microphysics of Clouds and Precipitation, D. Reidel Publishing Company

Putaud J.-P., Van Dingenen, R., Mangoni, M., Virkkula, A., Raes, F., Maring, H., Prospero, J., Swietlicki, E., Berg, O., Hillamo, R., and Makela, T., (2000) Chemical mass closure and assessment of the origin of the sub-micron aerosol in the marine boundary layer and the free troposphere at Tenerife during ACE-2, Tellus, 52B, 141-168.

Raes F., Wilson, J., and Van Dingenen, R., (1995) Aerosol dynamics and its implication for the global aerosol climatology, in "Aerosol Forcing of Climate", (Eds. R.J. Charson and J. Heintzenberg) , John Wiley & Sons.

Raes, F., (1995) Entrainment of free tropospheric aerosols as a regulating mechanism for cloud condensation nuclei in the remote marine boundary layer, Journal of Geophysical Research, 100, 2893-2903.

Rodhe H., (1983) Precipitation scavenging and tropospheric mixing, in Precipitation Scavenging, Dry Deposition, and Resuspension (Pruppacher et al., Eds), 719-728.

Schulz M., Balkanski, Y.J., Guelle, W., and Dulac F., (1998) Role of aerosol size distribution and source location in a three-dimensional simulation of a Saharan dust episode tested against satellite-derived optical thickness, Journal of Geophysical Research, 103, 10579- 10592.

Vogt R., Crutzen, P.J., and Sander, R., (1996) A mechanism for halogen release from sea-salt aerosol in the remote marine boundary layer, Nature, 383, 327-330.

Wang P.-H., Rind, D., Trepte, C.R., Kent, G.S., Yue, G.K., and Skeens., K.M., (1998) An empirical model study of the tropospheric meridional circulation based on SAGE II observations, Journal of Geophysical Research, 103, 13801-13818.

Wu Z., R.N. Newell, Y. Zhu, Y., Anderson, B.E., Browell, E.V., Gregory, G.L., Sachse, G.W., Collins Jr., J.E., (1997) Atmospheric layers measured from the NASA DC-8 during PEM-West B and comparison with PEM-West A., Journal of Geophysical Research, 102, 28353-28365.

Zimmermann P.H. (1984) Ein dreidimensionales numerisches Transportmodell fur atmospharische Spurenstoffe, Thesis, Univeristy of Mainz, FRG.

Active and Passive Sensing of Aerosol from Space

Matvienko G.G. and Belov V.V.

Institute of Atmospheric Optics SB RAS, Tomsk, 634055, Russia, 1, Akademicheskii ave.

It is well known that satellite methods are efficient means of remote monitoring of the atmospheric aerosol in real time. The algorithms of aerosol optical thickness (AOT) reconstruction from the data of NOAA/AVHRR radiometers recorded in the visible range of the spectrum have been developed and positive experience in their application for obtaining global information on the AOT of various regions of the global ocean has been accumulated (Afonin et al, 1997). On the other hand, there has been little and study on the applicability of these algorithms for monitoring of the aerosols above the Earth's underlying surface or interior water basins and also for aerosol classification. Undoubtedly, satellite measurements in the infrared range of the spectrum in combination with the data obtained in the visible range extend the capabilities of remote aerosol monitoring in the daytime and provide the basis for detecting local aerosol formations at night.

The principal feasibility of using the satellite measurements in the infrared wavelength range for detection of local regions with enhanced aerosol turbidity in the surface atmospheric layer and for the aerosol classification has been investigated at the Institute of Atmospheric Optics of the SB RAS. To this end, numerical experiments were carried out to simulate (for mid-latitudes in summer) night and daytime measurements of the intensities (radiative temperatures) of upwelling heat radiation in IR-channels (channel 3: 3.55–3.93 μm, channel 4: 10.3–11.3 μm, and channel 5: 11.5–12.5 μm) of the AVHRR radiometer above a homogeneous underlying surface versus the content of the aerosol of different types (rural and urban) in the surface atmospheric layer. Two situations were considered: a) standard vertical profiles of the meteorological parameters and the underlying surface temperature (surface temperature T=294 K) for the mid-latitudes in summer and b) nonstandard low values UST = 278 K (simulating the meteorological conditions above the Lake Baikal surface).

Recently the data from the IAO information base on meteorological and aerosol characteristics of the atmosphere over Tomsk have been statistically processed in order to estimate their typical values and variability ranges. The results have been used for computer simulation of the upwelling radiation intensity measured in the AVHRR visible and infrared channels for a wide range of optical and geometrical observational conditions.

Based on the results of computer simulation, the dependence of this intensity on the surface albedo, temperature, atmospheric humidity, aerosol content, and its optical characteristics (single scattering albedo and the asymmetry parameter of the scattering phase function) has been studied. The computed data for NOAA (12, 14, 15, and 16) satellites have been put in the basis of models of atmospheric haze characteristics and atmospheric transmittance that are needed for atmospheric correction of AVHRR data and reconstruction of the AOT. These results are exemplified below in Table 1:

The solution of this problem yields the following main results:

1) the sensitivity of measurements in AVHRR channels to variations of aerosol and meteorological parameters has been determined at various optical and geometrical observational conditions and the error of AOD reconstruction from satellite data has been estimated to be about 0.03-0.05;
2) the requirements for the accuracy of specification of the key atmospheric parameters have been formulated for AOD reconstruction with the preset accuracy;

I. Barnes (ed.), Global Atmospheric Change and its Impact on Regional Air Quality, 19–24.

3) the process of formation (due to aerosol scattering of solar radiation) of a local minimum of the signal in the first, second, and third AVHRR channels (0.63, 0.84, and 3.75 μm) at small sun elevation angle and small azimuth of observation has been studied. Under these conditions, the relative contribution of the solar radiation scattered by aerosol becomes dominant in the signal as compared to the fraction of radiation reflected from the surface.

Table 1. Continental aerosol. Observation at the angle of 30°.

	Sm = 5 km						Sm = 20 km					
Hs	A1	Dif	A2	Dif	T34	Dif	A1	Dif	A2	Dif	T34	Dif
75	7.9	-1.0	4.3	-0.6	1.1	-0.4	3.6	-0.8	1.7	-0.4	0.6	-0.1
50	7.7	-1.4	4.4	-0.8	1.1	-0.2	3.4	-1.2	1.7	-0.6	0.6	-0.1
30	7.9	0.2	5.0	0.6	1.4	0.1	3.8	-0.1	2.1	0.2	0.7	0.0
20	7.5	1.1	5.1	1.4	1.6	0.5	4.0	0.6	2.5	0.8	0.8	0.2
15	6.7	1.5	4.8	1.7	1.7	0.7	4.0	1.0	2.6	1.1	0.8	0.2
10	5.3	1.5	4.2	1.8	1.7	0.8	3.6	1.2	2.6	1.3	0.8	0.3
5	3.1	1.2	2.8	1.5	1.4	0.7	2.5	1.0	2.1	1.2	0.8	0.3
2	1.6	0.7	1.7	1.0	1.0	0.5	1.4	0.7	1.4	0.9	0.7	0.2

Sm is the meteorological range;
HS is the sun elevation angle, in degrees.;
A1 and A2 are the intensity of atmospheric haze (in units of albedo, %) in the 1st and 2nd AVHRR channels for the sun azimuth of 0°;
T34 is the difference of radiative temperature in the 3rd and 4th AVHRR channels for the sun azimuth of 0°;
Dif are the differences in A1, A2, and T34 for the azimuth of 0 and 180°.

Furthermore, we have analyzed the IAO information database on aerosol parameters of the atmosphere. These analyses have allowed us to reveal the periods of the highest aerosol and soot content in the atmosphere. For these periods we have carried out the comprehensive analysis of ground-based measurements and space measurements in the visible AVHRR channels. Processing of the satellite information (more than 2000 space images) included the following stages:
(a) calibration, geographic reference, visualization, and rejection of cloudy images;
(b) statistical analysis of the space-time variability of the surface albedo for the territory around Tomsk to find "dark" surface areas optimal for AOT measurements, which are characterized by low values of albedo (about 2-3%) and high spatial homogeneity;
(c) atmospheric (molecular) correction of measurements in the visible AVHRR channels (based on the models mentioned above) with allowance for the real state of the atmosphere at the time of satellite observations.

Upon solution of this problem, we have obtained the following results:
(1) the data of satellite and ground-based observations have confirmed the computed estimates of the sensitivity of AVHRR channels to variations of aerosol characteristics;
(2) a statistically significant correlation has been found between the satellite and ground-based observations (Figure 1), which is promising for the use of AVHRR data in monitoring of the properties of the tropospheric aerosol at the territory of Western Siberia.

Methodological aspects of the problem of laser sensing of the atmosphere from space have already been analyzed in (Zuev, 1974, Zakharov, 1988). The perspectives for

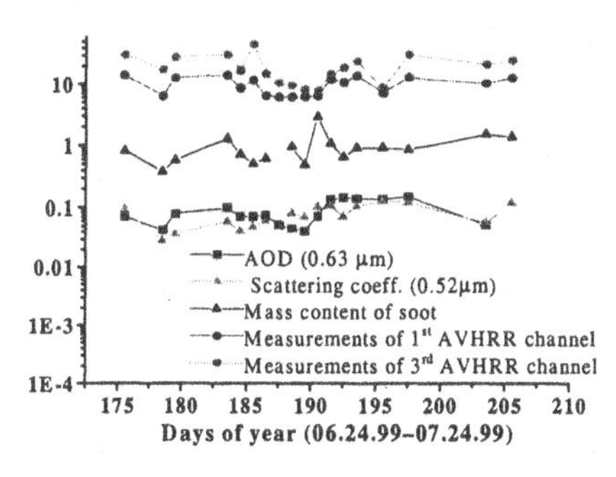

Figure 1. Comparison of measured aerosol characteristics and the data in the 1st and 3rd AVHRR channels (relative units).

engineering implementation of the nearest projects that have already been developed are discussed in many publications, for example, in Reading (1991) and McCormick (1993). By the present time the first cycles of investigations have already been carried out with the Russian space lidar BALKAN. The lidar was placed on the orbital station "Mir." Since August, 1995 it has been operated in the stationary mode (Balin et al., 1997). The Russian-French lidar ALISSA on the module "Priroda" of the same station was put into orbit in May, 1996. In September, 1994 the first successful experiment on multifrequency laser sensing of the entire depth of the Earth's atmosphere from a board of the Shuttle (Winker, 1996) was carried out. This orbital experiment was accompanied by synchronous sessions of ground-based and airborne lidar sensing (Renger et al., 1995).

High efficiency of first spaceborne lidar experiments on the study of global distributions of cloudiness, vertical profiles of the aerosol characteristics and some other parameters of the underlying surface in the next stage pose the expedient based on automatic specialized orbital complexes. Within the context of the Program "Etalon" of the Russian Space Agency (RSA) projects on the development of a network of small specialized satellites equipped with detectors of different types (electrical, magnetic, optical, etc.) have been initiated. It is planned to equip one of the satellites of the above-indicated network, namely, "TECTONICA-A," with a multifrequency lidar operating in a long-term automatic mode (Matvienko et al., 1998). The lidar specifications are given in Table 2.

Table 2. The lidar specifications.

Height of the orbit, km		400-600	
Working wavelengths, μ	1.064	0.532	0.355
Pulse energy, J	0.2	0.05	0.01
Pulse duration, ns	27	27	30
Pulse repetition frequency, Hz		10/100	
Angular beam divergence, mrad	1.0	0.4	0.4
Detection regime		Photon counting	
Aperture area of the mirror, m^2		2.0	
Field-of-view angle, mrad		1.0-20.0	
Spatial resolution, km		0.3-5.0/1.0-10.0	
Mass, kg		150	

As follows from the parameters listed in Table 2, three harmonics of a commercial garnet laser with diode pumping, for example, of an FP-50 laser of Fibertek, Inc. will be used.

A distinctive design feature of the aerosol lidar TECTONICA-A is the large aperture area of the receiving mirror (2 m^2). This is caused by the necessity of recording of energetically weak backscattered signals. This requirement follows from our preliminary estimates (Matvienko et al., 1998) and the experience on the space lidar LITE operation (Winker, 1996). Automatically this means a segmented mirror has to be deployed in orbit. The consequences of such a design are obvious difficulties with mutual alignment of the mirror segments and the laser transmitter and, as a result, the necessity of using large field-of-view angles. Definite possibilities for optimization of the receiving system and primarily for a decrease in the field-of-view angle are associated with the use of an adaptive secondary mirror in the receiving telescope.

Results of the adequate numerical experiment by the Monte Carlo method presented below deal with a study of the peculiarities of forming the active optical background illumination in receiving channels of the given promising lidar. It is well known, that the method allows one to estimate signals with different multiplicity of interaction and angles of photon arrival at the detector. This provides a basis for a detailed analysis of the lidar return signal waveform as a function of the optical-geometrical conditions of the experiment.

Obviously, the problem of separating the signals reflected from lower cloudiness has yet to be solved. As is well known, this cloudiness differs by large thickness. In addition, signals backscattered from liquid-drop and mineral formation noticeably differ in their polarization characteristics. However, in the case of sensing from space this problem calls for additional study. Design features of the receiving systems of multiwavelength lidar which is being developed, due to large (up to 20 mrad) field-of-view angles, imposes special requirements on the accuracy of estimates of the multiple scattering background under various conditions, including extremal ones. Just the multiple scattering background restricts the informational capabilities of laser sensing channels to obtain reliable quantitative data on the atmospheric parameters.

We note at once that signals coming from the middle atmosphere (for sensing in the nadir) are well described by the single scattering approximation. Rigorous calculations of signals from the troposphere without cloudiness are illustrated in Figure 2. These results are presented in the form of altitude dependence of the relative multiple scattering contribution $P_m(h)$ to the total signal $P(h)$. The value of this functional depends on the boundary conditions of the experiment and primarily on the angular receiving aperture. It can be seen that with an increase in reception angle up to the limiting values indicated above the contribution of $P_m(h)$ to the signal reflected from the atmospheric boundary layer reaches 30% (for the sensing parameters listed above). On the basis of the approach of Zuev et al. (1974) the values of multiple scattering may be estimated. The contribution of multiple scattering will not be the limitation in usage of segmented mirrors in aerosol space lidars.

In connection with the fast development of the oil and gas industries, the problem of monitoring the gaseous emissions from pipe-lines becomes increasingly urgent, especially in view of their ecological hazard. Therefore, it becomes of primary importance to develop the corresponding methods for analysis of the local gaseous composition of the atmosphere. These methods should ensure fast acquirement of data on the gases atmospheric composition on large spatial scales.

Laser methods of sensing of gaseous atmospheric constituents most completely satisfy the above requirements. Since resonance absorption has the largest interaction cross section, the DIAL method, which employs this effect, is highly sensitive.

We estimated numerically (based on numerical simulation) possibilities of remote detection of gas anomalies in the lower atmosphere and hydrocarbon emissions from pipe lines.

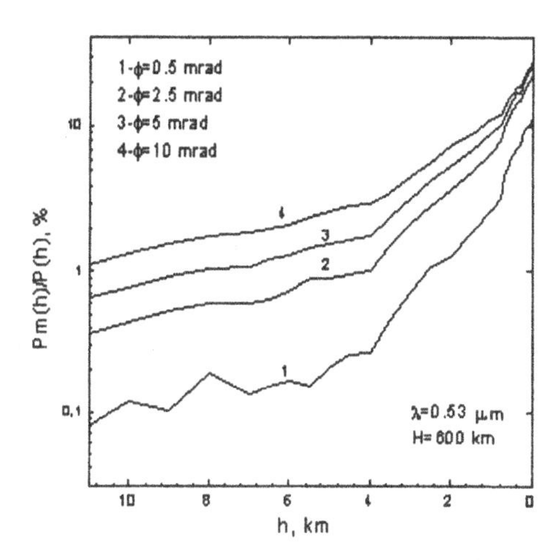

Figure 2. Effect of the lidar field-of-view angle on the formation of the multiple scattering background for $\phi = 0.5$, 2.5, 5.0, and 10 mrad.

Sensing of methane and ethane with the DIAL method allows one to use absorption bands from the near and mid-infrared spectral ranges. The spectroscopic parameters for methane and ethane bands are presented in Rothman (1992). The strongest band of methane is that centered near 3 μm (~ 3000 cm^{-1}, band with the total intensity of $1.08 \cdot 10^{-17}$ cm·$molecule^{-1}$). The only strongest band of ethane is centered near 3 μm (~ 2900 cm^{-1}, with the total intensity of $4.65 \cdot 10^{-17}$ cm·$molecule^{-1}$). This band is closely co-ordinated with the methane band. Radiation of the tunable laser in the spectral range from 3.1 to 3.6 μm can be obtained by efficient conversion of the frequency of the pulse-periodic TEA CO_2 laser into the third harmonic. Nowadays noticeable results are achieved in generation of harmonics and combination frequencies of CO_2-laser radiation (Karapuzikov et al., 2000).

In numerical modeling with the use of the software package SAGDAM (Firsov et al., 1999; Zuev et al., 1992) the following input parameters the space-borne lidar were used:

Orbit height	400 – 600 km
Wavelength range	3040 – 3645 nm
Emission line halfwidth	0.1 – 0.2 nm
Pulse repetition frequency	100 – 200 Hz
Duration of a pulse	30 ns
Energy per pulse	≥ 10 mJ
Frequency instability	0.02 nm
HWHM of emission line, nm	0.02 nm
Beam divergence	1 mrad
Time of signal accumulation	0.1 – 0.5 c
Spatial horizontal resolution	1 – 5 km
Receiving aperture of the telescope	≥ 1 m^2
Quantum efficiency of the receiver	0.2
Field-of-view angle of the receiving telescope	3 mrad
Transmission of the receiving and transmitting optics	0.5
NEP of the photodetector ($InSb|_{LN2}$)	$1 \cdot 10^{-12}$ $W/Hz^{1/2}$

The measurement of the concentration of methane is possible on wavelengths of the third harmonics TEA CO_2 - laser. For measurement of densities close to background the pair of analytical lengths of waves 3415.544 and 3420.130 nm is suitable, and the high densities (30-60 ppm) are troubleshot on lengths of waves 3429.601 and 3434.489 nm.

The problem of a choice of possible wavelengths for the analysis of C_2H_6 requires realization of additional research of the absorption spectrum of this molecule in the field of base frequencies and harmonicses CO_2 - laser. However, there are two wavelengths 3336.8 and 3333.0 nm, the usage of which will allow the detection C_2H_6 at the 0.1 ppm level.

The precomputations have shown, that the realization of space lidar is possible which will supply localization of anomalies of methane concentration at a level of background densities (~ 1.8 ppm) with an error of 25 % and ethane, at concentrations of 0.1 ppm with an error of 15-25%.

References

Afonin S.V., Belov V.V., and Makuschkina, Simulation of the upwelling thermal radiation scattered by aerosol, takig into account temperature inhomogeneities on a surface. Part 3. Small scale high-temperature anomalies, *Atmos. Oceanic Opt.* **10,** N. 2 (1997). 114-118.

Afonin S.V., Belov V.V., and Makuschkina, Ir images transfer through the atmophere, *Atmos. Oceanic Opt.* **10,** N. 4-5 (1997)278-288.

Balin Yu. S., Tikhomirov A. A., Samoilova S. V. Preliminary results of cloud and underlying surface sensing with the lidar BALKAN. *Atmos. Oceanic Opt.* **3.** (1997)333-352.

Firsov K. M., Kataev M. Yu., Mitsel' A. A., Ptashnik I. V., and Zuev V. V., *JQSTR*, **61,** No. 1, (1999)25-37.

Karapuzikov A.I., Malov A.N., Sherstov, I.V. *Infrared Phys. Technol.,* **41,** (2000) 77-85.

Matvienko G. G., Kokhanenko G. P., Shamanaev V. S., Alekseev V. A. Project of the spaceborne lidar TECTONICA-A, *Intern. Symp. Remote Sensing, Conf. Abstr., Barselona,* (1998)37-38.

McCormick M. P., Winker D. M., Browell E. V. Scientific investigations planned for the lidar in-space technology experiment (LITE), *Bull. Meteor. Soc.,* **74,** (1974)205-214.

Reading C, coordinator. *ESA SP-1143: Report of the Consultation Meeting, May 1991, ESA Publication Division, ESTEC, Noordwijk, The Netherlands,* (1991)32.

Renger W., Kieme C., Schreiber H.-G., Wirth M., Moerl P. Airborne backscatter lidar measurements at 3 wavelengths during ELITE. Final Results, *Workshop Proc. (IROE-CNR, Florence, Italy),* (1995)15-19.

Rothman L.S., Gamache R.R., Tipping R.N. et al, *J. Quant. Spectrosc. Radiat. Transfer* **48**, (1992)469-507 .

Zuev V. E., Krekov G. M., Naats I. E. Determination of aerosol parameters of the atmosphere by laser sounding from space., *Acta Astronautica,* **1,** (1974) 93-103.

Zakharov V. M., ed. *Laser Sensing of the Atmosphere from Space, Gidrometeoizdat, Leningrad,* (1988)213 .

Zuev V. V., Mitsel' A. A., and Ptashnik I. V., *Atmos. Oceanic Opt.,* **5,** No. 9, (1992)970-977.

Winker D. M. Multipple scattering effects observed in LITE data: the good, the bad, and the ugly, *Proc. MUSCLE,* **8,** (1996)1-5.

Global Self-Regulation of Ozone in the Earth Atmosphere

V.V. Lunin and A.B. Zosimov

Moscow State University, Department of Chemistry, Vorobjevi gori,
119899 Moscow, Russia, vvlunin@kge.msu.ru

This report is a generalization of the experimental and theoretical research in the field of physical chemistry of the Earth's atmosphere, which was made in the Catalysis and Gas Electrochemistry Laboratory of the Chemistry Department of MSU.

Approximately 5 years ago we suggested that processes, leading to ozone layer decomposition and consequently to increases in the CO_2 percentage in the Earth's atmosphere should be closely related and that chemical reactions with the involvement of both these gases should be included into natural oxygen cycle. After making such a suggestion we faced tough problems. It transpired that the natural oxygen cycle is not closed as supposed by modern ideas. We have could not find any full description of the general natural oxygen cycle in literature, so, we selected the data about all natural oxygen-containing substances cycles, presented in Figure 1.

It is well known that seasonal fluctuations of the CO_2 and ozone percentage are inharmonious. Increases of the CO_2 percentage are related to decreases of the O_3 concentration. Thus, year-by-year the increase of the CO_2 amount in the atmosphere is accompanied by a ozone layer decomposition process.

We proceeded from the well known assumption that the total amount of ozone in the atmosphere is the product of photosynthesis reactions. About $1.25 \cdot 10^{16}$-$1.5 \cdot 10^{16}$ mole of carbon is involved in these processes [1]. Biological oxidation is incomplete, consequently, every year about $3,3 \cdot 10^{14}$ moles of carbon is in oceanic sedimentations and soils [2]. The same amount of CO_2 is absorbed from the atmosphere with O_2 emission.

From the organic sediments in the ground and oceans approximately $1.5 \cdot 10^{13}$ moles of CH_4 are formed and liberated into the atmosphere. About $3 \cdot 10^{13}$ moles of oxygen oxidize that amount of aforementioned gas.

The natural cycles of N and S transfer about 10^{13} moles of oxygen to the ground.

The bivalent iron in the crusts can be oxidized by the water under the 10 km depth [3]. The hydrogen resulting from that process diffuses by means of earth splits into the atmosphere. About $3 \cdot 10^{13}$ moles of oxygen can oxidize this hydrogen.

Other outlets of oxygen in the lithosphere are unknown. Not more than $7 \cdot 10^{13}$ moles of O_2 can be eliminated from the atmosphere by means of the outlets shown in Figure 1. Thus, the total flow of O_2 in the atmosphere is not less than $2.4 \cdot 10^{14}$moles and the outlet of CO_2 is not less than $3 \cdot 10^{14}$ moles per year.

CO_2 can be consumed in 200 years, but it can be liberated from the ocean since the amount of this gas in the ocean is approximately 60 times more than in atmosphere. All the CO_2 from the oceans can be spent in the processes of photosynthesis in no less than 10^4years. When the percentage of CO_2 is less than 0.01volume %, photosynthesis almost stops [4]; nowadays its value in the atmosphere is near by this limit.

Emission of O_2 from photosynthesis should increase its concnetration in atmosphere at about $7 \cdot 10^{-6}$ volume parts per year.

Oxygen retards the photosynthesis processes. When its amount is more than 25 – 26 volume %, photosynthesis also almost stops [4]. In order to increase the percentage of O_2 from 21% to 26 volume % one needs about $6 \cdot 10^3$ years. Thus, the characteristic time for a O_2 or CO_2 catastrophe for biosphere is $6 \cdot 10^3$-10^4 years.

I. Barnes (ed.), Global Atmospheric Change and its Impact on Regional Air Quality, 25–30.

However, it is well known than during the all Neozoic period (about $50 \cdot 10^6$ years) the amount of O_2 was changing very slowly [3]. So, one can assume that there are outlets of O_2 in the crusts, in the form of oxygen-containing compounds. The capacity of these outlets should be about 10^{14} moles of O_2 per year. The solubility of O_2 in water is very low, consequently, it can penetrate into the lithosphere in the form of its compounds.

We assumed that one of the oxygen outlets can be the flow of ozone from the stratosphere to the Earth's surface, which is approximately $6 \cdot 10^{13}$ moles of ozone per year. This amount is equivalent to $9 \cdot 10^{13}$ moles of molecular oxygen per year.

Ozone dissolves in water approximately 30 times better than O_2, therefore, it can penetrate into the ocean and into the Earth's crusts as aqueous solution. However, ozone is a very powerful oxidizing agent and the time of its existence in such solutions is very short. So, it was necessary to investigate the decomposition mechanisms of ozone in marine water.

It is known, that the ozone layer absorbs all UV-radiation with the wavelength 200-300 nm. Thus it serves as a protector of all living organisms from UV-rays. The absorption spectra of nucleic acids (max 260 nm) and proteins (max 280 nm) are located in this region. Actually, molecular oxygen absorbs the radiation with the wavelength less than 254 nm, and the role of ozone in this region can be neglected.

Calculations have been made of the value of absorption of UV-radiation with the wavelength 260 and 280 nm for the case that all stratospheric ozone will be decomposed. In this instance molecular dissipation of the light in the atmosphere and absorption of the UV-radiation by nitrogen and sulfur oxides and, of course, troposphere ozone shall take place. It was found that radiation with the wavelength 260 nm, which is more dangerous for the living organisms due to the interruption of the work of genetic mechanisms, will be absorbed in the troposphere. The intensity of the radiation with the wavelength 280 nm will be decreased 10,000 times. Thus, in the case of destruction of all the stratospheric ozone, probably, a biological catastrophe will not occur, because of the fact that dangerous UV-radiation will be absorbed in the troposphere.

So, the positive role of the ozone for the biosphere is not only absorption of UV-radiation, but also decrease of the photosynthesis rate due to absorption of the light with the wavelength 450-740 nm (which is essential for photosynthesis) and inhibition of chlorophyll [5]. Because of this fact, the speed of the processes of emission of O_2 and absorption of CO_2 are decreasing too. So, the probability of the appearance of CO_2 and O_2 catastrophes is declining.

An estimation of the amounts of the ozone-destroying compound, which is necessary for to yield the modern stratospheric ozone concentration, has also been made.

The first theoretical calculation of the stationary concentration of stratospheric ozone was made by Hrgian [6]. He took into account only oxygen, nitrogen and solar UV-radiation. Other authors then made similar calculations with addition of supplementary compounds, which exist in the stratosphere. We used the original calculation scheme with several added equations to represent the presence of the ozone-destroying compound.

Vertical profiles of the ozone distribution were obtained, which were similar to the real ones, however, the total amount of ozone was hundreds of times in excess of the existing amount. This implies that large amounts of ozone-destroying compounds must exist in the atmosphere in order to explain the current status of the ozone layer.

It is well known that there are a lot of ozone destruction mechanisms, which were proposed by the different authors. There are hydrogen, chlorine and nitrogen cycles [7, 8].

The hydrogen decomposition cycle can affect only the higher part of the ozone layer. UV-radiation with the wavelength 180 nm, which is able to decompose water molecules, exists there. The total number of NO_x in the whole atmosphere is below the amount necessary for ozone decomposition. So, we assumed, that the main ozone-decomposing substance in the stratosphere is chlorine in the form of Cl_2 and ClO_x and the source of these species is the oxidation of the chlorine ion by ozone.

We assumed that in any catalytic ozone destruction cycle in the stratosphere photochemical reactions should occur, e.g., for active chlorine formation from CFC UV-radiation of less than 250 nm is necessary. The same applies to the hydrogen cycle. Molecular oxygen and ozone absorb UV-radiation and, indeed, can block the production of the active species, which can decompose ozone.

If active chlorine is a product of photochemical reactions, then ozone concentration oscillations of considerable amplitude are possible. In the random decreases of the ozone concentration and increasing of UV-radiation intensity the rate of Cl production and ozone decomposition will increase. And vise versa, in the random increasing of ozone concentration the Cl production will decrease. In this case the ozone concentration can have two stable values. One of these is near zero, another – close to the theoretical value, calculated from oxygen photochemical cycle without taking into account any ozone destructing compounds.

In order to verify our assumption under laboratory conditions a series of experiments were made, i.e. oxidation of Cl^- by ozone in the presence of CO_2. Liberation of Cl_2 was observed (A.V. Levanov report). On the basis of the experimental data, calculations of the equilibrium composition of the solution, with Cl^- and compounds where Cl exhibits a +1 oxidation state, and gas phase over such solutions were made. These values are a function of the pH. On the basis of the experiments it was established for the first time, that the yield of the process of oxidation of Cl^- by ozone depends on the CO_2 concentration in the air.

At pH levels below 5.5 Cl_2 emission take places and at values over 5.5 ClO^- ions are collected in the solution. They can oxidize some organic compounds in solution: oceans or other water areas. The pH level 5.5-5.7 is produced in the water drops when they come in contact with air because of the CO_2 dissolving in water (at present-day concentrations) [9]. Consequently, when the CO_2 percentage in the air is higher then it modern level, Cl_2, a powerful ozone destroying agent, will be liberated from the chloride solutions, such as drops of sea water, oxidized by air with ozone. Should the CO_2 concentration in the air decrease, with an associated pH level increase, Cl_2 will not be liberated, and the processes of chemical oxidation of organic compounds will be accelerated and CO_2 concentrations in the solutions, and, hence, in the air will increase. Therefore, one can uphold that the natural mechanism of the self-regulation of the ozone, CO_2 and O_2 concentrations in the Earth's atmosphere exist.

On the basis of such a mechanism, perhaps, one will be able to explain the connection between global trends and seasonal oscillations of the CO_2 and O_3 concentrations in the atmosphere. However, the local interruption of the ozone layer in the form of ozone holes over the poles, requires more detailed discussion. For the explanation of this phenomenon we assumed, that chlorine oxides can be formed on the ice particles of PSC as a result of HCl oxidation by ozone. The possible scheme of the HCl flow from the oceans to the stratosphere is shown on the Figure 2. Together with it, it was known, that the composition of atmospheric condensations changes when humid air masses move from seaside to inland continent changes.

First of all, let us deal with the level of chloride ions in atmospheric condensations [6, 9]. The molar ratio Cl/Na in atmospheric condensations changes from 1 over the ocean to 0.5 over a continent [6]. Chlorine ions can be partly displaced from chloride solutions in the form of HCl, if solutions absorb non-metal oxides, e.g. SO_2, NO_2 etc.

It is well known that H_2CO_3 acidity is weaker than HCl. However, indeed this process can occur. Only one condition is necessary: multiple vaporization of the drops of marine water with subsequent condensation on the salt crystals. CO_2 should dissolve in these drops; on evaporation of the drop to the solid phase, in the first instance, carbonates will be isolated since they have lower solubility than chlorides. This will be accompanied by HCl liberation to the gas phase.

Since air around drops in clouds is mixed by eddies, in the condensation of the new drop on the nuclei, arising from the evaporated drop, all of the HCl is not able to return back to the new drop. The possibility of this mechanism of HCl emission from the NaCl solutions was tested experimentally. In multiple evaporating of up to 50% from the starting value, of the 5% solution of NaCl by the CO_2-containing gas flow, followed by watering, carbonates were collected in the solution. This experiment proved that Cl^- is removed from the solution by the air flow, like eddies in clouds, in the form of HCl.

The emission of the marine salt from oceans is a known process. We assumed that variation of the Cl/Na molar ratio is linked with displacement of HCl from the drops by CO_2 [6], and maybe some other gases, such as SO_2, NO_2.We obtained an estimation of the HCl emission to the gas phase – around 10^{13} moles per year. This is about 2 times greater than HCl emission from volcanoes.

In the lower part of the troposphere rains will probably elutriate HCl from the gas phase to the ocean. However, some amount of HCl can be transferred through the tropopause to the stratosphere by the rising atmospheric flows. Such a process becomes possible, since there is no water in liquid and vapour phases in the polar vortex. Consequently, HCl can be collected in the polar stratosphere.

We tried to simulate in our laboratory the further fate of HCl adsorbed on the ice. Under the low temperature, ice with adsorbed HCl was affected by pure ozone. Thus, without any light treatment, chlorine oxides were formed. Further they can participate in different photochemical processes, which lead to the ozone hole formation (S.V. Savilov report). Consequently, the chlorine cycle of the ozone decomposition is more effective, than we thought before, since there is additional source of active chlorine – HCl. On the basis of our experimental work and estimations, we can vindicate, that natural mechanisms of self-regulation of ozone, O_2 and CO_2 levels in the atmosphere exist. They prevent the biosphere from catastrophic variations in the percentages of O_2 and CO_2.

Reactions of the oxidation of the chlorine ions by ozone play the dominant role in these processes. The products of these reactions are ozone-destroying compounds or water-soluble oxygen-chlorine substances, which can oxidize organic matter in the Earth's crust.

Acknowledgements

This work has been supported by FCP "Integratsia" 467. The authors thank Serguei V. Savilov for helpful discussions and preparation of this text.

References

1. Wayne R.P. Chemistry of the atmospheres. Oxford University Press. 2000.
2. Global warming. Greenpeace report. // Ed. Leggett J. Oxford University Press, 1990.
3. Sorohtin O.G., Ushakov S.A. Globalnaya evolutsia Zemli. Moscow: MGU, 1991.
4. Polevoy V.V. Fiziologiya rasteniy. Moscow: Vishaya shkola, 1989.
5. Semenov S.M., Kunina I.M., Kuhta B.A. Troposferniy ozon I rost rasteniy v Evrope. Moscow: Meteorologiya i Gidrologiya, 1999.
6. Hrdian A.H. Fizika atmosfery. Leningrad: Gidrometeoizdat, 1979.
7. Molina M.J. *Pure & Appl. Chem.* **68** (1996) 1749.
8. van Loon G. W., Duffy S.J. Environmental Chemistry. Oxford University Press. 2000.
9. Batcher S., Garlson R. Vvedeniye v himiyu atmosfery. Moscow: Mir, 1977

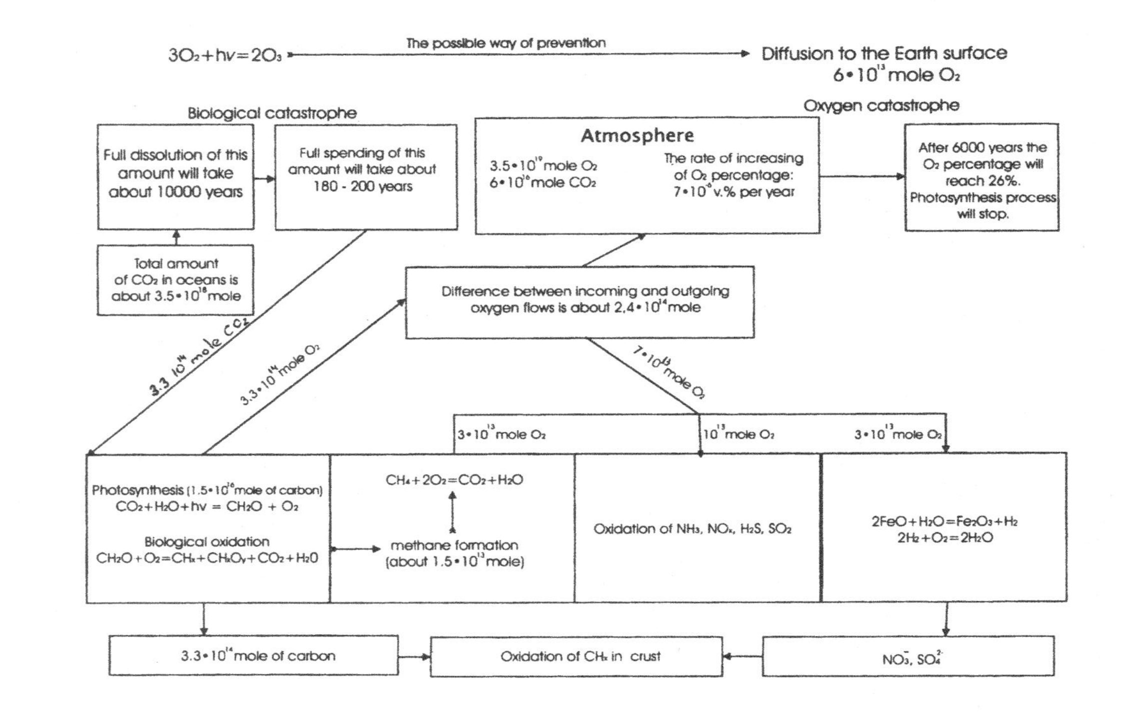

Figure 1._Oxygen cycle in the Earth atmosphere.

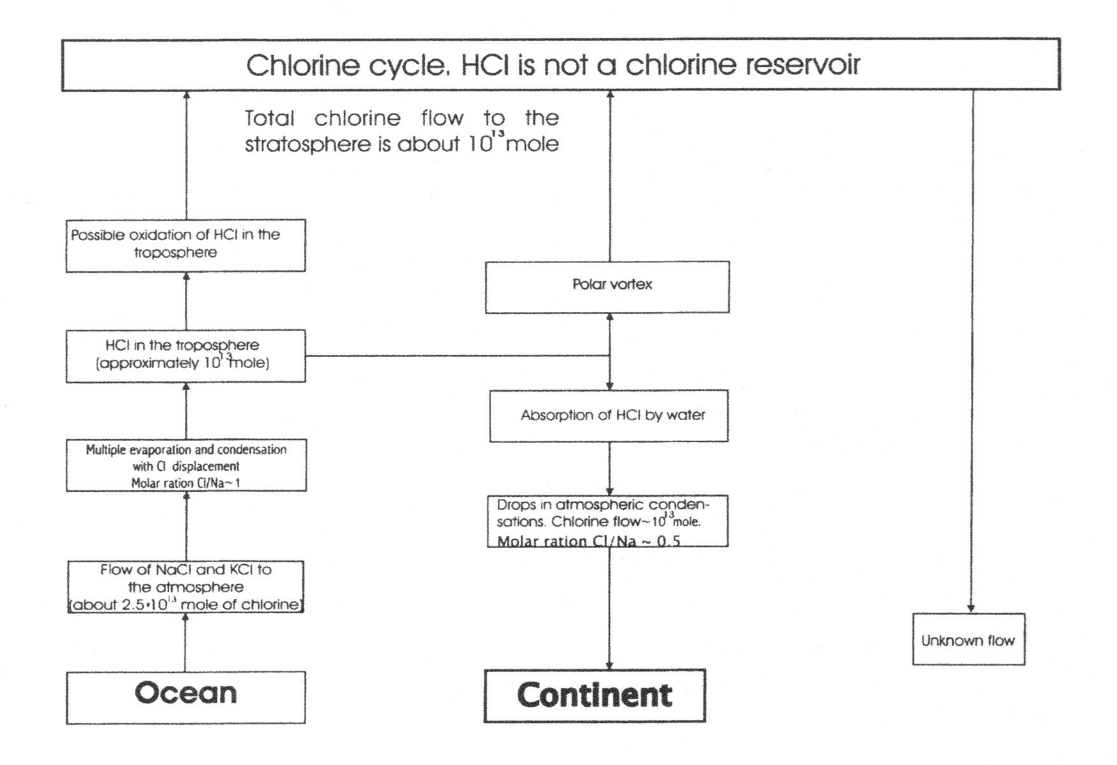

Figure 2. The main chlorine flows in the Earth atmosphere.

Tropospheric Multiphase Chemistry in Field, Modelling and Laboratory Studies

Hartmut Herrmann, Paolo Barzaghi, Olaf Böge,Erika Brüggemann, Barbara Ervens, Andreas Donati, Johannes Hesper, Thomas Gnauk, Yoshiteru Iinuma, Konrad Müller, Christian Neusüß and Antje Plewka

Institut für Troposphärenforschung, Permoserstr. 15
D-04318 Leipzig, Germany

Introduction

Aerosol field characterization experiments have been performed within different field campaigns and main findings are presented here. Recent campaigns include (a) the campaigns of the 'Stadtaerosol'-project within the German AFS programme and (b) the Saxonian 'SLUG' project of our group. Tropospheric aerosols are necessary for cloud formation and clouds, on the other hand, are processing the chemical constituents of the multiphase system including gases, cloudwater and aerosol constituents. Therefore, both aerosol-gas as well as cloudwater chemical systems are studied in laboratory experiments which are briefly described. Finally, a multiphase model is described which is used to simulate cloud chemical conversions and, in a first series of numerical experiments, aerosol organic composition changes.

Results and Discussion

1.) Aerosol Field Measurements: EC/OC Quantifications

The quantification of organic material in tropospheric aerosol particles is an important research topic. Generally, it may consist of organic material of either primary or secondary origin

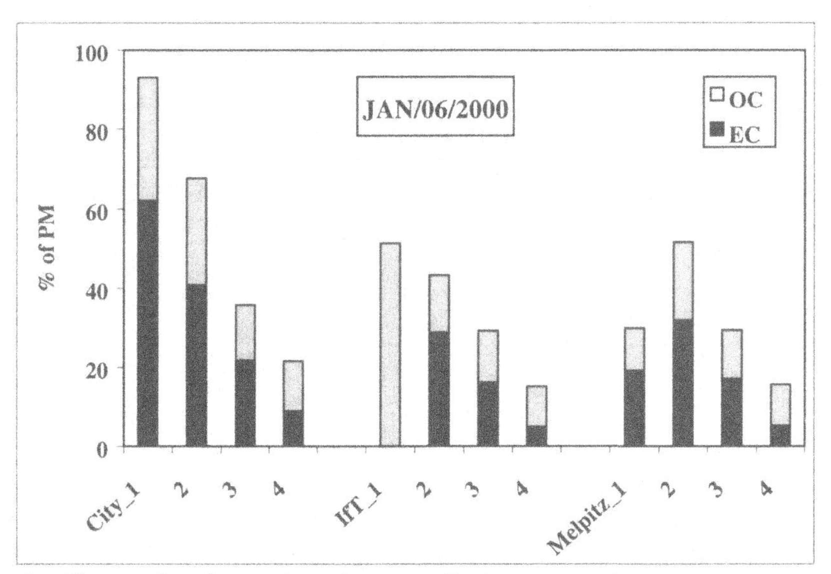

Figure 1:. Elemental and organic carbon contributions to particle mass as from Berner impactor measurements, winter campaign 1999/2000, 'SLUG'-project

Primary organic material may consist of soot and, possibly, organic matter of biogenic origin. Soot again is produced by incomplete combustion of carbon-containing materials. It is

I. Barnes (ed.), Global Atmospheric Change and its Impact on Regional Air Quality, 31–36.

a heterogeneous material with carbon content possibly reduced to 50% and an oxygen content of up to 30%. It includes an organic component addresses as OC (soluble in organic solvents and/or volatile) and a component insoluble (in organic solvents and/or resistant to oxidation at T < 400°C) which is addressed as EC (elemental carbon, sometimes 'BC = black carbon' or 'GC = graphitic carbon'. OC in tropospheric particles may contain contributions of secondary origin, i.e. organic substances which have been formed in the atmosphere by conversion of organics leading to condensable products (low vapor pressures) in the course of tropospheric oxidation. EC and OC in our laboratory is always separated by a gas evolution method (see Neusüß et al. (2000) for more details.

Results for the size-resolved quantification of OC and EC in the 'SLUG' campaign in the center of Leipzig in winter 1999/2000 are shown in Figure 1. It can clearly be seen that under these winterly conditions almost all of the particle mass on impactor stage 1 (particle radii 50 –120 nm) is due to organic and elemental carbon. For bigger particles, the fraction of EC is continuously decreasing, especially in the inner city where both OC and EC can be regarded to be mainly caused by traffic. At the other locations of aerosol measurements, i.e. at the IfT in 10 m height and at the IfT research station Melpitz theses patterns are less clear because other influences rather then just traffic emissions appear to influence the measured particle composition. At the IfT this may be caused by a higher contribution of aged particles transported to the sampling location mixed by fresh emissions from both traffic and house-heating.

2.) Aerosol Field Measurements: INDOEX

Finally, our group within an IfT effort has contributed to the INDOEX campaign in the year 1999. With a simplified two stage impactor aerosol was sampled onboard the research vessel R/V Ron Brown. The OC/EC determination has then been performed in our laboratory. For the method and overview on INDOEX aerosol measurements Quinn et al. (2002). For detailed results on carbon analysis, see Neusüß et al. (2002). In Figure 2 results are given from this series of measurements where not only INDOEX data but also data for the preceding AEROSOL campaign are shown. Circles in this Figure indicate the TC ≡ OC + EC mass per cubic meter of air. As can be seen, there is a considerable bearing of particle organics and elemental carbon south of the Indian subcontinent. Particulate organics here are suspected to be caused by biomass burning, here mostly used for cooking purposes, emissions from traffic and industrial activities with no adequate exhaust treatments.

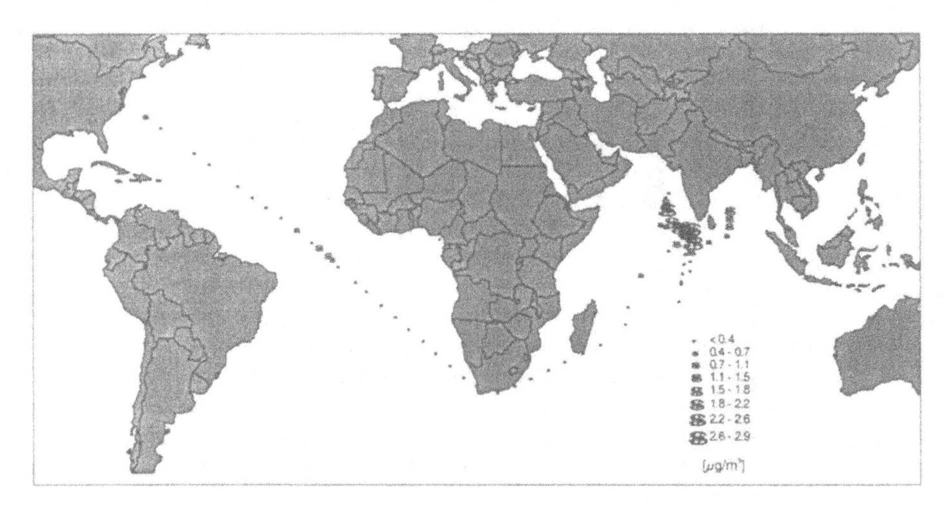

Figure 2: Results for TC during the AEROSOLS and INDOEX campaigns in 1999.

3.) Aerosol Reactor

In order to better understand the interaction of gas phase processes and particles a gas phase aerosol reactor is currently set up at the IfT chemistry department. A schematic of this experiment is shown in Figure 3.

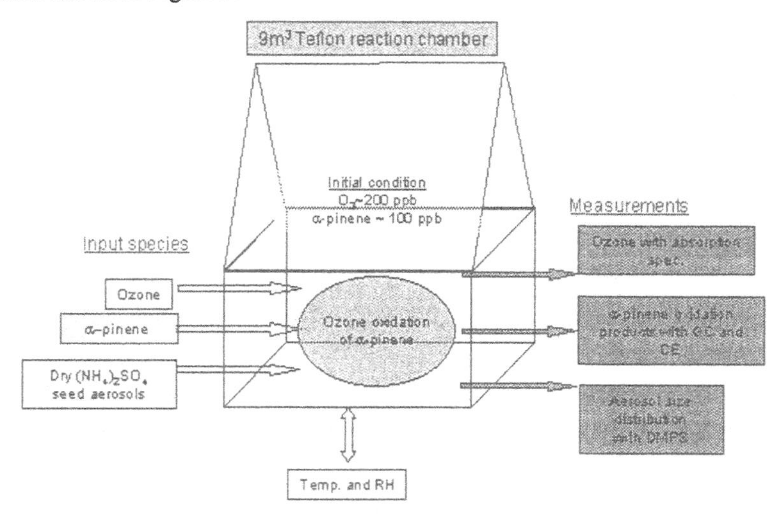

Figure 3: IfT chemistry department aerosol reactor

Currently, α-pinene is reacted with ozone in the presence of inorganic seed aerosols and the particle phase products are identified and quantified by means of derivatisation-GC/MS, LC/MS, CE and, upcoming CE/MS. The focus of future studies will be extensive chemical analytics of the particle phase which is modified by gas phase processes involving terpenes, aromatics and oxygenates and the oxidants ozone, OH and NO_3.

4.) Aqueous Phase Chemistry

Chemical reactions of the free radicals OH·, NO_3· and Br· with compounds of importance for the tropospheric aqueous phase were investigated by means of direct, laser-based methods. The radicals were generated by laser photolysis of adequate precursors and the absorption signals of these radicals were detected. From these measurements the rate coefficients for the reactions were derived. Reactions of Br atom with organic compounds were studied at different temperature by UV absorption. A method to study the reactivity of the OH radical with organic compounds was developed. The concentration and temporal behavior of NO_3 radical, produced by few pulses into an acidic aqueous solution, were compared with the concentration of the reaction products identified by HPLC.

Reactions of Br atoms with organic compounds

Bromine atoms were generated by excimer-laser-photolysis at 248 nm of aqueous solutions containing bromoacetone as a precursor. For the kinetic studies a laser-photolysis-long path-laser-absorbance apparatus (LP-LPA) was used. The Br atom absorption was measured directly at 296 nm using the 200W output of a Hg(Xe) lamp and a monochromator coupled to a photomultiplier. All the experiments were carried out under pseudo-first order conditions. The time resolved Br atom decay traces were captured and averaged by a digital oscilloscope and then transferred to the PC to extract the first order rate constants. The second order rate constants were extrapolated from the plot of measured k_{1st} vs concentration of the reactant (Figure. 4).

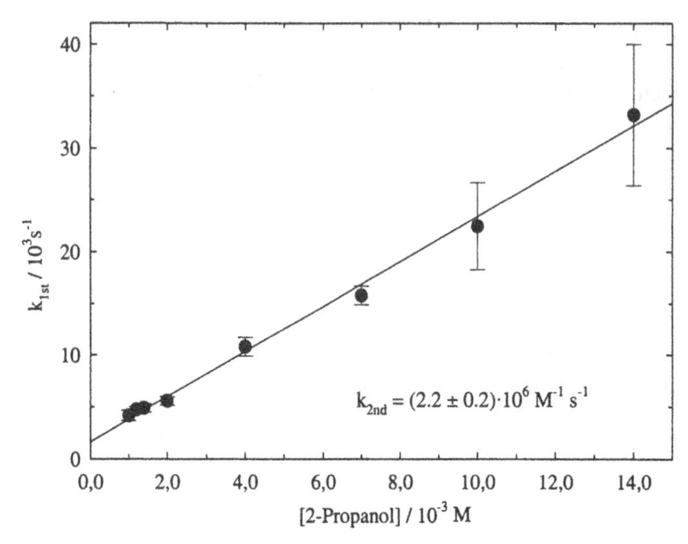

Figure 4: k_{1st} = f ([2-Propanol]) for the reaction of Br atom with 2-Propanol; T = 298K

Reactions of OH radicals and Peroxyl Radicals

Rate constants of OH radical reactions with organic compounds have been investigated at different temperatures. The intention of this work is to increase the knowledge in this area by using a new technique. Usually competition kinetic methods were used to determine rate constants of the OH radical with organic compounds, because of the difficulty of the direct detection approach (low extinction coefficient of OH, overlapping absorption with organic peroxyl radicals). The new method follows the absorption/time profile at 244 nm taking the interference of RO_2 radicals into account. A new method was developed to determine the time dependence of OH radical absorption in the aqueous phase. Two different ·OH-radical sources were used. The first one is the direct Excimer laser flash photolysis of water at 193 nm:

$$H_2O + h\nu \text{ (193 nm, 400 mJ)} \longrightarrow H\cdot + \cdot OH$$

The second source used for kinetic investigations is the excimer laser flash photolysis of hydrogen peroxide at 248 nm:

$$H_2O_2 + h\nu \text{ (248 nm, 650 mJ)} \longrightarrow 2\ \cdot OH$$

The spectroscopic properties of the peroxyl radicals ($RO_2\cdot$) (e.g. the absorption band, the extinction coefficient) are needed for the kinetic investigations to utilize the absorption/time profiles recorded at 244 nm.

$$\cdot OH + RH \longrightarrow H_2O + R\cdot$$

In the presence of oxygen (solution is saturated with oxygen) the alkyl-radical (R·) reacts quickly to form the corresponding peroxyl-radical:

$$R\cdot + O_2 \longrightarrow RO_2\cdot$$

This reaction pathway is the rate determining step. Peroxyl-radicals investigated with this system absorb light in the UV-range. Therefore the absorption/time profile measured at 244 nm can be used to calculate the rate constant of the OH-radical reaction by following the build up of the peroxyl-radical and by taking the absorption of the OH-radical into account.

Reactions of OH / NO_2 / NO_3 with Phenol

The flash photolysis of nitrate anions at λ = 248 nm and peroxodisulphate anions at λ = 351 nm was used to study the oxidation process of phenol by OH / NO_2 and NO_3 / NO_2 in aqueous solution under different experimental conditions. Two different mononitrophenols (isomer *ortho* - and *para* -) and a dihydroxy derivative (catechol) were identified as the main reaction products by means of HPLC-DAD-ED technique and their yields of formation where directly compared with the initial radical concentrations of OH, NO_2 and NO_3, respectively (Figure. 5).

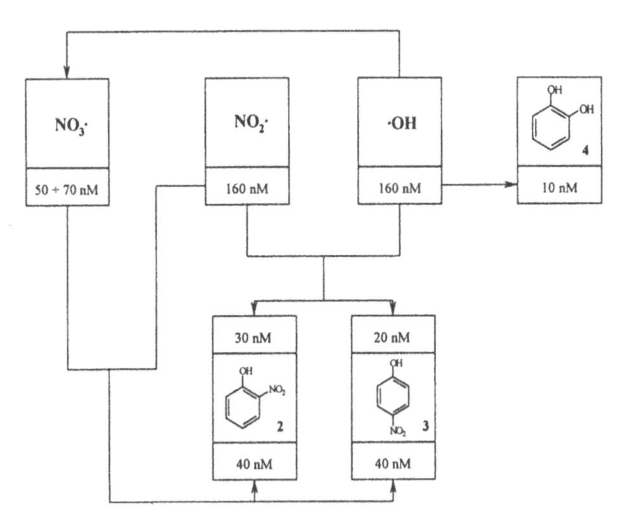

Figure 5: Schematic of radical and product yields with the possible reaction pathways in the oxidation of phenols by OH/NO_2 and NO_3/NO_2 in aqueous solution.

The obtained results show that the agreement between initial radicals yield and product concentrations is generally reasonable. The obtained results will be implemented into a module for tropospheric multiphase chemistry modelling for a better description of nitrophenol formation.

5.) Multiphase Modelling with CAPRAM

Multiphase Modelling is performed with a chemical mechanism comprising RACM and the Chemical Aqueous Phase Radical Mechanism (CAPRAM) its version 2.4 (MODAC mechanism). In Figure 6 the composition of the mass is shown obtained from the model after one simulation day. It becomes evident that the organic mass (i.e. the mass of low volatile organics which can be expected to remain in the residual particle) contributes to about 1% to the total mass.

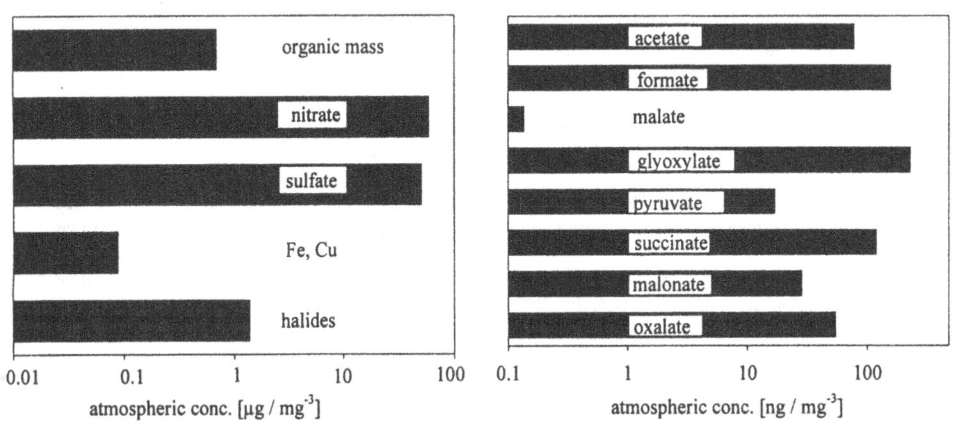

Figure 6: Mass distribution in a cloud droplet (after 24 h simulation time); left: total mass [µg m^{-3}], right: distribution of organic mass [ng m^{-3}]

In the right Figure this bar is further resolved into single compounds showing that oxalate, pyruvate and malonate are the main contributors. Obviously sources of malic acid are underestimated in the model. In field studies its concentration is comparable to those of the smaller dicarboxylic acids.

The analysis of formation pathways shows that, in fact, the formation both of oxalic and pyruvic acid is connected strongly to the amount of aromatics in the gas phase. More than 90% of glyoxal and methylglyoxal are produced by aromatics. Other pathways such as the oxidation of glycolic acid as oxalate precursor are much less important maybe due to the fact that currently no sources of hydroxyacetaldehyde are implemented in the model. However, the oxidation of hydroxy ketones, which are mainly formed in the gas phase in the oxidaton of alkanes and methacrolein, contributes only less to 5% to the methylglyoxal formation. In general, the formation of oxygenated compounds, e.g. acetone, hydroxy ketone and methylglyoxal takes mainly place in the gas phase, but they are further oxidized in clouds.

In the literature it is suggested that the high amount of oxalic acid in aerosol particles can be partly explained by decay processes from higher acids. The intermediate products in the succinate decay such as oxalacetic or mesoxalic acid are stabilized by decarboxylation so that the carbon chain becomes shorter. It seems reasaonable that finally glyoxylic acid is formed. Therefore, the presented reaction scheme shows such correlations between long and small dicarboxylic acids.

Summary and Outlook

As can be seen from this contribution tropospheric multiphase chemistry needs to be studied by all three approaches used in the atmospheric sciences. Field measurements, laboratory experiments and modelling studies are to be applied. Upon further development based on laboratory measurements future tropospheric multiphase models should be able to be applied for the interpretation of complex field campaigns characterizing gases, particles as well as cloud droplets and their respective interaction.

Acknowledgements

The studies results of which have been reported here were supported by (i) the 'Bundesministerium für Bildung und Forschung (BMBF)' and (ii) by the ‚Sächsisches Landesamt für Umwelt und Geologie' within the 'SLUG'-project (‚Korngrößendifferenzierte Identifikation der Anteile verschiedener Quellgruppen an der Feinstaubbelastung'). HH also acknowledges support by the Fonds der Chemischen Industrie.

References

Herrmann, H., B. Ervens, H. W. Jacobi, R. Wolke, P. Nowacki and R. Zellner, CAPRAM2.3: A chemical aqueous phase radical mechanism for tropospheric chemistry, *J. Atmos. Chem.* **36** (2000) 231-284.

Müller, K., A 3-year study of the aerosol in northwest saxonia (Germany), *Atmos. Environ.* **33** (1999) 1679-1685.

Neusüß, C., M. Pelzing, A. Plewka and H. Herrmann, A new analytical approach for size-resolved speciation of organic compounds in atmospheric aerosol particles: Methods and first results. *J. Geophys. Res.* **105** (2000) 4513-4527.

Neusüß C., T. Gnauk, A. Plewka, H. Herrmann, P. K. Quinn, Carbonaceous aerosol over the Indian Ocean: OC/EC fractions and selected specifications from size-segregated samples taken onboard the R/V Ron Brown, *J. Geophys. Res.*, in press (2002).

Quinn P.K., D. J. Coffmann, T. S. Bates, T. L. Miller, J. E. Johnson, E. J. Welton, C. Neusüß, M. Miller, P. J. Sheridan, 2000, Aerosol optical properties during INDOEX 1999: Means, variability, and controlling factors, *J. Geophys. Res.*, in press (2002).

Stable Carbon Isotope Ratio Measurements: A New Tool to Understand Atmospheric Processing of Volatile Organic Compounds

J. Rudolph

Centre for Atmospheric Chemistry, York University, 4700 Keele Street, Toronto, Ontario, M3J 1P3 Canada

Introduction

Measurements of stable isotope ratios have been extremely valuable to better understand the atmospheric cycles of a number of important trace gases, such as ozone, carbon monoxide and - dioxide, nitrous oxide, methane and hydrogen (c.f. Lowe et al., 1994; Conny and Currie, 1996; Conny et al., 1997; Manning et al., 1997; Brenninkmeijer and Röckmann, 1997; Manning, 1999; Lowe et al., 1999). Due to experimental limitations, applications of stable isotope ratio measurements for gas phase components generally have been limited to studies of compounds that are present in the atmosphere at levels of at least several ten ppb. Only recently it has been demonstrated that the stable carbon isotope ratio measurements of volatile organic compounds (VOC) in the atmosphere are possible at sub-ppb levels (Rudolph et al., 1997). Traditionally, applications of stable isotope ratio measurements focus on constraining atmospheric trace gas budgets and source apportionment. However, it has been shown that stable carbon isotope ratios of atmospheric VOC can also be used to determine the extent of chemical processing ("photochemical age") of the studied compound since its emission into the atmosphere (Rudolph and Czuba, 1999).

Knowledge of the isotopic composition of emissions, as well as understanding the isotope fractionation associated with atmospheric removal are essential for the interpretation of ambient measurements. Unfortunately, the amount of information presently available for understanding the isotopic composition of atmospheric VOC is presently very limited. In this paper an overview of our present level of knowledge is given, including a brief description of the state of experimental methods and discussion of the conceptual framework for understanding the stable carbon isotope ratios of VOC in the atmosphere.

Measurement methodology

Conventional ("dual-inlet") isotope ratio mass spectrometry (IRMS) methods for stable carbon isotope ratio measurements require sample sizes in the range of at least some 10 μg of carbon. Typically mixing ratios of VOC in air are in the range of a few ppb or less. Consequently obtaining a sample of sufficient size for stable carbon isotope analysis of VOC would require processing several m^3 of ambient air and isolating an individual VOC from the sample. It is therefore not surprising that no acceptable method based on "dual inlet" IRMS has been developed. However, in the seventies a method was developed that allows to couple gas-chromatographic separation on-line to IRMS (Matthews and Hayes, 1978). The procedure is based on a continuous oxidation of organic compounds in the effluent of the gas-chromatographic separation column and on-line determination of the $^{12}C/^{13}C$ in the formed carbon-dioxide by IRMS. Combined with the increased sensitivity of recently developed IRMS instruments this allows to determine the stable carbon isotope ratio for samples containing only a few ng of carbon. Due to the separation power of gas chromatography this method is also suitable for the analysis of complex mixtures such as atmospheric VOC.

All presently published methods for analysis of VOC in the atmosphere use a very similar procedure (Rudolph et al., 1997 and 2000, Tsunogai et al., 1999). To some extent this method is a modification of established procedures for measurements of atmospheric VOC by gas chromatography. The main modification is the increased sample volume. Typically the VOC are pre-concentrated from sample volumes in the range of several dm^3 (at standard

I. Barnes (ed.), Global Atmospheric Change and its Impact on Regional Air Quality, 37–42.

temperature and pressure), about an order of magnitude more than required for concentration measurements.

Prior to enrichment of the VOC from the air samples, carbon dioxide is removed by passing the samples through a small trap packed with a chemisorbent, e.g. potassium carbonate or Ascarite. Similarly water is removed either cryogenically or by a selective adsorbent. The large sample volumes generally require a two-step cryogenic enrichment procedure. The first trap typically consists of a 20 cm long stainless steel tube with an i.D. of 6-10 mm packed with glass beads. The second trap consists of a small capillary with approximately 0.3-0.5 mm i.D. Rapidly (within a few seconds) heating the capillary trap (40K/s) from a temperature close to that of liquid nitrogen to approximately 500K allows injection of the trapped VOC into the chromatographic system, which is fast enough for a subsequent separation on high resolution capillary columns.

The column effluent passes through a commercial combustion interface that converts hydrocarbons to carbon dioxide and water. The design of the interface generally follows that described by Matthews and Hayes (1978). About 0.5 ml/min of the carrier gas is transferred via an open split and a fused silica restriction capillary to the ion source of the isotope ratio mass spectrometer. Masses 44 ($^{12}C^{16}O_2$), 45 ($^{13}C^{16}O_2$ and $^{12}C^{17}O^{16}O$) and 46 ($^{12}C^{18}O^{16}O$) are monitored continuously and stored in digitized form for subsequent evaluation of the chromatograms. The areas of the chromatographic peaks for masses 44, 45 and 46 are integrated and the $^{13}C/^{12}C$ ratio is calculated from the mass 45/44 ratios after applying a small correction for the ^{17}O contribution to mass 45 following carbon isotope ratio measuring procedures.

Stable carbon isotope ratios are nearly always measured relative to a standard and expressed as relative difference in ($\delta^{13}C$) in per mil (‰). It is common practice to report $\delta^{13}C$ relative to the internationally accepted Vienna Peedee belemnite reference point (VPDB). Tests with artificial mixtures (Rudolph et al. 1997 and 2000) demonstrated that an accuracy of better than 0.5‰ could be achieved.

The reproducibility of the measurements depends on the available amount of carbon and the quality of the separation of the studied compound. For complex ambient samples repeat measurements resulted in a reproducibility of 0.5‰ to 1‰ for well-separated peaks and samples containing at least several ng of carbon.

Stable carbon isotope ratios of VOC emissions

Presently most of the information available on the stable carbon isotope ratios of VOC emitted into the atmosphere is indirect, based on data for crude oil, natural gas, and other types of parent material. Obviously, using such indirect information requires some understanding of the potential isotope fractionation associated with the emission processes. There are a few studies of the stable carbon isotope composition of the most important incomplete combustion sources, specifically biomass burning and transportation related emissions (Czapiewski et al., 2001; Rudolph et al., 1997 and 2001). Figure 1 shows the frequency of observations of $\delta^{13}C$ values for hydrocarbons emitted from transportation related sources and biomass burning. On average the emissions are very close to the average stable carbon isotope ratios of the burnt fuel. The transportation derived emissions average around –27‰, nearly identical to the average stable carbon isotope ratio of crude oil. Similarly, the biomass burning emissions have an average $\delta^{13}C$ value of –24.7‰, which is very close to the average composition of the different woods burnt in these studies, –25.1‰. The spread of the $\delta^{13}C$ values from both source categories is only small, for the transportation related sources the width of the distribution is less than 2‰, for biomass burning it is slightly wider, nearly 3‰. This somewhat larger variability of the stable carbon isotope ratios for

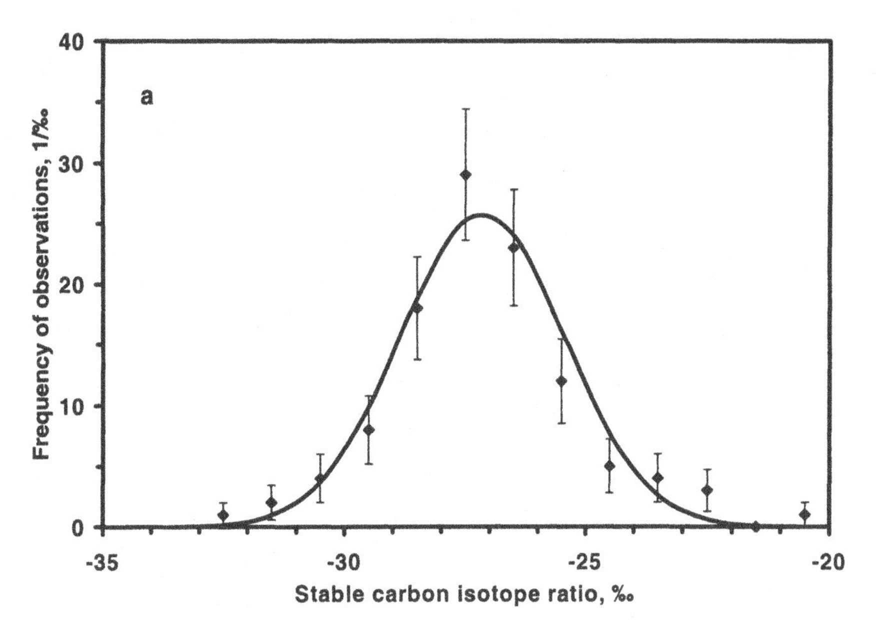

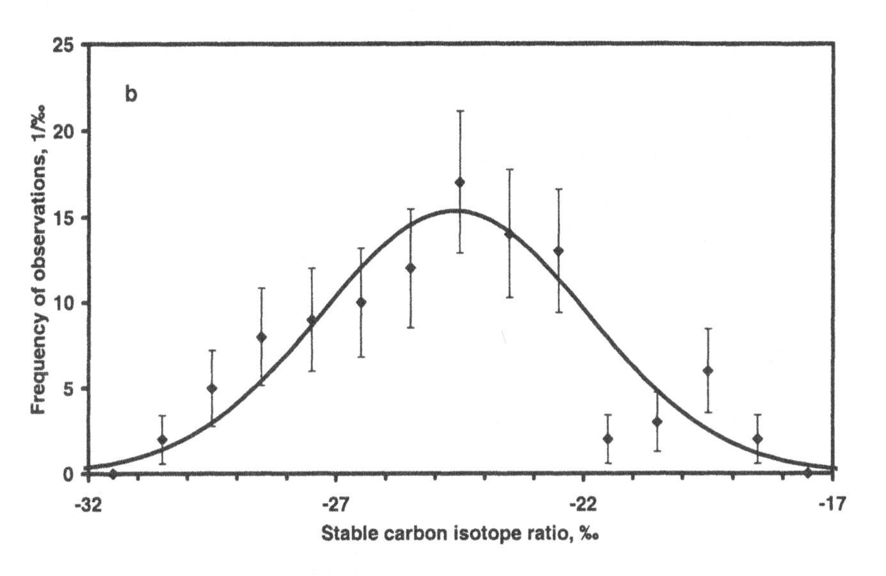

Figure 1. Frequency distribution for $\delta^{13}C$ of hydrocarbons in samples characteristic for transportation derived emissions in the greater Toronto area (a) and biomass burning (b). Data are taken from Czapiewski et al. (2001) and Rudolph et al. (2001). Width of the intervals is 1‰. The error bars are calculated from counting statistics for the given interval. For comparison normal distributions centered at –27‰ (a) and –24.7‰ (b) with a standard deviation of 1.7‰ (a) and 2.7‰ (b) are also shown (solid lines).

biomass burning can be explained by the variability of the $\delta^{13}C$ values of the fuel wood as well as a small fractionation depending on the burning conditions.

Although the vast majority of compounds emitted from biomass burning and transportation sources have stable carbon isotope ratio very close to that of the fuel, for two compounds, methyl chloride and ethyne, large fractionations have been reported. Ethyne is enriched in ^{13}C by some 20‰ to 30‰, methyl chloride emitted form biomass burning is on average depleted by 25‰.

The isotopic fractionation during removal of atmospheric VOC

For many VOC the reaction with OH-radicals is the dominant removal mechanism. The kinetic isotope effect (KIE) for this type loss process has been studied, although only for a limited number of compounds, specifically hydrocarbons. Table 1 summarizes the carbon KIEs for the reaction of a number of hydrocarbons with OH-radicals. In spite of the limited data available, we can already recognize some systematic features that allow first order estimates of KIEs for OH-reactions of hydrocarbons that have not yet been measured.

Table 1:. Carbon kinetic isotope effects for the reaction of non-methane hydrocarbons with OH-radicals

Compound	Average KIE[1], ‰	Error ‰	Predicted[2] KIE, ‰	Difference, ‰
Propane	3.44	0.26	3.46	0.02
n-Butane	2.84	0.17	2.60	0.24
Hexane	1.41	0.92	1.73	0.32
Propene	11.70	0.19	11.12	0.58
1-Butene	7.40	0.32	8.34	0.94
1-Hexene	5.22	0.41	5.56	0.34
(z)-2-Butene	9.34	1.98	8.34	1.00
1,3-Butadiene	7.55	0.88	8.34	0.79
Isoprene	6.94	0.80	6.67	0.27
Ethyne[‡]	15.80	0.61	16.68	0.88
Benzene	8.13	0.80	5.56	2.57

[1] For convenience the KIE is given as relative difference in per mil between the rate constants for reaction of the compound containing no ^{13}C atoms and the ^{13}C labeled compound.
[2] Calculated from an inverse dependence of the KIE on number of carbon atoms (n_c), KIE= a/n_c with a constant of 10.4‰ for alkanes and 33.4‰ for unsaturated compounds.

The most striking observation is that for each of the n-alkanes the KIE is approximately a factor three lower than for unsaturated hydrocarbons with the same carbon number. This difference can be explained by the different reaction mechanisms. The reaction of alkanes with OH-radicals is initialized by the abstraction of a hydrogen atom whereas at ambient temperature the reaction with unsaturated hydrocarbons proceeds via an addition of the OH-radical. Furthermore, within each group of compounds the KIE is decreasing with increasing number of carbon atoms. Thus for each of the groups of compounds we can describe the observations by a simple inverse dependence between the KIE and the number of carbon atoms (see Table 1). There are some minor differences between the observations and the KIE predicted from such a simple model, which points towards a dependence of the KIE on details of the chemical structure, e.g. for benzene there is a difference of 2.6‰. Although this difference is significantly higher than uncertainty of the measurements, the presently available data set is too limited to establish a systematic relation between such differences and specific chemical structures.

Chemical processing in the atmosphere and the stable carbon isotope ratio of VOC

The dependence between OH-radical processing for an atmospheric VOC and its stable carbon isotope ratio can be described by the following equation (Rudolph and Czuba, 2000):

$$\delta_z = (t_{av} * [OH] * {}^{OH}k_z) * {}^{OH}\varepsilon_{KIE} + {}^{0}\delta_z \quad (1)$$

δ_z: Observed ambient stable carbon isotope ratio for compound z
t_{av}: Average age of compound z
[OH]: Average OH-radical concentration
${}^{OH}k_z$: Rate constant for the reaction of compound z with OH-radicals
${}^{OH}\varepsilon_{KIE}$: Kinetic isotope effect for the reaction of compound z with OH-radicals in per mil
${}^{0}\delta_z$: Stable carbon isotope ratio of the emissions for compound z

If attack by OH-radicals is the main atmospheric reaction of z, t_{av} * [OH] represents the average of the extent of photochemical processing of compound z, often referred to as its photochemical age. Based on equation 1 it is possible to use isotope data for differentiating between chemical reaction and dilution as the reason for changes in ambient concentrations of VOC. An example is given in Figure 2.

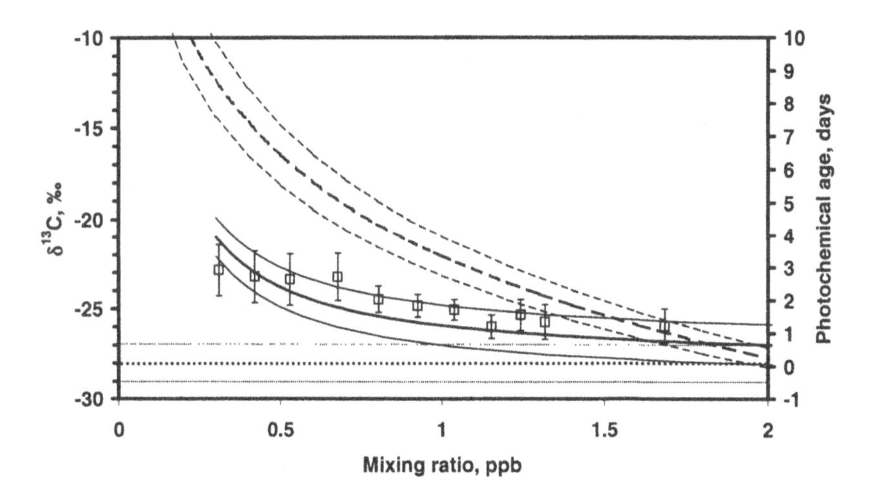

Figure 2. Plot of stable carbon isotope ratios of benzene versus mixing ratio for measurements at a suburban location in the greater Toronto area in summer (squares). The data point represents the average of three measurements; the error bars indicate the standard deviation. The photochemical age axis is derived from equation 1, using an OH-radical concentration of $2*10^6$ molecules cm^{-3} and the KIE and rate constant for the reaction of OH-radicals with benzene from laboratory measurements. Also shown are the theoretical curves for dilution of emissions by air masses free of benzene (dotted line), dilution of emissions with air containing 0.3 ppb of benzene with a stable carbon isotope ratio of –21‰ (solid line), and the dependence expected for a hypothetical isolated air mass. The thin lines indicate the estimated one-sigma uncertainties of the theoretical curves based on the uncertainties of the isotopic composition of the emissions and the error of the KIE.

In the case of dilution of the emissions with air masses that are free of benzene, the isotopic composition of benzene is not going to change with changing concentrations. It should be remembered that a variation in source strength would have the same effect. If we assume that the change in concentration is exclusively due to photochemical removal, there would be a systematic increase in the stable carbon isotope ratio with decreasing concentration. The steepness of this change will depend on the KIE of the loss process. For the example presented in Figure 2 it is obvious that the observations are not compatible with either of these two extreme cases. A possible explanation of the observations is a limited amount of photochemical processing equivalent to an average photochemical age of a few days with a dilution process. But also a dilution with air masses that have significant concentrations of benzene can explain the observations. In this case the stable carbon isotope ratio of the "mixture" will vary between the isotopic compositions of the different air masses. The difference in stable carbon isotope ratio between the air masses may be due to variations in the source composition, but can also be the result of different photochemical aging of the air masses. This is one of the limitations of a photochemical age derived from the stable carbon isotope ratios; we can determine the average age of the studied compound, but without additional information not the individual contributions of different air masses.

References

Brenninkmeijer, C. A. M. and T. T. Röckmann; Principal factors determining the 18O/16O ratio of atmospheric CO as derived from observations in the southern hemispheric troposphere and lowermost stratosphere, *J. Geophys. Res.*, **102** (1997) 25,477-25,485.

Czapiewski, K. v., E. Czuba, L. Huang, D. Ernst, A.L. Norman, R. Koppmann, and J. Rudolph; Isotopic composition of non-methane hydrocarbons in emissions from biomass burning, *J. Atmos. Chem.*, submitted, (2001).

Conny, J. M. and L. A. Currie; The isotopic characterization of methane, non-methane hydrocarbons and formaldehyde in the troposphere, *Atmos. Environ.*, **30** (1996) 621-638.

Conny, J. M., Verkouteren R. Michael, and L. A. Currie; Carbon 13 composition of tropospheric CO in Brazil: A model scenario during the biomass burn season, *J. Geophys. Res.*, **102** (1997) 10,683-10,693.

Lowe, D. C., C. A. M. Brenninkmeijer, G. W. Brailsford, K. R. Lassey, and A. J. Gomez; Concentration and ^{13}C records of atmospheric methane in New Zealand and Antarctica: Evidence for changes in methane sources, *J. Geophys. Res.*, **99** (1994) 16,913-16,925.

Lowe, D. C., W. Allan, M. R. Manning, T. Bromley, G. Brailsford, D. Ferretti, A. Gomez, R. Knobben, R. Martin, Z. Mei, R. Moss, K. Koshy, and M. Maata; Shipboard determinations of the distribution of ^{13}C in atmospheric methane in the Pacific, *J. Geophys. Res.*, **104** (1999) 26,125-26,135.

Manning, M. R., Characteristic modes of isotopic variations in atmospheric chemistry, Geophys. Res. Lett., **26** (1999) 1263-1266.

Manning, M. R., C. A. M. Brenninkmeijer, and W. Allan; Atmospheric carbon monoxide budget of the southern hemisphere: Implications of $^{13}C/^{12}C$ measurements, *J. Geophys. Res.*, **102** (1997) 10,673-10,682.

Matthews, D. E., and J. M. Hayes; Isotope ratio monitoring gas chromatography-mass spectrometry, *Anal. Chem.*, **50** (1978) 1465-1473.

Rudolph, J., E. Czuba, A.L. Norman, L. Huang, and D. Ernst; Stable carbon isotope composition of non-methane hydrocarbons in emissions from transportation related sources and atmospheric observations in an urban atmosphere, *Atmos. Environ.*, submitted (2001).

Rudolph, J. and E. Czuba; On the use of isotopic composition measurements of volatile organic compounds to determine the "photochemical age" of an air mass, *Geophys. Res. Lett.*, **27** (2000) 3865-3868.

Rudolph J., E. Czuba, and L. Huang L; The stable carbon isotope fractionation for reactions of selected hydrocarbons with OH-radicals and its relevance for atmospheric chemistry, *J. Geophys. Res.*, **105** (2000), 29,329-29,346.

Rudolph, J., D. C. Lowe, R. J. Martin, and T. S. Clarkson; A novel method for compound specific determination of C13 in volatile organic compounds at ppt levels in ambient air, *Geophy. Res. Lett.*, **24** (1997) 659-662.

Tsunogai U., N. Yoshida, and T. Gamo; Carbon isotopic composition of C_2-C_5 hydrocarbons and methyl chloride in urban, coastal, and maritime atmospheres over the western North Pacific, *J. Geophys. Res.*, **104**, (1999) 16,033-16,039.

Air Pollution Monitoring in Romania

Raluca Mocanu[1] and Eiliv Steinnes[2]

[1]*Department of Analytical Chemistry, Faculty of Chemistry, "Al. I. Cuza" University, 6600 Iasi, Romania*
[2]*Department of Chemistry, Norwegian University of Science and Technology, 749 Trondheim, Norway*

In most European countries environmental protection and limitation of pollution problems are the subject of numerous research and monitoring programs. In Romania, environmental protection issues connected to intense local pollution sources have been considered of particularly great importance, especially during the last decade.

The most important institutions responsible for environmental protection and monitoring in Romania are the following: The Ministry of Waters, Forests and Environmental Protection (MWFEP), the Environmental Protection Agency, the Sanitary Police, and the Romanian Waters Company. According to the laws of environmental protection every industrial plant is required to have a laboratory for controlling its own emissions. After 1990 numerous non-governmental organisations with an environmental profile were established. Moreover, many universities, independent of their main profile (chemistry, biology, agriculture, engineering etc) have established ecological or environmental modules which contribute scientifically to environmental issues.

Basically the aims of all these institutions are:

- to assess the state of the environment;
- to monitor the evolution of the environment;
- to improve the conditions of the environment.

Within the MWFEP network (Ministry of Waters, Forests and Environment Protection, 1996) the air quality is followed up by:

- *Background pollution monitoring* comprising of 4 stations located at heights exceeding 1000 m, where the average daily concentrations of ozone, sulphur dioxide and nitrogen dioxide in air as well as physical and chemical parameters of precipitation are measured.
- *Impact pollution monitoring* carried out at the national level. This includes a supervising network for the environmental quality and for the collection of emission data as well as processing and reporting the data. A monitoring network covering all the country's counties performs routine measurements of sulphur dioxide, nitrogen dioxide, ammonia, and suspended matter.

Figure 1 shows the distribution of specific polluting substances in the highly polluted zones of the country: H_2S, CS_2, phenol, formaldehyde, HCl, Cl_2, heavy metals, etc. The environmental quality is strongly affected by a wide range of activities carried out in the country, and in addition by transboundary pollution. Still, according to CORINAIR data (European Atmospheric Emissions Information by Country. European Topic Center on Air Emissions, 1990), the quantities of polluting substances produced in Romania in 1989 (a reference year for the country) was smaller per capita than the average reported for other European countries. In recent years, due to a general decrease in industrial activity, and partly to certain measures for environment protection, the present emissions of air pollutants are generally significantly below the 1989 level.

In addition to the above mentioned pollutants the following are also the subject of great concern (Ministry of Waters, Forests and Environment Protection, 1999):

- *Suspended matter*: both annual mean values and maximum concentrations per 24 hours exceed the MPC (maximum permitted concentration) values in some highly industrialised towns in Romania.

I. Barnes (ed.), Global Atmospheric Change and its Impact on Regional Air Quality, 43–48.

- *Specific polluting substances* (Cl_2, H_2S, phenols, heavy metals etc.) also exceed the permitted norms in several industrial towns. The magnitude of this problem may be illustrated with the following example: in 1999, S.C. "Romplumb" Baia Mare released to the atmosphere about 36t of lead as suspended matter - this is an enormous quantity.
- *Stratospheric ozone*: Systematic measurements of ozone concentrations (from the Earth´s surface up to the upper stratosphere) have been made in Romania since 1980. The data library allowed determination of the long-term variation of ozone with a high degree of accuracy. According to these data the total global decrease of ozone was 9.14%. This is in agreement with the decrease for medium latitudes of Europe estimated at 10-15 % for the same period.
- *Emission of "Greenhouse" gases*: After 1989 the „greenhouse" gas emissions decreased mainly due to the decrease of the economic activity, but also due to the application of different programs for emission reductions.
- *Acid and alkaline precipitation.* The acid precipitation is mainly due to SO_2 and NO_x.

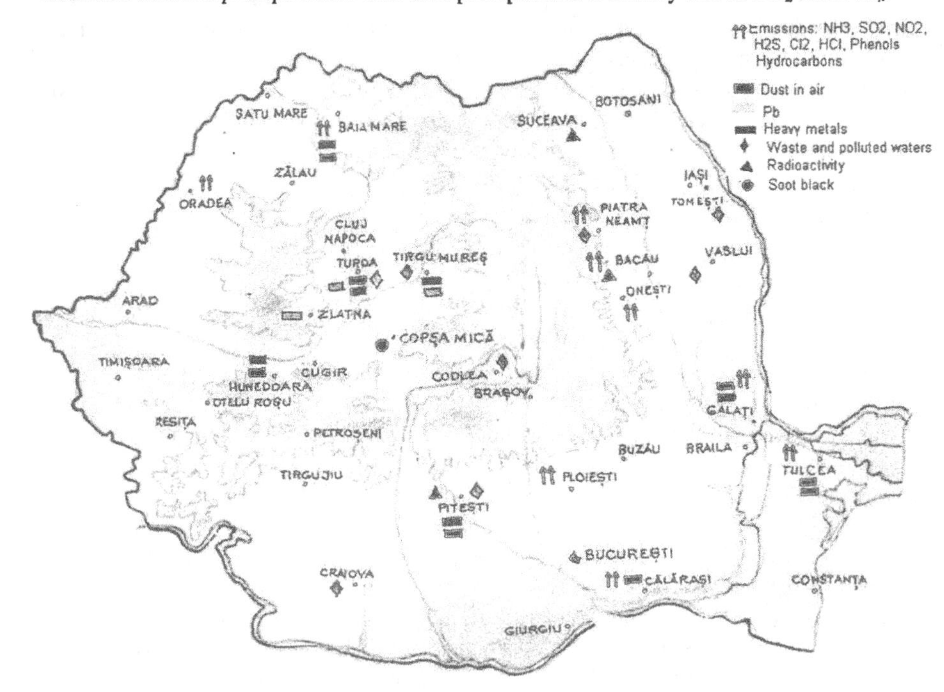

Figure 1: Pollutant distributions in the highly polluted zones of Romania.

The contribution of "Al. I. Cuza" University to the national efforts for environmental monitoring and protection was initiated a few years ago. During this time several collaboration projects were established between the research groups of the authors, and recently also with a third group from Dubna, Russia (Dr. Marina V. Frontasyeva). The Norwegian as well as the Russian group has particular experience in monitoring of atmospheric heavy metals (Steinnes, 1984; Steinnes, 1995; Frontasyeva *et al.*, 1994). Air pollution control is a key factor in a comprehensive environmental policy. In order to assess properly the atmospheric deposition of air pollutants only well proven and standardised methods should be used. Moreover these methods should be generally applicable.

The conventional methods of investigating the atmospheric deposition of heavy metals are based on precipitation samplers. For financial reasons this approach to monitoring is restricted to a limited number of stations. Moreover these measurements are often associated with difficulties due to low concentrations, contamination problems, and the inhomogeneous nature of precipitation. All this considered, the use of naturally growing mosses for monitoring of heavy metal deposition has been developed as a complementary method, and this approach has had considerable success.

The moss technique to study the regional distribution of heavy metal deposition was first proposed by the Swedish scientists Å. Rühling and G. Tyler (Ruhling and Tyler, 1968; Ruhling and Tyler, 1971) and is now used on a regular basis in many European countries. According to a program comprising more than 20 nations the moss method is now used every five years to monitor heavy metal deposition in Europe, and the UN program ICP Vegetation is co-ordinating this work. This approach has been shown to be simple, comprehensive and representative, and at the same time possible to carry out at a moderate cost.

Moss monitoring: Why?

Mosses have a great potential as biomonitors of air pollution. They can be used because they have a great capacity to sorb and retain many heavy metals from precipitation and dry deposition. Moreover, unlike higher plants the mosses have no roots or cuticle layer (Steinnes *et al.*, 1994). Hence sorption processes occur over the entire surface.

Moss surveys have several advantages over conventional deposition monitoring based on precipitation sampling:

- The moss sampling is simple, allowing a large number of sites to be included in the survey.
- The chemical analyses are much easier to perform due to the higher concentrations in moss, and the contamination problems often inherent in precipitation sampling are avoided.
- Moss analysis normally does not allow the estimation of absolute metal deposition rates in the way precipitation analysis does. A quantitative approach however is to calibrate moss data versus atmospheric deposition values from analysed precipitation samples by using linear regression (Berg and Steinnes, 1997). The calculated regression equation can be used to transform moss concentration data to absolute deposition rates.
- On the basis of such regression equations, the relative uptake efficiencies of elements in moss (E_x %) may be estimated for those elements exhibiting statistically significant relationships between moss concentrations and wet deposition. In laboratory studies it has been indicated that the efficiency factor of Pb is very close to 100%. Based on this assumption E_x (%) is calculated by the equation:

$$E_x\% = (K_x/K_{Pb}) \times 100$$

where: K_x = slope of the regression line of element x in moss versus atmospheric deposition

K_{Pb}= slope of the regression line of Pb in moss versus atmospheric deposition.

This method is independent of possible baseline values for the studied elements.

Atmospheric heavy metal deposition studies in Romania by moss bio-monitoring

The feather moss species *Hylocomium Splendens* and *Pleurozium Schreberi* were used in the large–scale bio-monitoring studies carried out in the northern half of Europe (Ruhling *et al.*, 1987; Ross, 1990). When moving this activity farther south these mosses may have to be replaced by other species of moss or lichen. In order to do so it is necessary to calibrate the new species against the present ones with respect to collection efficiency of important trace elements. It is also advantageous to carry out calibrations against bulk deposition. A new biomonitor species could be introduced for local or regional studies in order to be able to estimate absolute deposition rates provided that such calibration is done.

The first results concerning heavy metal pollution in Romania obtained by using the moss monitoring method was published in 1994 (Nordic Council of Ministers, 1994) and was restricted to part of Transilvania. A second survey based on sampling in 1995-1996 covered the central and northern part of the country (Lucaciu *et al.*, 1999). According to that study the most polluted area with respect to As, Cd, Cr, and Fe is the Romanian Carpathians and with Pb the north-western part of Transilvania.

Until recently no attempt had been made to characterise the industrial and local sources of pollution in the province of Moldavia using the moss technique. That would require the use of new moss species since the commonly used *Hylocomium Splendens* and *Pleurozium Schreberi* are not present in most of the region. Two alternative moss species, *Hypnum Cupressiforme* and *Brachytecium Salebrosum* more frequently found in this part of Romania were selected (Cucu-Man *et al.*, 2000). The latter species has not been described before in the literature as biomonitor. *Hypnum Cupressiforme* and *Brachytecium Salebrosum* were selected taking into account that they are widely distributed at the given climatic conditions, easily distinguished from other mosses and their structure is similar to Hylocomium Splendens.

Moss samples from 23 sites in Moldavia were analysed by ENAA (epithermal neutron activation) and FAAS (flame atomic absorption spectrometry). The obtained data showed a higher concentration for all elements in *Hypnum Cupressiforme* than in *Brachytecium Salebrosum*. The highest correlation between samples of the two species collected at the same site was found for Cu as can be observed in Figure 2.

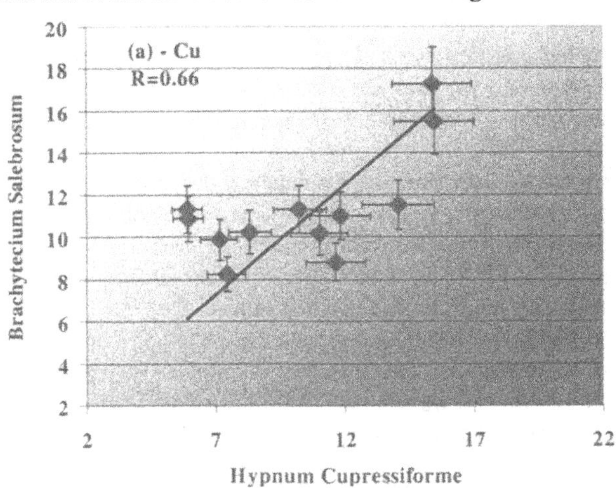

Figure 2: Concentration of Cu in the *Hypnum Cupressiforme* and *Brachytecium Salebrosum* investigated moss species obtained using the FAAS technique. The given concentrations are in μg retained element per g moss.

The high correlation between crustal elements (Sc, Al, Fe, Na, La, Sm, Th) indicates possible contamination of some moss samples with soil particles (Table 1).

Table 1: Correlation coefficient between crustal elements.

	Na	Al	Sc	Cr	V	Fe	Co	La	Sm	Th
Na	1.00									
Al	0.94	1.00								
Sc	0.93	0.96	1.00							
Cr	0.89	0.91	0.96	1.00						
V	0.91	0.97	0.95	0.90	1.00					
Fe	0.88	0.92	0.96	0.93	0.89	1.00				
Co	0.80	0.85	0.93	0.89	0.86	0.88	1.00			
La	0.92	0.96	0.94	0.93	0.94	0.90	0.84	1.00		
Sm	0.92	0.94	0.97	0.91	0.95	0.94	0.88	0.93	1.00	
Th	0.93	0.95	0.98	0.98	0.93	0.96	0.88	0.96	0.96	1.00

Higher Pb concentrations than elsewhere were found in the densely populated and industrialised parts of Moldavia (Iasi, Botosani, Vaslui). It is likely to be a relationship between these higher values and high automobile traffic, to a large extent with old vehicles.

The distribution of Cu identifies a number of local sources (Iasi, Vaslui counties). This can be explained by the usage of $CuSO_4$ pesticides in vineyards; these counties are big wine producers.

The highest levels of Zn were found in Iasi the region. It is probably due to specific pesticides used in orchards. A high and relatively constant Zn level of 20-40 $\mu g\ g^{-1}$ is generally present in all regions. This may be explained by contribution from the tree canopy, considering the fact that Zn is an essential element in vascular plants and *Hypnum Cupressiforme* and *Brachytecium Salebrosum* are both epiphytic mosses growing on trees (see Figure 3).

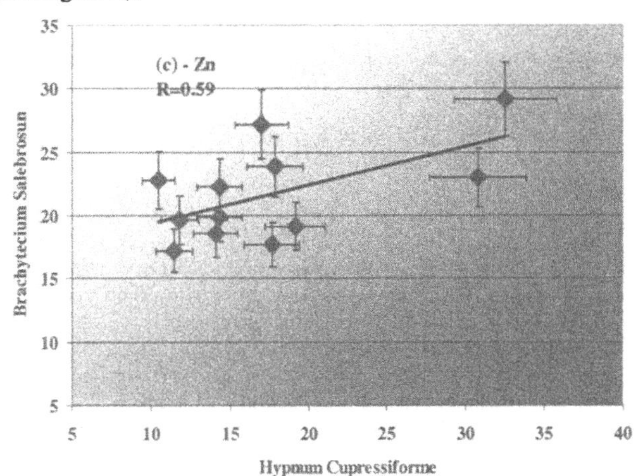

Figure 3: Concentration of Zn in the *Hypnum Cupressiforme* and *Brachytecium Salebrosum* investigated moss species by using FAAS technique. The given concentrations are in μg retained element per g moss.

The positive correlations observed between *Hypnum Cupressiforme* and *Brachytecium Salebrosum* encourage us to use them as biomonitors. An intercalibration with *Hylocomium Splendens* (which was already transplanted in the area of interest) will offer more useful information about this opportunity.

Conclusions

We consider that moss bio-monitoring may play an important role in future studies of trace element (heavy metal) deposition in the environment. This requires a generally accepted procedure to be used, including the selection of the sampling sites, and intercalibration between different moss species will be necessary in many cases. Samples can be stored for many years for further studies. The results from experience developed in Norway and elsewhere for a long period of time proved the feasibility, reliability, and utility of the method. Its extension almost all over Europe has facilitated the establishment of a European Network. The transfer of the organising of the European Heavy Metals in Mosses Project from the Nordic Council of Ministers to the United Nations Economic Commission for Europe International Co-operative Programme on the Effects of Air Pollution on Natural Vegetation and Crops (ICP Vegetation) offers some promise for extension of the moss bio-monitoring technique also to other continents.

References

Berg T. and E. Steinnes; Use of mosses (Hylocomium Splendens and Pleurozium Schreberi) as biomonitors of heavy metal deposition: from relative to absolute deposition values, *Environ.Poll.* **98**, (1997) 61-71.

Cucu-Man S., I. Bejan, R. Mocanu and E. Steinnes; The use of mosses to study heavy metal atmospheric deposition in Moldavia (Romania), 1-st International Conference on Analytical Chemistry, Brasov (2000), Romania.

European Atmospheric Emissions Information by Country; European Topic Center on Air Emissions, *Corinair* **90**, 1-11.

Frontasyeva, M.V., V.M. Nazarov, E. Steinnes; Moss as monitor of heavy metal deposition: comparison of different multi-element analytical techniques, *J. Radioanal. Nucl. Chem.*, **181**, (1994) 363-371.

Lucaciu A., M.V. Frontasyeva, E. Steinnes, Ye.N. Cheremisina, C. Oprea, T.B. Progulova, L. Stanciu, L. Timofte; Atmospheric deposition of heavy metals in Romania studied by the moss biomonitoring technique employing nuclear and related analytical techniques and GIS technology, *J. Radioanal. Nucl. Chem.*, **240(2)**, (1999) 457-458.

Ministry of Waters, Forests and Environment Protection; Environment Protection Strategy, Report - 1996, Bucuresti (1996), Romania.

Ministry of Waters, Forests and Environment Protection; Environment Protection Strategy (1999), Bucuresti, Romania

Report NORD Nordic Council of Ministers, Copenhagen, 1994.

Ross H.B.; On the use of mosses (Hylocomium Splendens and Pleurosium Schreberi) for estimating atmospheric trace metal deposition, *Water, Air and Soil Poll.* **50**, (1990) 63-76.

Ruhling A. and G. Tyler; An ecological approach to the lead problem, *Bot. Notiser* **12**, (1968) 321-342.

Ruhling A. and G. Tyler; Regional differencies in the heavy metal deposition over Scandinavia, *J. Appl. Ecol.* **8**, (1971) 497-507.

Ruhling E., L. Rasmussen, K. Pilegaard, A. Makinen and E. Steinnes; Survey of atmospheric heavy metal deposition in Nordic Countries in 1985. Report NORD **21**, Nordic Council of Ministers, Copenhagen (1987) 44.

Steinnes E.; Monitoring of trace element distribution by means of mosses, *Fresenius J. Anal. Chem.* **317**, (1994) 350-356.

Steinnes E., J.E. Hanssen, J.P. Rambaek and N.B. Vogt; Atmospheric deposition of trace elements in Norway: temporal and spatial trends studied by moss analysis, *Water, Air and Soil Poll.* **74**, (1994) 121-140.

Steinnes E.; A critical evaluation of the use of naturally growing moss to monitor the heavy metal deposition, *Sci. Total Environ.*, **160/161**, (1995) 243-249.

Modelling

Modelling of Ozone and Secondary Organic Aerosol across Europe: Results from the EMEP models

David Simpson and Yvonne Andersson-Sköld
EMEP MSC-W, The Norwegian Meteorological Institute
P.B. 43, N-0313 Oslo, Norway (david.simpson@dnmi.no)

Introduction

For over two decades, models from EMEP MSC-W[1] have been used to aid governments in their design of control strategies to reduce air pollution in Europe. Indeed, EMEP long-range transport models (e.g. Eliassen and Saltbones, 1983) played a key role in even the first UN-ECE[2] Protocol on sulphur, signed in Helsinki in 1985. Modelling demonstrated that sulphur could be transported many 100s of km from the source, and was used to quantify the contribution of each country to acid deposition across Europe. The sulphur model developed in this early work has been steadily extended and improved to give today's EMEP models for acidification, ozone and most recently aerosols.

In recent years the EMEP models have been used to develop complex and ambitious international agreements, notably the UN-ECE "Gothenburg Protocol" and the European Union's National Emissions Ceiling (NEC) Directive. It is anticipated that the next round of negotiations for both of these organisations will also make heavy use of the EMEP models; with much more focus on particulate matter and hence both inorganic and organic aerosols. Given the importance of the EMEP models in this context, it is important that these models are thoroughly evaluated, and that they reflect the scientific consensus on air pollution physics and chemistry. This paper summarises some work on the capabilities of the latest photo-oxidant models, and presents some findings of the most recent explorations concerning secondary organic aerosol formation.

The EMEP Photo-oxidant models

The Gothenburg Protocol and NEC Directive made use of ozone results from the MSC-W Lagrangian photo-oxidant model, which has been described in detail by Simpson (1992, 1993, 1995) and Simpson et al. (1997). This model has been thoroughly evaluated against measurements, not only of ozone but of species which are more difficult to predict, such as HCHO and PAN (Simpson, 1992, 1993, Malik et al., 1996, Solberg et al., 1995, MSC-W, 1998). As an illustration, Figure 1 shows the cumulative frequency distribution of modelled versus observed noontime ozone obtained at two very different locations, Birkenes in southern Norway, and Schauinsland in southern Germany. These data are based upon 5-summers (April-September) of simulations, and suggest that the Lagrangian model does a good job of reproducing observed values on a statistical basis. Extensive time-series comparisons can be found in Malik et al. (1996) and MSC-W (1998).

---

[1] EMEP: The Cooperative Programme for Monitoring and Evaluation of the Long-Range Transmission of Air Pollutants in Europe; MSC-W: Meteorological Synthesizing Centre –West. See www.emep.int.
[2] United Nations Economic Comission for Europe

I. Barnes (ed.), Global Atmospheric Change and its Impact on Regional Air Quality, 51–56.

In a comparison of the EMEP model with 3 Eulerian models (LOTOS, EUMAC and REM3), Hass et al. (1997) concluded that the models were comparable in terms of performance against observations. Similarly the chemical mechanism of this model has been shown to be comparable to other recent mechanisms (Derwent, 1993, Kuhn et al., 1998 and Andersson-Sköld and Simpson, 1999). In the latter study, the EMEP model was shown to give similar values for ozone, HCHO and even radicals to the much more extensive (1800 reaction) scheme of Andersson-Sköld (1995).

Eulerian models are now replacing Lagrangian models at MSC-W. Their superior physical formulation allows improved spatial resolution and the models can cope with pollutants which are transported both within and above the atmospheric boundary layer. The first such model at MSC-W was the sulphur model of Berge and Jakobsen (1998). This has been extended to nitrogen pollutants (Jonson et al., 1998) and ozone (MSC-W, 1998, Jonson et al. 1999). Efforts in EMEP now focus on the development of a new generation of Eulerian model that unifies all these modelling approaches into one flexible system (Simpson et al. 2001).

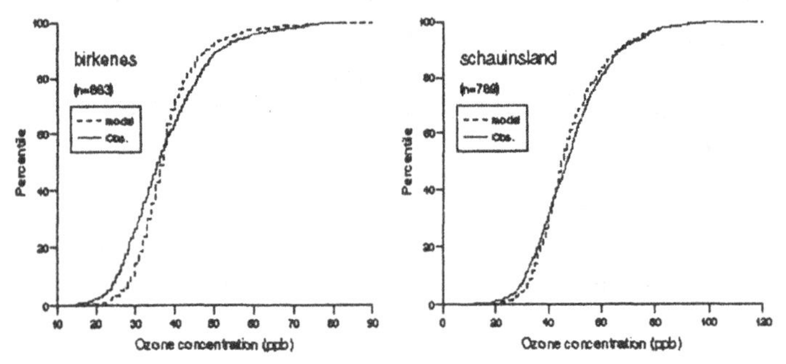

Figure 1: Frequency distributions of 12 GMT ozone concentrations from the EMEP Lagrangian model and observations, at a site in Norway (Birkenes) and Germany (Schauinsland). From EMEP Report 2/98 (MSC-W, 1998)

Future scenarios with the Photo-oxidant Models

The EMEP models have been increasingly used to predict changes in ozone concentrations (and related exposure indices) for future years. Reis et al. (2001) looked in detail at road-traffic emissions in the year 2010. Jonson et al. (2001) used the Eulerian model in conjunction with a global model to investigate the effect of changing free tropospheric ozone on European ozone levels, again for the year 2010. This study found that over Europe, an increase of free tropospheric ozone of the order of 10 ppb could be expected in summer, resulting in an increase of boundary layer ozone of about 2 ppb over present levels. However, despite this increase in the background ozone, exceedance of ozone threshold levels is expected to be reduced significantly in Europe as a result of the foreseen reductions in precursor emissions from this region. Finally, Tuovinen et al. (2001) used the Lagrangian model to compare the effects of increasing background ozone against the effects of changing climate (temperature change) for scenarios for the year 2050. This study showed that the increasing background ozone values had a much greater effect on ozone concentrations than any effects caused by temperature increases.

Secondary Organic Aerosol Formation

Our understanding of secondary organic aerosol is in a state of great flux at present. Important new results appear at regular intervals. The work of Pankow (1994), Odum et al (1996) and others provides both a theoretical and experimental basis for describing SOA yields in smog chambers, based upon gas-particle partitioning theory. This theory provides a quantitative explanation for the wide range of aerosol yields previously observed by different experiments, although without giving any details of the actual compounds involved in the aerosol formation. More recently, as new analytical techniques have become available, chemists have succeeded in identifying a significant proportion of the actual compounds in both the gas and aerosol phases of smog-chamber reactants (e.g. Kamens and Jaoui, 2001). Kinetic models can now be constructed which show rather good success in predicting the extent and composition of aerosol formation from such important precursor compounds as α-pinene under a wide range of conditions (Kamens et al., 1999, Kamens and Jaoui, 2001, Andersson-Sköld and Simpson, 2001).

However, these successes apply mainly to smog-chambers (SCs), where conditions are usually very far removed from those of the ambient atmosphere. Differences include temperature and relative humidity in SCs, which are often far from ambient levels. More importantly, SCs usually involve the reactions of very high concentrations of a single substance (e.g. several 10s to 100s of ppb of α -pinene, although sometimes with other HCs added as radical scavengers) with an inorganic seed aerosol. A consequence of this is that the reaction products in SCs condense onto an aerosol phase, which is derived from this reactant. In the ambient atmosphere it is very likely that the SOA will condense onto pre-existing aerosol. The latter will be a very complex mixture of compounds which are derived from primary emissions and SOA condensation, and where aerosol-phase chemical reactions among these species may well have generated complex non-volatile compounds.

It is also notable that when the products of terpene degradation have been identified in ambient aerosol samples, the concentrations found have of the order of a few ng/m^3 – a factor of 1000 less than the known amount of organic compounds in the aerosol (e.g. Kavouras et al., 1999, Yu et al., 1999). Clearly many other compounds are present in the atmospheric aerosol, but we do not know if these compounds are mainly anthropogenic or biogenic in origin, primary or secondary, or indeed some reacted mixture of all of these.

Further, there are serious problems associated with the use of simple partitioning coefficients (K-values), as discussed in Andersson-Sköld and Simpson (2001). Even in simple conditions, such K values should be very sensitive to temperature because of the controlling role of vapour pressure in establishing K. Unfortunately, since we do not know the species which are actually contributing to the aerosol we cannot specify this temperature dependence with any accuracy. (It is difficult even when we do know the species. Activity coefficients are also an important added complication for these calculations that we can hardly begin to deal with for atmospheric conditions.) K-values are also only appropriate as long as the aerosol-phase compound does not react further. The success of this approach in smog-chambers does not ensure that it is suitable for the more complex atmospheric aerosol.

Results for SOA formation

From the above, it is obvious that attempting to use models to predict the formation and composition of SOA in ambient atmospheres is a very uncertain activity. Nevertheless, it is instructive and important to apply our best-available theories in order to begin the iterative process of matching model-results and observations, a process, which will ultimately lead to better models and which should generate further ideas for the type of measurements which can be used to decide between competing theories.

Andersson-Sköld and Simpson (2001) conducted a two-part study in order to make an estimate of the organic aerosol formed over Nordic Europe. Firstly, the kinetic-aerosol mechanism, which Kamens et al (1999) developed for the nighttime chemistry of α-pinene with ozone was extended, to include daytime reactions and the dimer formation between pinic acid and nor-pinic acid, which Hoffman et al (1998) postulated. The new mechanism, denoted Kam-2, was tested against a series of smog-chamber data, which covered α-pinene concentrations from 20-900 ppb and NOx concentrations of between zero and 240 ppb, and shown to work well over this wide range. The inclusion of dimer formation was shown to be essential in order to match the experimental results, lending more credibility to the postulate that either this particular dimer is formed, or that some corresponding mechanism for generating low-volatility products is operating in the SC.

In the second part of this study, three different schemes for predicting SOA formation were tested in the EMEP Lagrangian oxidant model. The three schemes tested were (i) The α-K model which uses the parameters for the gas-particle partitioning largely derived from Griffin et al. (1999a,b); (ii) An "α-K-T" model in which the partitioning coefficients K were modified to reflect the effects of temperature. In this model, lower temperature leads to greater condensation rates, an important factor when modelling aerosol formation over the Nordic area; (iii) The Kam-2 model. This model also makes use of partitioning coefficients, with a temperature variation suggested by Kamens et al. (1999). In fact this temperature variation was also that used for the "α-K-T" model above.

These three models have different advantages and disadvantages, as discussed further in Andersson-Sköld and Simpson (2001). However, despite their very different origins, models (ii) and (iii) actually gave rather similar predictions for SOA – they differed by "only" a factor of two and correlated very well with each other. Model (i) showed a much poorer correlation with model (iii), because of its lack of temperature control on SOA formation. Whichever method was used, this study suggested that biogenic sources contributed much more to SOA formation than anthropogenic sources, at least in this Nordic area (Figure 2).

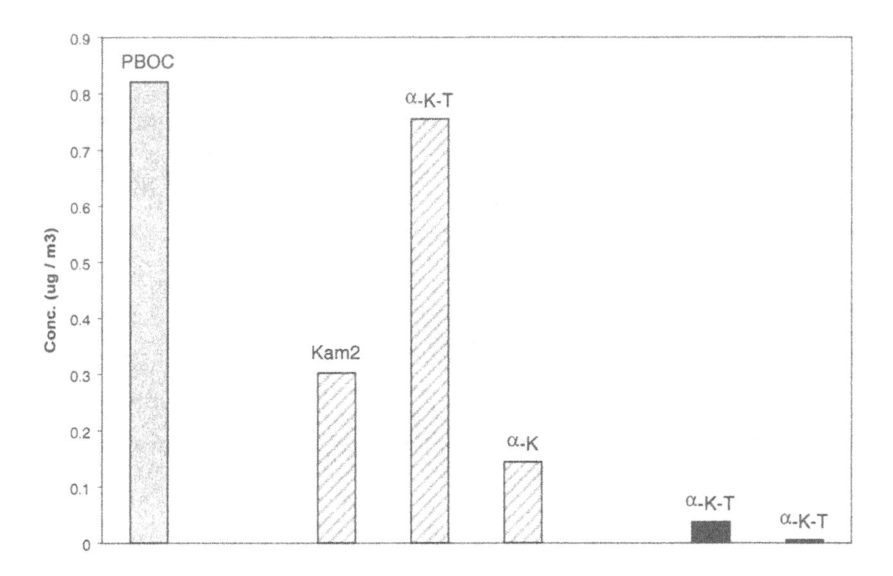

Figure 2: Calculated concentrations of organic carbon at Aspvreten, Sweden, for July 1994 (See Andersson-Sköld and Simpson, 2001). Dark shading: PBBOC (primary and background, assumed non-reactive); Striped shading: biogenic SOA from 3 methods (Kam-2, α-K-T, α-K, see text); Black shading: anthropogenic SOA from two methods.

This study was limited to some extent by the use of the Lagrangian EMEP model with its coarse vertical resolution. Work is currently underway to perform a similar analysis with the new Eulerian model. Still, the uncertainties highlighted above and in the work of Andersson-Sköld and Simpson strongly suggest that model work alone is too speculative for much progress to be made. Our understanding will only improve substantially when we have a better idea of the composition of the atmospheric aerosol.

References

Andersson-Sköld, Y. and Simpson, D., Comparison of the chemical schemes of the EMEP MSC-W and IVL photochemical trajectory models, *Atmos.Environ.*, **33** (1999) 1111-1129.

Andersson-Sköld, Y. Updating the chemical scheme for the IVL photochemical trajectory model, Swedish Environmental Research Institute (IVL), Report B 1151, Gothenburg, Sweden (1995).

Andersson-Sköld, Y. and Simpson; Secondary organic aerosol formation in northern Europe. A model study. *J. Geophys. Res.*, **106** (2001) 7357-7374.

Berge, E. and Jakobsen, H.A., A regional scale multi-layer model for the calculation of long-term transport and deposition of air pollution in Europe, *Tellus* B **50**, (1998) 205-223.

Eliassen, A. and Saltbones, J. , Modelling of long-range transport of sulphur over Europe: a two year model run and some experiments, *Atmos. Environ.*, **17** (1983) 1457-1473.

Griffin, R., Cocker, D., Flagan, R., and Seinfeld, J., Organic aerosol formation from the oxidation of biogenic hydrocarbons, *J. Geophys. Res.*, **104** (1999a) 3555-3568

Griffin, R., Cocker, D., Seinfeld, J., and Dabdub, D, Estimate of global organic aerosol from oxidation of biogenic hydrocarbons, *Geophys. Res. Lett.*, **26** (1999b) 2721-2724

Hass, H., Builtjes, P.J.H., Simpson, D., and Stern, R., Comparison of model results obtained with several European regional air quality models, *Atmos. Environ.*, **31** (1997) 3259-3279.

Hoffmann, T: et al., Molecular composition of organic aerosols formed in the alpha-pinene

Jonson, J.E., Bartnicki, J., Olendrzynski, K., Jakabson, H.A., and Berge, E., EMEP Eulerian model for atmospheric transport and deposition of nitrogen species over Europe. *Env. Poll.*, **102** (1998), 289-298.

Jonson, J.E., Sundet, J.K., and Tarrason, L., Model calculations of present and future levels of ozone and ozone precursors with a global and a regional model, *Atmos. Environ.*, **34** (2001), 525-537.

Kamens, R.M., Jang, C., Chien, C., and Leach, K., Aerosol formation from the reaction of α-pinene and ozone using a gas-phase kinetics-aerosol partitioning model, *Environ. Sci. Tech.*, **33** (1999) 1430-1438.

Kamens, R.M. and Jaoui, M., Modeling aerosol formation from α-pinene + NOx in the presence of natural sunlight using gas-phase kinetics and gas-particle partitioning theory, *Environ. Sci. Technol.*, **35** (2001) 1394-1405.

Kuhn, M.., et al.., Intercomparison of the gas-phase chemistry in several chemistry and transport models, *Atmos. Environ.*, **32** (1998) 693-709.

Malik, S., Simpson, D., Hjellbrekke, A.-G. , and ApSimon, H.., Photochemical model calculations over Europe for summer 1990. Model results and comparison with observations, Norwegian Meteorological Institute, Oslo, Norway, EMEP MSC-W Report 2/96 (1996).

MSC-W, Transboundary photooxidant air pollution in Europe. Calculations of tropospheric ozone and comparison with observations, Norwegian Meteorological Institute, Oslo, Norway, EMEP MSC-W Report 2/98 (1998).

Odum, J.R., Hoffmann, T., Bowman, F., Collins, D., Flagan, R., and Seinfeld, J., Gas/particle partitioning and secondary organic aerosol yields, *Environ. Sci. Tech.*, **30** (1996) 2580-2585.

Pankow, J. An absorption model of the gas/aerosol partitioning involved in the formation of secondary organic aerosol, *Atmos. Environ.*, **28** (1994) 189-193.

Reis, S., Simpson, D., Friedrich, R., Jonson, J.E., Unger, S., and Obermeier, A., Road traffic emissions - predictions of future contributions to regional ozone levels in Europe, *Atmos. Environ.*, **34** (2000) 4701-4710.

Simpson, D., Long-period modelling of photochemical oxidants in Europe. Model calculations for July 1985, *Armos. Env.*, **26** (1992) 1609-1634.

Simpson, D., Long-period modelling of photochemical oxidants in Europe for two extended periods: 1985 and 1989. Model results and comparisons with. *Atmos. Env.*, **27** (1993) 921-943.

Simpson, D. ,Biogenic emissions in Europe 2: Implications for ozone control strategies, J. Geophys. Res., 100, (1995) 22891-22906.

Simpson, D., Jonson, J.E., and Unger, S., Progress with the "Unified" EMEP model, in EMEP MSC-W Report 1/2001, Norwegian Meteorological, Institute, Oslo, Norway, (2001) 33-38.

Simpson, D. , Olendrzynski, K. , Semb, A. , Storen, E. , and Unger, S. , 1997, Photochemical oxidant modelling in Europe: multi-annual modelling and source-receptor relationships, Norwegian Meteorological Institute, EMEP MSC-W Report 3/97.

Solberg, S. , Dye, C. , Schmidbauer, N. , and Simpson, D. , 1995, Evaluation of the VOC measurement programme within EMEP, Kjeller, Norway, Norwegian Institute for Air Research (NILU) Report 5/95.

Tuovinen, J.-P., Simpson, D., Mayerhof, P., Lindfors, V: and Laurila, T., Surface ozone exposures in Northern Europe in changing environmental conditions, Presented at "Changing Atmosphere: 8th European Symposium on the Photo-Chemical Behaviour of Air Pollutants, Torino Italy, 17-20th Sept. 2001.

Yu, J. D.,Griffin, R, Cocker, D., Flagan, R. and Seinfeld, J., Observations of gaseous and particulate products of monoterpene oxidation in forest atmospheres, *Geophys. Res. Lett.*, **26** (1999) 1145-1148.

Emissions

Emission of Atmospheric Pollutants from Aero Engines

Peter Wiesen, J. Kleffmann and R. Kurtenbach

Bergische Universität Wuppertal, Phys. Chemie / FB 9, D-42097 Wuppertal, Germany

Introduction

Air traffic experienced rapid expansion during the last 40 years and became an integral and vital part of modern society. Since 1960 passenger traffic increased at nearly 9% per year (expressed as passenger kilometres). Recent scenarios have estimated a continuous increase during the next 15 years of 5% per year whereas an average annual increase of air traffic fuel consumption of only 3% per year is estimated caused by increased efficiency of aircraft engines. Aviation currently consumes 2-3% of the fossil fuels used worldwide, of which the majority (~80%) is used by civil aviation and by military aircraft (~20%). Although aviation is still a relatively small user of fossil fuel at this time, aircraft are the main anthropogenic emitters of gases and particles in the upper troposphere and lower stratosphere (UT/LS), where they have an impact on the atmospheric composition and where the lifetime of the pollutants is much higher than in the lower regions of the atmosphere.

The concerns about emissions from aircraft were originally focused on their contribution to local air quality in the vicinity of airports and led to the introduction of legislation in the U.S.A. and subsequently to international standards for control of fuel venting and emissions of certain species below 915 m.

Triggered by the development of the first civil supersonic aircraft in the 1960s, subsequent research was undertaken in the 1970s to quantify the impact of NO_x emissions from high-flying aircraft on the stratospheric ozone layer (CIAP, COMESA, COVOS).

Later on the emphasis shifted to addressing the role and contribution of aviation emissions to the global issues of ozone loss and potential climate change through various research activities such as the European projects e.g. AERONOX, POLINAT, MOZAIC, AEROCHEM, AEROTRACE, AEROCONTRAIL, AEROPROFILE, AEROJET, AEROJET II, AERONET, PARTEMIS, AERONET II[1], which led to a significant improvement of the current understanding of the impact of aircraft emissions on these phenomena.

Air traffic, engine combustion and aircraft emissions

Aero engines emit a large variety of gaseous pollutants and particles in the UT/LS (altitude 10 - 12 km), leading to increasing levels of CO_2, CO, NO_x, ozone, water vapour and non-methane hydrocarbons and particles. Engine emissions are studied either by:

- engine tests in a ground level test bed,
- engine tests in an altitude simulation test facilities and
- in-flight measurements (e.g. plane chasing).

Currently, emissions are only regulated for some species, namely CO, total hydrocarbons (HC), NO_x and smoke, over a so-called LTO (landing/take-off) cycle, see Figure 1, in which the emission values are quoted as an emission index with units of grams per kilogram of fuel burnt. Although the LTO part of a flight mission is a very small proportion, currently cruise emissions are not controlled and not included in engine certification testing. It is not possible to achieve flight operations conditions on the test facilities used for certification testing. Currently, controlled measurements of emissions from a very limited number of engines operating at flight conditions have been made in altitude simulation test facilities and from in-flight measurements. Therefore, engine manufacturers have developed methods for

[1] Further information on these projects can be obtained from http://www.cordis.lu/en/home.html

I. Barnes (ed.), Global Atmospheric Change and its Impact on Regional Air Quality, 59–64.

calculation of in-flight emissions from the ground-level certification data using typical combustor inlet parameters, which are unfortunately regarded as commercially sensitive and consequently they are not normally available. In Figure 2 a comparison of all available in situ measured NO_x emission index values with corresponding predicted values is shown.

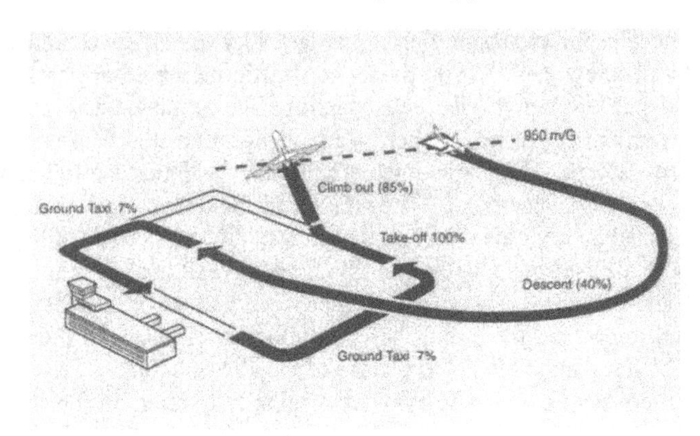

Figure 1: A typical LTO cycle for aero engine certification.

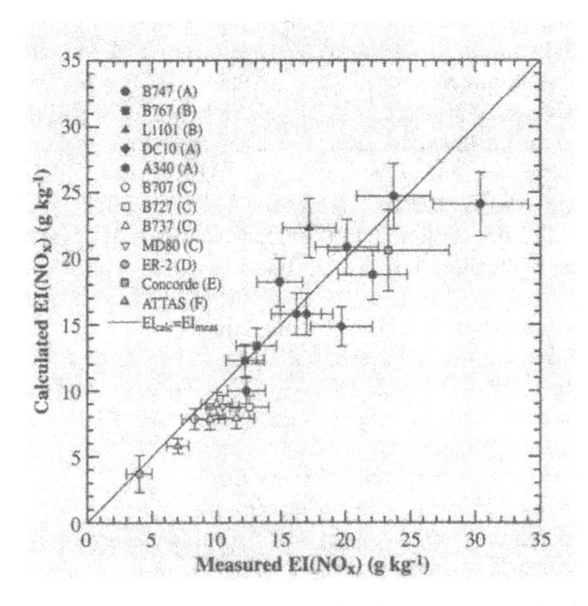

Figure 2: Comparison of all available in situ measured NOx emission index values with corresponding predicted values: (A) Schulte et al. 1997, (B) Schlager et al. 1997, (C) Schulte and Schlager (1996), (D) Fahey et al. 1995a, (E) Fahey et al. 1995b, and (F) Haschberger and Lindermeir (1996) (from Penner et al., 1999).

Historical trends in emissions show that very substantial decreases in HC and CO emissions were achieved due to improvements in engine design and greater fuel efficiency, the last of which is shown in Figure 3. Along with this only a relatively small increase in NO_x emissions has been observed due to improvements in combustor technology, rather than the much larger increase, which would be predicted due to the much higher combustion temperatures and pressures associated with the more fuel efficient engine cycles.

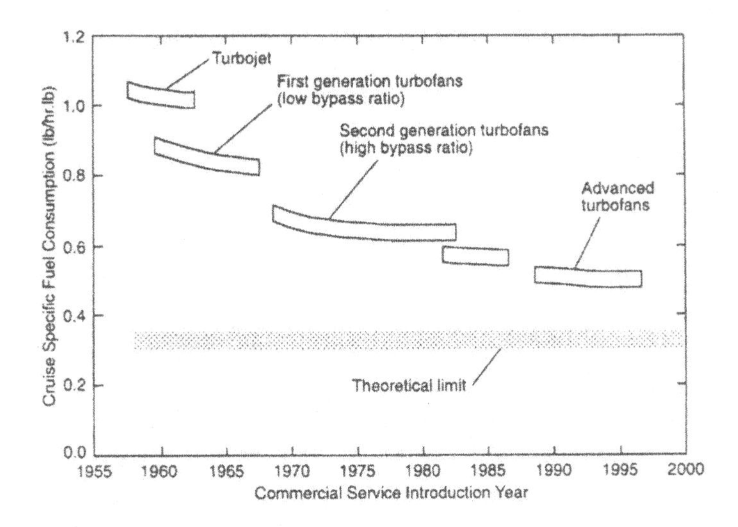

Figure 3: Fuel consumption improvements as engine technology has advanced from straight turbojets to turbofan engines with increasing bypass ration (from Brasseur et al., 1998).

Unregulated emissions

Whereas the formation and emission of nitrogen oxides NO_x (NO and NO_2) has been the object of numerous investigations (e.g. Brasseur et al., 1998; Penner et al., 1999) there is still only limited information available on the emission of other reactive nitrogen species (NO_y) such as HONO, the exact composition of the hydrocarbons emitted and their partially oxidised products (ketones, aldehydes, acids) and particles. A better knowledge of direct NO_y emissions from aircraft and the formation of NO_y species in the aircraft plume is needed because of the potential impact of these species on stratospheric and tropospheric ozone. A detailed knowledge of the emission of non-methane hydrocarbons and partially oxidised hydrocarbons is of particular importance because of the reactivity of these species, which may lead to chemical reactions forming, e.g. organic nitrates and peroxynitrates in the immediate wake of aircraft. A substantially better understanding of particulate emission is required since these species may contribute directly to climate change due to their scattering and absorption properties, their impact on contrail formation and also because of heterogeneous reactions that may occur.

NO_y species measurements

With respect to NO_y species, other than NO and NO_2, most experiments which are currently available dealt with the detection of HONO and HNO_3. From global three-dimensional model studies on NO_y emissions from aircraft it was concluded that aircraft sources appear to have only a minimal impact on lower tropospheric NO_y mixing ratios in the Northern Hemisphere and in much of the Southern Hemisphere (Kasibhatla, 1993). Arnold et al. (1994) studied the emission of HONO and HNO_3 at cruise altitude using an AAMAS instrument, which was flown aboard the DLR research aircraft FALCON. Figure 4 shows the result of an experiment in which the plume of a DC-9 airliner was monitored at a flight altitude of 9.5 km at a distance of 2 km, which corresponds to a plume age of 9 s. The authors observed an increase of the HONO concentration from a background level of about 5 pptV to maximum values of about 520 pptV when the FALCON flew within the plume of the DC-9 airliner.
Experiments carried out in the plume of the Concorde supersonic aircraft (Fahey et al., 1995b)

showed high NO_x/NO_y ratios indicating that HNO_3 was not abundant in the plume in the first minutes to hours after emission. HNO_3 formation is probably controlled by the reaction of OH with NO_2. The observed OH peak in the plume was attributed to the rapid photolysis of HONO which is probably formed in the reaction OH + NO within microseconds of their emission from the engine.
Further experiments, e.g. with different types of combustors or engines such as low-NO_x combustors are necessary in order to set up a reliable database for the emission of NO_y species from aircraft.

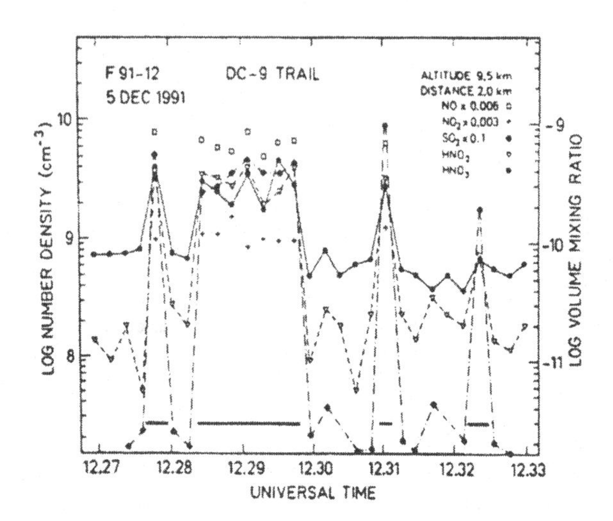

Figure 4: Measured concentration of various trace gases and water vapour in the plume of a DC-9 aircraft (from Arnold et al., 1994). The thick lines indicate the times when the measuring aircraft was within the plume of the DC-9.

Volatile organic compound measurements
The chemical composition of the exhaust from aircraft turbine engines with respect to the hydrocarbons has been investigated in several studies during the last 20 years. In these studies commercial as well as military engines were investigated (see Brasseur et al, 1998 and references therein). From these experiments it was concluded that the emissions of the species investigated are strongly dependent on the operating conditions of the engines. Hydrocarbon emissions decrease with increasing engine power and are generally small for cruise conditions. Although a large number of different species were detected in the exhaust, it was observed that under specific conditions only a small number, 8 - 10 species, in particular unsaturated HC, contribute up to 80% of the total emission of organic species. However, it should be pointed out that different aero engines show significantly different emission characteristics and that a possible impact of jet fuel formulation on the emission of organic species is only poorly understood at present. Therefore, more experiments with different types of combustors or engines and different fuel types are still necessary to set up a reliable database for the emission of these trace species from aero engines.

Particulates
Traditionally, particulates from jet engines have been considered as essentially carbon (soot) particles, with some fuel or fuel breakdown products absorbed onto them. For most engines currently in operation certification data, i.e. smoke number data, for the LTO cycle are available. However, these data provide an indication of the mass of the particle emissions, but

not of mean particle size or particle distribution. More recently, particle size measurements at the engine exhaust became available, which show that aero engines emit ultra-fine particles (diameter: <30 nm). It should be pointed out that while there are prediction equations for calculation of gaseous emissions at altitude conditions from sea-level data available, such relationships have not yet been established for particulates. However, more recently a unified model for ultrafine aircraft particle emissions has been developed (K rcher et al., 2000).
For atmospheric issues, sulphur compounds, arising from the fuel sulphur, are also important as precursors of atmospheric aerosols and contrails. Figure 5 shows as an example how the fuel sulphur content influences the emission of particulates from jet engines.

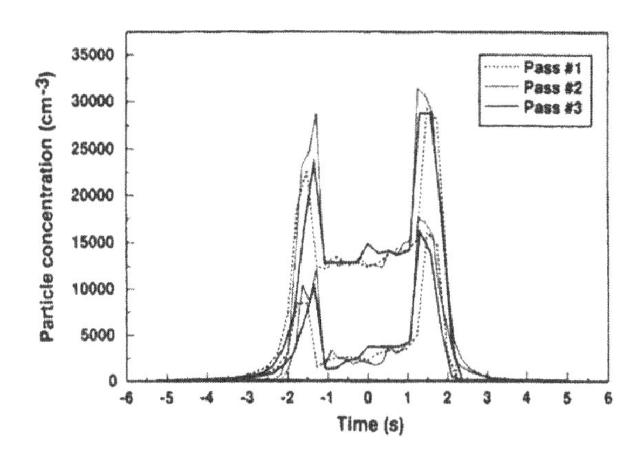

Figure 5: Aerosol number concentration vs. time for particles larger than 7 nm (top curves) and larger than 18 nm (bottom curves) in diameter from three passes through the plumes of an ATTAS twin-engine jet research aircraft, while using fuel with elevated sulphur content in the right engine (from Schumann et al., 1996).

During operation of the DLR ATTAS twin-engine jet research aircraft, different high- and low-sulphur fuels were burned at the same time on the two engines. The observations showed visible and measurable differences between contrails caused by the different sulphur levels. However, it is worth mentioning that even the combustion of synthetic sulphur-free fuels such as SASOL leads to the emission of particles possibly caused by the emission of condensable organic species.
For the USA a relationship between aircraft fuel usage above 7 km altitude and contrail frequency implies that contrail coverage is limited mainly by the number of aircraft flights, not by atmospheric conditions at cruise altitude (Minnis et al., 1997). In 1992, aircraft line-shaped contrails are estimated to cover about 0.1% of the earth s surface on an annually averaged basis. The contrail cover is projected to grow to 0.5% by 2050 at a rate, which is faster than the rate of growth in aviation fuel consumption (Penner et al., 1999).

Conclusions and future emission reduction

Air traffic is currently of only minor importance for the greenhouse effect and the ozone budget of the atmosphere. Presently, NO_x emissions from subsonic aircraft in 1992 are estimated to have increased ozone concentrations at cruise altitudes in northern mid-latitudes by up to 6%, compared to an atmosphere without aircraft emissions.
The best estimate of radiative forcing by air traffic in 1992 is 0.05 Wm^{-2}, which corresponds to 3.5% of man-made radiative forcing. For 2050 reference scenarios predict an increase to 0.19 Wm^{-2} (5%) by air traffic (Penner et al., 1999).
There is a range of options to further reduce the impact of aviation emissions including

changes in aircraft and engine technology (e.g. low NO_x combustor/engine), fuel, operational practices and regulatory and economic measures. Flight safety, operational and environmental performance and costs are the dominant factors for aviation industry when assessing any new aircraft purchase or engineering or operational changes. Since the typical lifetime of an aircraft is 25 – 35 years, this factor will to a large extent determine the rate at which technology advances and policy options related to technology can further reduce aircraft emissions during the next decades.

References

Arnold, F., J. Schneider, M. Klemm, J. Scheid, T. Stilp, H. Schlager, P. Schulte, and M.E. Reinhardt, 1994: Mass spectrometer measurements of SO_2 and reactive nitrogen gases in exhaust plumes of commercial jet airliners at cruise altitude. In: Impact of Emissions from Aircraft and Spacecraft upon the Atmosphere [Schumann, U. and D. Wurzel (eds.)]. Proceedings of an international scientific colloquium, Cologne, Germany, 18–20 April 1994. DLR-Mitteilung 94-06, Deutsches Zentrum für Luft- und Raumfahrt (German Aerospace Center), Oberpfaffenhofen and Cologne, Germany, pp. 323–328.

Brasseur, G., R.A. Cox, D. Hauglustaine, I. Isaksen, J. Lelieveld, D.H. Lister, R. Sausen, U. Schumann, A. Wahner and P. Wiesen; European scientific assessment of the atmospheric effects of aircraft emissions, *Atmos. Environ.* **32** (1998) 2329--2418.

CIAP (1975) Climate Impact Assessment Program. Final Report. U.S. Department of Transportation. DOT-TST-75-51 to 53 (8 volumes).

COMESA (1976) The report of the committee on meteorological effects of stratospheric aircraft. Parts 1 and 2. Meteorological Office, Bracknell, U.K.

COVOS, Final Report of Comité d'Etudes sur les Conséquences des Vols Stratosphériques.

Fahey, D.W., E.R. Keim, E.L. Woodbridge, R.S. Gao, K.A. Boering, B.C. Danube, S.C. Wofsy, R.P. Lohmann, E.J. Hintsa, A.E. Dessler, C.R. Webster, R.D. May, C.A. Brock, J.C. Wilson, R.C. Miake-Lye , R.C. Brown, J.M. Rodriguez, M. Loewenstein, M.H. Proffitt, R.M. Stimpfle, S.W. Bowen, and K.R. Chan, In situ observations in aircraft exhaust plumes in the lower stratosphere at midlatitudes. J. Geophys. Res. **100** (1995a) 3065–3074.

Fahey, D.W., E.R. Keim, K.A. Boering, C.A. Brock, J.C. Wilson, H.H. Jonsson, S. Anthony, T.F. Hanisco, P.O. Wennberg, R.C. Miake-Lye, R.J. Salawitch, N. Louisnard, E.L. Woodbridge, R.-S. Gao, S.G. Donelly, R. Wamsley, L.A. Del Negro, B.C. Daube, S.C. Wofsy, C.R. Webster, R.D. May, K.K. Kelly, M. Loewenstein, J.R. Podolske, and K.R. Chan, Emission measurements of the Concorde supersonic aircraft in the lower stratosphere. Science **270** (1995b) 70–74.

Haschberger, P. and E. Lindermeir, Spectrometric in-flight measurement of aircraft exhaust emissions: first results of the June 1995 campaign, *J. Geophys. Res.* **101** (1996) 25995–26006.

Kärcher, B., R.P. Turco, F. Yu. M.Y. Danilin, D.K. Weisenstein, R.C. Miake-Lye and R. Busen, A unified model for ultrafine aircraft particle emissions, J. Geophys. Res. **105** (2000) 29379 – 29386.

Kasibhatla, P.S., NO_y from subsonic aircraft emissions: a global 3-dimensional study, Geophys. Res. Lett. 21 (1993) 1707–1710.

Minnis, P., J.K. Ayers and S.P. Weaver, Surface-based observations of contrail occurrence frequency over the U.S., April 1993 – April 1994, NASA Reference Publication 1404, National Aeronautics and Space Administration, Hampton, VA, USA, 79 pp. (1997).

Penner, J., D.H. Lister, D.J. Griggs, D.J. Dokken and M. McFarland (eds.), Aviation and the global atmosphere: a special report of IPCC working groups I and III, Cambridge University Press, ISBN 0-521-66404-79 (1999).

Schlager, H., P. Schulte and H. Ziereis, In situ measurements in aircraft exhaust plumes and in the North Atlantic flight corridor, in: U. Schumann (ed.), *Pollutants from Air Traffic: Results of Atmospheric Research 1992-1997, DLR Mitteilung 97-04* (1997), ISBN 1434-8462, pp. 57–66.

Schulte, P. and H. Schlager, Flight measurements of cruise altitude nitric oxide emission indices of commercial jet aircraft, *Geophys. Res. Lett.* **23** (1996) 165–168.

Schulte, P., H. Schlager, H. Ziereis, U. Schumann, S.L. Baughcum, and F. Deidewig, NO_x emission indices of subsonic long-range jet aircraft at cruise altitude: in situ measurements and predictions, *J Geophys. Res.* **102** (1997) 21431–21442.

Schumann, U., J. Ström, R. Busen, R. Baumann, K. Gierens, M. Krautstrunk, F.P. Schröder and J. Stingl, In situ observations of particles in jet aircraft exhaust and contrails for different sulphur containing fuels, J. Geophys. Res. **101** (1996) 6853–6869.

Schumann, U. and G. Amanatidis, Aviation, aerosols, contrails and cirrus clouds (A^2C^3), *Air Pollution Research Report* 74, ISBN 92-894-0461-2, CEC, Brussels (2001).

Arctic Regions

The Impact of Halogen Chemistry on the Oxidation Capacity of the Troposphere

Ulrich Platt

Institut für Umweltphysik, University of Heidelberg, INF 229, D-69120 Heidelberg

Introduction

The role of reactive halogen species in the destruction of stratospheric ozone is well known and largely understood [e.g. Solomon 1990] and it is clear that reactive halogen species (RHS = X, X_2, XY, XO, HOX, where X, Y denotes Cl, Br and possibly I) contribute considerably to the loss of stratospheric ozone. Only in recent years has it become clear that RHS can also play a significant role in the troposphere. For instance, recently in the tropospheric boundary layer (BL) significant amounts of BrO and IO could be directly determined by Differential Optical Absorption Spectroscopy (DOAS), also indirect evidence for Cl and Br atoms were found under certain conditions. These observations were made at a variety of sites (see Table 1):

(1) Bromine monoxide was found in the Arctic and Antarctic troposphere by ground based and satellite observations (see Table 1 for references). These episodes of BL BrO were only found in springtime. It could be shown that BrO (at levels up to 30 ppt) is always connected to episodes of BL ozone destruction [e.g. Platt and Lehrer 1995].

(2) Measurements by chemical amplification [Perner et al. 1999], DOAS [Tuckermann et al. 1997], and Hydrocarbon Clock data [e.g. Jobson et al. 1994] suggest ClO levels in the ppt range in the Arctic BL.

(3) In addition IO and OIO was found to reach several ppt at coastal sites in Ireland and at the Canary Islands (see Table 1), and also in the Arctic [Hönninger and Platt 2001] and Antarctica [Friess et al. 2001]. Model calculation suggest that O_3 destruction due to these IO levels is comparable to the classic photochemical O_3 loss in the tropospheric boundary layer [Stutz et al. 1999].

(4) Even higher BrO mixing ratios (up to 90 ppt) were found in the Dead Sea basin [Hebestreit et al. 1999, Matveev et al. 2001], where also complete BL ozone destruction is associated with episodes of high BrO. Also at other salt lakes e.g. near Salt Lake City, USA, elevated levels of BrO (up to 6 ppt) and also ClO (around 15 ppt) were reported [Stutz et al. 2001].

(5) Recently evidence is accumulating that there exist significant amounts of BrO (of the order of 1-2 ppt) also in the free troposphere. Measurements of the BrO column density from the ground [e.g. Eisinger et al. 1997, Kreher et al. 1997, Otten et al. 1998, Frieß et al. 1999] and from satellite [Richter et al. 1998, Wagner and Platt 1998, Wagner et al. 2001] consistently yield higher values than obtained by integrating stratospheric in-situ profiles or profiles from balloon-borne spectroscopic measurements [e.g. Harder et al. 2000, Ferlemann et al. 1998]. Additional evidence comes from the analysis of diurnal profiles of the BrO – 'slant column' density (SCD). BrO SCD data measured by ground based ZSL-DOAS fit model calculations of the SCD much better if a tropospheric (vertical) BrO column density of $(0.5 - 2)\cdot 10^{13}$ cm^{-2} is assumed corresponding to 1-2 ppt BrO throughout the troposphere [Friess et al. 1999, van Roozendael et al. 2001]. The remarkable similarity in the tropospheric residual at various measurement sites suggests that BrO could be a rather ubiquitous constituent of the global troposphere.

Tropospheric Sources of Inorganic Halogens Species

Inorganic halogen species (in the following denoted as InX = X, X_2, XY, XO, HOX, $XONO_2$, HX, where X, Y = Cl, Br, I) are released to the atmosphere either by degradation of organic halogen compounds or by oxidation of sea salt halogenides (X^-).

I. Barnes (ed.), Global Atmospheric Change and its Impact on Regional Air Quality, 67–75.

Table 1: Observation of reactive halogen species in the troposphere and their probable source mechanism

Species	Found at	Technique	Conc. Level, Typ. Rate of O_3 Destruction	Reference
HOCl (?)	Marine BL	Mist Chamber	?	Pszenny et al. 1993
BrO	Arctic and Antarctic BL	DOAS (ground based)	Up to 30 ppt 1-2 ppb/h	Hausmann and Platt 1994, Tuckermann et al. 1997, Hegels et al. 1998, Martínez et al. 1999
BrO	Arctic and Antarctic BL	DOAS (satellite)	around 30 ppt (assuming 1000m layer)	Wagner and Platt 1998, Richter et al. 1998, Hegels et al. 1998, Wagner et al. 2001
BrO	Salt lakes (Dead Sea, Salt Lake City, Caspian Lake)	DOAS	Up to 100 ppt 10-20 ppb/h	Hebestreit et al. 1999, Matveev et al. 2001, Stutz et al. 2001, Wagner et al. 2001
Br	Arctic BL	Hydrocarbon Clock	$(1\text{-}10)\times10^7 cm^{-3}$	Jobson et al. 1994, Ramacher et al. 1997, 1999
BrO	Mid-Lat. Free Troposphere	DOAS (difference)	1-2 ppt ≈0.05 ppb/h	Frieß et al. 1999, v. Roozendael et al. 2001
BrO	Polar free Troposphere	Airborne DOAS		McElroy et al. 1999
Cl	Arctic BL	Hydrocarbon Clock	$(1\text{-}10)\times10^4 cm^{-3}$?	Jobson et al. 1994, Ramacher et al. 1997, 1999
Cl	Remote Marine BL	Hydrocarbon Clock	$(1\text{-}15)\times10^3 cm^{-3}$?	Wingenter et al. 1996
Br_2, BrCl				Foster et al. 2001
IO, OIO	Coastal Areas	DOAS	Up to 6 ppt ≈0.2 ppb/h	Alicke et al. 1999, Allan et al. 2000, Frieß et al. 2001

The dominating source of InX in the stratosphere is the photochemical degradation of fully halogenated compounds (like CF_2Cl_2 or CF_2ClBr). In contrast, tropospheric InX can only be released from less stable, partially halogenated organic compounds such as methyl bromide, CH_3Br, $CHBr_3$, CH_2Br_2, CH_2I_2, or CH_2BrI [e.g. Cicerone 1981, Carpenter et al. 1999]. Some of these species (e.g. CH_3Br) originate to about 50% from anthropogenic sources, others, in particular polybrominated species, are only emitted from biological sources, for instance in the ocean or in coastal areas. The lifetime of CH_3Br is of the order of one year, $CHBr_3$, has several days lifetime, while CH_2I_2 is photodegraded in minutes. [Wayne et al. 1995, Davis et al. 1996, Carpenter et al. 1999]. In addition, even less stable halocarbon species may also be emitted, which – probably due to their instability - escape detection. Another source only of importance to the troposphere is release of InX from sea salt (e.g. sea salt aerosol or sea salt deposits) which appears to proceed by three main pathways:

(1) Strong acids can release HX from sea-salt halides. Under certain conditions (see below) HX can be heterogeneously converted to reactive halogen species (RHS).

(2) Oxidising agents may convert (sea salt) Br^- or Cl^- to Br_2 or BrCl. In particular HOX (i.e.

HOBr and HOCl) is likely to be such an oxidator. Thus the halogen release mechanisms are autocatalytic (see below) as proposed by Tang and McConnel [1996] and Vogt et al. [1996]. Since the development of a radical-chain and the chain branching mechanism is similar to that of chemical explosions these processes became known as 'Bromine Explosion' [Platt and Lehrer 1995, Platt and Janssen 1996, Wennberg 1999]. But direct photochemical [Oum et al. 1998] oxidation of halides by O_3 may also occur, though probably quite slowly.

3) Oxidised nitrogen species, in particular N_2O_5 and NO_3, perhaps even NO_2, can react with sea-salt bromide or chloride to release either HBr, HCl or photolabile species, in particular $BrNO_2$ [Finlayson-Pitts et al. 1990, Behnke et al. 1997, Schweitzer et al. 1999].

Table 2: Sources of Reactive Halogen Species found in various parts of the Troposphere

Species, Site	**Likely Source Mechanism**
ClO_X in the Polar Boundary Layer	By-Product of the 'Bromine Explosion'
BrO_X in the Polar Boundary Layer	Autocatalytic release from Sea Salt on ice ('Bromine Explosion' mechanism)
BrO_X in the Dead Sea Basin	Bromine Explosion (Salt Pans)
BrO_X in the Free Troposphere	(1) Photo – Degradation of hydrogen-containing organo halogen species (e.g. CH_3Br) (2) "Spill-out" from the Boundary Layer (3) Transport from Stratosphere (?)
IO_X in the Marine Boundary Layer	Photo – Degradation of short-lived organo-halogen species (e.g. CH_2I_2)

Tropospheric Cycles of Inorganic Halogen Species

The following gives a brief account of the reaction cycles of inorganic halogen species in the troposphere (see Figure 1), more information can be found for instance in the review of Wayne et al. [1995], Platt and Janssen [1996], Platt and Lehrer [1995], and Lary et al. [1996]. Following release by the mechanisms described above inorganic halogen species will rapidly be photolysed to form halogen atoms, which, in turn, are most likely to react with ozone:

$$X + O_3 \rightarrow XO + O_2 \qquad (R1)$$

Typical conversion time constants via R1 for Cl are around 0.1 s and for Br and I of the order of 1 s at tropospheric background O_3 levels. Halogen atoms are regenerated in a series of reactions including photolysis of XO:

$$XO + h\nu \rightarrow X + O \qquad (J2)$$

Where $J_2 \approx 3\times10^{-5}\ s^{-1}$, $4\times10^{-2}\ s^{-1}$, $0.2\ s^{-1}$ for X = Cl, Br, I, respectively. In the parts of the troposphere affected by pollution reaction with NO is another source of halogen atoms:

$$XO + NO \rightarrow X + NO_2 \qquad (R3)$$

Where $k_3 = 1.7\times10^{-11}\ cm^3 molec.^{-1}s^{-1}$ (X = Cl) and $k_3 = 2.1\times10^{-11}\ cm^3 molec.^{-1}s^{-1}$ (X = Br, I). In addition to that the self reactions of XO (or reaction with another halogen oxide YO) can play an important role:

$$XO + YO \rightarrow X + Y + O_2 \quad (\text{or } XY + O_2) \qquad (R4)$$

In the case of XO = YO = BrO the rate constant $k_4 = 3.2\cdot10^{-12} cm^3 molec.^{-1}s^{-1}$ (overall rate constant of the self-reaction of BrO including the channel to Br_2, accounting for ≈20% at 250 K). Reactions of ClO + BrO [LeBras and Platt 1995] and reactions involving IO proceed considerably faster. Reaction 4 is the rate determining step in an ozone-destruction cycle involving R1, R4, (and photolysis of X_2 or XY – species).

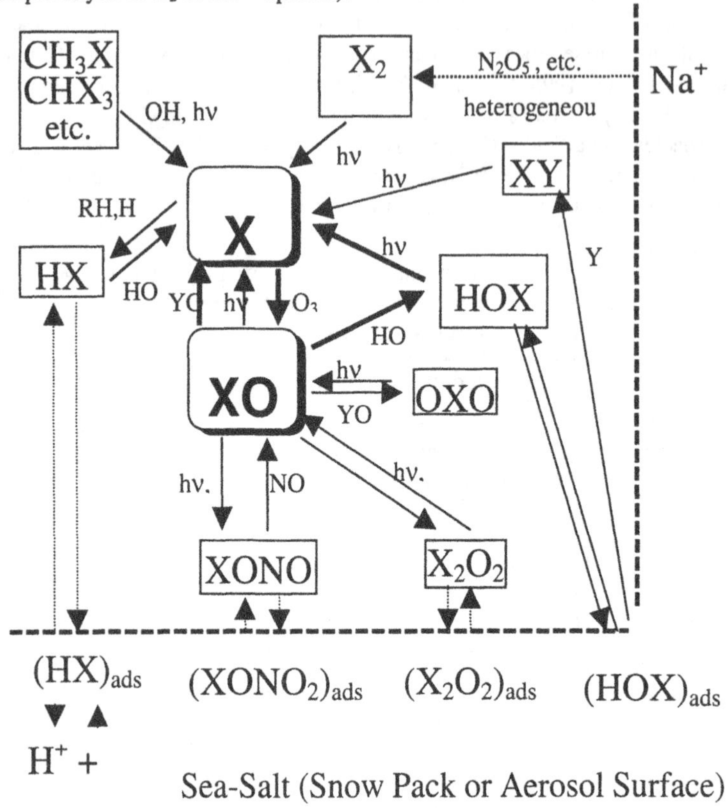

Figure 1: Simplified scheme of the inorganic halogen reactions in the boundary layer (X = Cl, Br, I) after Wayne et al. [1995] and Platt and Janssen [1996]. Sources are release from sea-salt (aerosol or deposits) or photochemical degradation of organohalogen species. Important questions concern the XO/X interconversion and recycling from hydrogen halides (HX). Fat lines: ozone destruction cycles.

This type of halogen catalysed ozone destruction has been identified as the prime cause for polar boundary layer ozone destruction [Oltmanns et al. 1986, Barrie et al. 1988, Barrie and Platt 1997]. The channels of R1 leading to OXO would not contribute to the ozone loss.

Ultimately the XO/X ratio is, therefore, largely determined by reactions R1, R3, R4, and (during daytime) J_2. Overall a small fraction (less than 0.1% for Cl, about 0.1 - 1 % for Br and several 10% in the case of I, see e.g. Platt and Janssen [1996]) of the inorganic halogen will be present in the form of halogen atoms. In addition, halogen atoms can react with saturated- (Cl) or unsaturated hydrocarbons (Cl, Br) to form hydrogen halides, e.g.:

$$RH + Cl \rightarrow R + HCl \qquad (R5)$$

Here R denotes an organic radical. An alternative is the reaction of Cl and Br atoms with HCHO (or higher aldehydes) or Cl, Br, I with HO_2 also leading to conversion of halogen atoms to hydrogen halides. Bromine atoms can add to the C=C double bond of olefins leading to organic Br species, but, at least partly, also to HBr. While in the stratosphere with its low hydrocarbon HCHO and HO_2 concentration, the fraction of Br atoms being converted to BrO by reaction with

O_3 (R1) is close to 100%, in the troposphere roughly 99% of the Br atoms react with O_3 [Wayne et al. 1995, Platt and Janssen 1996]. HI is hardly formed in the atmosphere, however the short photolytic lifetime (around minute) of BrO leads to noticeable levels of Br atoms, which reaction with HCHO and HO_2 leading to conversion of reactive Br into HBr within a few hours [e.g. Platt and Lehrer 1995]. In the case of chlorine around 50% of the Cl atoms react with hydrocarbon (e.g. CH_4) thus the conversion of reactive Cl to HCl is even faster. In the lower troposphere hydrogen halides may be irreversibly lost by wet or dry deposition due to their high water solubility. In the free troposphere recycling of Br to the BrO_X reservoir is more likely, as will be discussed below. The only relevant gas phase 'reactivation' mechanism of HX is reaction with OH:

$$HX + OH \rightarrow X + H_2O \qquad (R6)$$

Assuming OH levels typical for the free troposphere and gas-phase chemistry only [HBr]/[BrO] ratios should typically reach 10 – 20, thus making HBr the most abundant InBr species there. Other important reservoirs for inorganic halogen species besides HX are HOX and $XONO_2$, where HOX is formed via the reaction of peroxy radicals with XO:

$$XO + HO_2 \rightarrow HOX + O_2 \qquad (R7)$$

(with k_7 is several times 10^{-11} cm^3s^{-1}). Formation of HOX is followed by its photolysis:

$$HOX + h\nu \rightarrow X + OH \qquad (R8)$$

Thus in the sunlit atmosphere a photo-stationary state between XO and HOX is established with [HOX]/[XO] ranging from roughly 1 to 10 (if heterogeneous reactions are neglected), depending on the oxidation capacity of the atmosphere. Important side effects of HOX formation are O_3 destruction [e.g. Barrie et al. 1988] as well as a reduction of the $[HO_2]/[OH]$ ratio as will be discussed below.

An interesting reaction cycle involving HOBr [Fan and Jacob 1992, Tang and McConnel 1996, Vogt et al. 1996] is the liberation of gaseous bromine species (and to a lesser extent chlorine species) from (sea-salt) halides by the Bromine Explosion mechanism:

$$HOBr + (X^-)_{Surface} + H^+ \rightarrow XBr + H_2O \qquad (R9)$$

Where XBr denotes BrCl or Br_2. The required H^+ (the reaction appears to occur at appreciable rates only at $pH < 6.5$ [Fickert et al. 1999]) could be supplied by strong acids, such as H_2SO_4 and HNO_3 originating from man made or natural sources.

Finally, in situations where elevated NO_X concentration prevail halogen nitrates, $XONO_2$ can form:

$$XO + NO_2 \overset{M}{\longleftrightarrow} XONO_2 \qquad (R10)$$

In contrast to $ClONO_2$ Bromine nitrate is assumed to be quite stable against thermal decay, but is readily photolysed and may be converted to HOX by heterogeneous hydrolysis:

$$XONO_2 + H_2O \xrightarrow{Surface} HNO_3 + HOX \qquad (R11)$$

or to Br_2 or BrCl by reaction with HY in solution:

$$XONO_2 + HY \xrightarrow{Surface} HNO_3 + XY \qquad (R12)$$

Overall $XONO_2$ is probably of minor importance at the low NO_X levels typically found in the free troposphere, but can play an important role in polluted air (for instance at polluted coastlines). Under conditions of high NO_X halogen release can also occur via the reaction of sodium halides at surfaces:

$$N_2O_5 + NaX(s) \rightarrow NaNO_3(s) + XNO_2 \qquad (R13)$$

[e.g. Finlayson-Pitts et al. 1989]. The XNO_2 formed in the above reaction returns to the gas phase and may photolyse to release a halogen atom or possibly further react with sea salt [Schweitzer et al. 1999]:

$$XNO_2 + NaX(s) \rightarrow NaNO_2(s) + X_2 \quad (R14)$$

The above reaction sequence would constitute a dark source of halogen molecules.

Recent laboratory investigations have shown that in the troposphere heterogeneous reactions (on aerosol or cloud particle surfaces, as originally suggested by Fan and Jacob [1992]) can not only produce but also efficiently recycle BrO_X (and possibly ClO_X) from HBr (or HCl) [e.g. Kirchner et al. 1997]. As pointed out by Abbat [1994] process is reaction 9. While HBr and HOBr are the most likely forms of InBr in the troposphere as outlined above, HCl is probably more abundant than total Br in the upper troposphere (≤100 ppt). The Br_2 or BrCl formed in reaction 9 will be rapidly photolysed to release Br (and Cl) atoms, which then react as described above.

Observations of Reactive Halogen Species in the Troposphere

Reactive halogen species, in particular Cl- and Br- atoms, ClO, BrO, IO, and OIO could be detected by several direct or indirect techniques (see Table 1) in the troposphere:

- Differential Optical Absorption Spectroscopy (DOAS) of BrO, IO, and OIO (and upper limits for ClO). This technique is being applied in a series of ground based measurements as well as from aircraft, balloons, and from satellite.
- 'Hydrocarbon Clock' observations of time - integrated amounts of Cl- and Br- atoms, where the term Hydrocarbon Clock refers to the observation of hydrocarbon ratios in airmasses of known chemical age.
- Detection of ClO by chemical amplification [Perner et al. 1999].
- Sampling (mist chamber) and wet chemical detection of reactive chlorine.
- Mass spectrometric detection of Br_2 and BrCl [Foster et al. 2001]

The Overall Picture of Inorganic Halogen Species in the Troposphere

From Table 2 which summarises the various regions of the troposphere, where reactive chlorine, bromine, or iodine species are found as well as their likely sources, it becomes clear that the occurrence of noticeable levels of RHS is a widespread phenomenon in many parts of the troposphere. Clearly spectacular changes in tropospheric chemistry, including frequently complete loss of O_3, occur in polar regions as well as in the vicinity of salt lakes due to release of reactive bromine and, to a minor extent, chlorine, but also other regions are affected. Reactive iodine appears to be abundant at most coastlines, but data are still sparse.

Particularly notable is the likely presence of ppt levels of BrO in the free troposphere, which can have a profound effect on the global budget of tropospheric ozone and other chemical key parameters of the troposphere (see below).

Overall we can now describe the following scenario for reactive bromine species in the free troposphere: From degradation of the ubiquitous CH_3Br (tropospheric mixing ratios 11 – 12 ppt in the northern hemisphere, 8 – 10 ppt in the southern hemisphere) by reaction with OH (lifetime from 10 to 18 months) a level of inorganic bromine approaching roughly 10% of the CH_3Br mixing ratio, or slightly more than 1 ppt, is expected (assuming about one month residence time of air masses in the upper troposphere). According to model calculations including heterogeneous chemistry and the observations in the polar boundary layer a large fraction (around 50 %) of this inorganic Br would be present in the form of BrO. In addition to CH_3Br other, shorter lived, organic bromine species like CH_2Br_2 or $CHBr_3$ might also contribute [e.g. Dvortsov et al. 1999]. Moreover, stratospheric air descending to the upper troposphere will contain ≈20 ppt of total InBr [e.g. Harder et al. 2000].

At the levels suggested by the available measurements tropospheric BrO can have a noticeable effect on several aspects of tropospheric chemistry, these include:

(1) As already discussed by several authors [e.g. Platt and Janssen 1996, Davis et al. 1996]

reactive halogen species, and in particular reactive bromine and iodine can readily destroy tropospheric ozone. While the halogen oxide self- and cross reactions are of negligible influence at 1-2 ppt BrO the XO – HO_2 cycle - reactions 7, 8 followed by conversion of OH to HO_2 (e.g. by reaction with CO) with the net result of converting O_3 to O_2 will lead to a rate of ozone loss comparable to photochemical O_3 destruction associated with HO_X reactions (due to $O(^1D) + H_2O$ and $HO_2 + O_3$). In addition the reaction of BrO – CH_3O_2 (rate and products of the reaction IO – CH_3O_2 is presently unknown) may have a similar effect on O_3.
(2) Reactions 7, 8 will lead to the conversion of HO_2 to OH and thus reduce the HO_2/OH ratio, in particular at low NO_X levels. At a given photochemical situation (O_3-, H_2O levels, insolation) the presence of BrO or IO will therefore increase the OH concentration.
(3) The reaction of XO with NO, (reaction 3) leads to conversion of NO to NO_2 and thus to an increase of the Leighton ratio ($L = [NO_2]/[NO]$). An increase in L is thus usually regarded as an indicator for photochemical ozone production (due to the presence of RO_2 (R = organic radical)) in the troposphere. However, as already noted by Platt and Janssen [1996] an increase in L due to halogen oxides will not lead to O_3 production.
(4) A more subtle consequence of reactive halogen species for the ozone levels in the free troposphere is due to the combination of the above effects (2) and (3) as pointed out by Stutz et al. [1999]: Photochemical O_3 production in the troposphere is limited by the reaction of NO with HO_2 (or CH_3O_2). However, the presence of reactive bromine will reduce the concentrations of both educts of this reaction and thus reduce the NO_X – catalysed O_3 production.
(5) Heterogeneous reactions with HX = HCl would lead to BrCl, which is readily photolysed to release Cl-atoms. This process would constitute a Br catalysed chlorine activation, which directly enhances the oxidation capacity of the troposphere.
(6) In addition tropospheric reactive bromine species could be responsible or contribute to the observed cases of upper tropospheric ozone loss [e.g. Kley et al. 1996].
(7) Gas-phase iodine species (like IO or HOI) may facilitate transport of I from the coast to inland areas and thus contribute to our iodine supply [Cauer 1939].
Clearly, at present more measurements are needed to ascertain the distribution and levels of free tropospheric reactive bromine. However it is already obvious that inorganic halogen species are likely to play a major – and yet not fully recognised – role in tropospheric chemistry.

References

Abbat J.P.D. (1994), Heterogeneous reaction of HOBr with HBr and HCl on ice surfaces at 228 K, Geophys. Res. Lett. 21, 665-668.

Alicke B., Hebestreit K., Stutz J., and Platt U. (1999), Detection of Iodine Oxide in the Marine Boundary Layer, Scientific Correspondence to Nature, 397, 572-573.

Allan B.J., McFiggans G., and Plane J.M.C. (2000), Observation of Iodine monoxide in the remote marine boundary layer, J. Geophys. Res. 105, 14363-14369.

Barrie L.A., Bottenheim J.W., Schnell R.C., Crutzen P.J. and Rasmussen R.A. (1988), Ozone destruction and photochemical reactions at polar sunrise in the lower Arctic atmosphere, Nature 334, 138-141.

Barrie L.A., and Platt U. (1997), Arctic tropospheric chemistry: overview to Tellus special issue, 49B, 450-454.

Behnke W., George C., Scheer V., and Zetzsch C. (1997), Production and decay of $ClNO_2$ from the reaction of gaseous N_2O_5 with NaCl solution: bulk and aerosol experiments, J. Geophys. Res. 102D, 3795-3804.

Carpenter L.J., Sturges W.T., Penkett S.A., Liss P.S., Alicke B., Hebestreit K., and Platt U. (1999), Short-lived alkyl iodides and bromides at Mace Head, Ireland: Links to biogenic sources and halogen oxide production, J. Geophys. Res., 104, 1679-1689.

Cauer H. (1939), Schwankungen der Jodmenge der Luft in Mitteleuropa, deren Ursachen und deren Bedeutung für den Jodgehalt unserer Nahrung (Auszug), Angewandte Chemie 52, Nr. 11, 625-628.

Cicerone R.J. (1981), Halogens in the Atmosphere, Rev. Geophys. and Space Phys. 19, 123-139.

Davis D., Crawford J., Liu S., McKeen S., Bandy A., Thornton D., Rowland F. and Blake D. (1996), Potential impact of iodine on tropospheric levels of ozone and other critical oxidants, J. Geophys. Res., 101, 2135-2147.

Dvortsov V.L., Geller M.A., Solomon S., Schauffler S.M., Atlas E.L., and Blake D.R. (1999), Rethinking reactive halogen budgets in the midlatitude lower stratosphere, Geophys. Res. Lett. 26, 1699-1702.

Eisinger M., Richter A., Ladstätter-Weißenmayer A., and Burrows J.P. (1997), DOAS Zenith sky observations: 1.BrO measurements over Bremen (53°N) 1993-1994, J. Atmos. Chem. 26, 93-108.

Fan S.-M. and Jacob D.J. (1992), Surface ozone depletion in the Arctic spring sustained by bromine reactions on aerosols, Nature 359, 522-524.

Ferlemann F., Camy-Peyret C., Harder H., Fitzenberger R., Hawat T., Osterkamp H., Perner D., Platt U., Schneider M., Vradelis P. and Pfeilsticker K. (1998), Stratospheric BrO Profile measured at Different Latitudes and Seasons. Measurement Technique, Geophys. Res. Lett., 25, 3847-3850.

Fickert S., Adams J.W., and Crowley J.N. (1999), Activation of Br_2 and BrCl via uptake of HOBr onto aqueous salt solutions, J. Geophys. Res. 104, 23719-23728.

Finlayson-Pitts B.J. (1993), Indications of Photochemical Histories of Pacific Air Masses from Measurements of Atmospheric Trace Species at Point-Arena, California - Comment, J. Geophys. Res. 98, 14991-14993.

Finlayson-Pitts, B. J.; Ezell, M. J.; Pitts, J. N. (1989), Formation of chemically active chlorine compounds by reactions of atmospheric NaCl particles with gaseous N_2O_5 and $ClONO_2$, Nature 337, 241-244.

Fitzenberger, R., H. Bösch, C. Camy-Peyret, M.P. Chipperfield, H. Harder, et al. (2000), First Profile Measurements of Tropospheric BrO, Geophys. Res. Lett. 27, 2921-2924.

Foster K.L., Plastridge R.A., Bottenheim J.W., Shepson P.B., Finlayson-Pitts B.J., Spicer C.W. (2001), The Role of Br_2 and BrCl in Surface Ozone Destruction at Polar Sunrise, Science 291, 471-474.

Frieß U., Otten C., M. Chipperfield, Wagner T., Pfeilsticker K., and Platt U. (1999), Intercomparison of measured and modelled BrO slant column amounts for the Arctic winter and spring 1994/95, Geophys. Res. Lett. 26, 1861-1864.

Frieß U., Wagner T., Pundt I., Pfeilsticker K. and Platt U. (2001), Spectroscopic measurements of tropospheric iodine oxide at Neumayer station, Antarctica, Geophys. Res. Lett. 28, 1941-1944.

Harder H., Camy-Peyret C., Ferlemann F., Fitzenberger R., Hawat T., Osterkamp H., Perner D., Platt U., Schneider M., Vradelis P. and Pfeilsticker K. (1998), Stratospheric BrO profile measured at different latitudes and seasons: Atmospheric observations, Geophys. Res. Lett. 25, 3843-3846.

Harder H., Bösch H., Camy-Peyret C., Chipperfield M., Fitzenberger R., Payan S., Perner D., Platt U., Sinnhuber B., and Pfeilsticker K. (2000), Comparison of measured and modelled stratospheric BrO: Implications for the total amount of stratospheric bromine, Geophys. Res. Lett. 27 3695-3698.

Hausmann M., and Platt U. (1994), Spectroscopic measurement of bromine oxide and ozone in the high Arctic during Polar Sunrise Experiment 1992, J. Geophys. Res. 99, 25,399-25,414.

Hebestreit K., Stutz J., Rosen D., Matveev V, Peleg M., Luria M., and Platt U. (1999), First DOAS Measurements of Tropospheric BrO in Mid Latitudes, Science 283, 55-57.

Hegels, E. et al., Global distribution of atmospheric bromine monoxide from GOME on earth observing satellite ERS-2, Geophys. Res. Lett. 25, 3127-3130, 1998.

Hönninger G. and Platt U. (2001), The Role of BrO and its Vertical Distribution during Surface Ozone Depletion at Alert, Atmos. Environ. in press.

Jobson B.T., Niki H., Yokouchi Y., Bottenheim J., Hopper F. Leaitch R. (1994), Measurements of C_2-C_6 hydrocarbons during Polar Sunrise Experiment 1992, J. Geophys. Res. 99, 25,355-25,368.

Kirchner U., Benter Th., and Schindler R.N. (1997), Experimental verification of gas phase bromine enrichment in reactions of HOBr with sea salt doped ice surfaces, Ber. Bunsenges. Phys. Chem. 101, 975-977.

Kley D., Smit H.G. J., Crutzen P.J., Vömel H., Oltmans S., Grassl H., Ramanathan V. (1996), Observation of near-zero ozone concentrations over the convective Pacific: effects on air chemistry. Science 274, 230-233.

Kreher K., Johnston P.V., Wood S.W., and Platt U. (1997), Ground-based measurements of tropospheric and stratospheric BrO at Arrival Heights (78°S), Antarctica, Geophys. Res. Lett. 24, 3021-3024.

Lary D.J., Chipperfield M.P., Toumi R., and Lenton T. (1996), Heterogeneous atmospheric bromine chemistry, J. Geophys. Res. 101, 1489-1504.

LeBras G., and Platt U. (1995), A Possible mechanism for combined chlorine and bromine catalysed destruction of tropospheric ozone in the Arctic, Geophys. Res. Lett., 22, 599-602.

Martínez, M., Arnold T. and Perner D. (1999), The role of bromine and chlorine chemistry for arctic ozone depletion events in Ny Alesund and comparison with model calculations, Ann. Geophys. 17, 941-956.

Matveev V., Peleg M., Rosen D., Tov-Alper D.S., Stutz J., Hebestreit K., Platt U., Blake D. and Luria M (2001), Bromine Oxide – Ozone interaction over the Dead Sea, J. Geophys. Res. 106, 10375-10378.

McElroy C.T. McLinden C.A., and McConnell J.C. (1999), Evidence for bromine monoxide in the free troposphere during Arctic polar sunrise, Nature 397, 338-340.

Oltmans S.J. and Komhyr W.D. (1986), Surface ozone distributions and variations from 1973-1984 measurements at the NOAA Geophysical Monitoring for Climate Change baseline observatories, J. Geophys. Res. 91, 5229-5236.

Otten C., Ferlemann F., Platt U., Wagner T., Pfeilsticker K. (1998), Groundbased DOAS UV/visible measurements at Kiruna (Sweden) during the SESAME winters 1993/94 and 1994/95, J. Atmos. Chem. 30, 141-162.

Oum, K.W., Lakin M.J., DeHaan D.O., Brauers T., and Finlayson-Pitts B.J. (1998a), Formation of molecular chlorine from the photolysis of ozone and aqueous Sea-Salt particles, Science 279, 74-77.

Perner D., Arnold T., Crowley J., Klüpfel T., Martinez M., and Seuwen R. (1999), The measurements of active chlorine in the atmosphere by chemical amplification, J. Atmos. Chem., in press.

Platt U., and Hausmann M. (1994), Spectroscopic measurement of the free radicals NO_3, BrO, IO, and OH in the troposphere, Res. Chem. Intermed. 20, 557-578.

Platt U., and Lehrer E., (1995), ARCTOC, Final Report to the European Union.

Platt. U., and Janssen C. (1996), Observation and role of the free radicals NO_3, ClO, BrO and IO in the Troposphere, Faraday Discuss. 100, 175-198.

Pszenny A.A.P., Keene W.C., Jacob D.J., Fan S., Maben J.R., Zetwo M.P., Springeryoung M., and Galloway J.N. (1993), Evidence of Inorganic Chlorine Gases other than Hydrogen Chloride in Marine Surface Air, Geophys. Res. Lett. 20, 699-702.

Ramacher B., Rudolph J., Koppmann R. (1999), Hydrocarbon measurements during tropospheric ozone depletion events, J. Geophys. Res. C104, 3633-3653.

Richter A., Wittrock F., Eisinger M., and Burrows J.P. (1998), GOME observations of tropospheric BrO in northern hemispheric spring and summer 1997, Geophys. Res. Lett. 25, 2683-2686.

Solberg S., Schmidtbauer N., Semb A., Stordal F. (1996), Boundary-Layer Ozone Depletion as Seen in the Norwegian Arctic in Spring, J. Atmos. Chem. 23, 301-332.

Solomon S. (1990), Progress towards a quantitative understanding of Antarctic ozone depletion, Nature 347, 347-354.

Stutz J., Hebestreit K., Alicke B., and Platt U. (1999), Chemistry of halogen oxides in the troposphere: comparison of model calculations with recent field data, J. Atmos. Chem. 34, 65-85.

Stutz J., Ackermann R., Fast J.D., and Barrie L. (2001), Atmospheric Reactive Chlorine and Bromine at the Great Salt Lake, Utah, Science, submitted.

Tang T. and McConnel J.C. (1996), Autocatalytic release of bromine from arctic snow pack during polar sunrise, Geophys. Res. Lett. 23, 2633-2636.

Tuckermann M., Ackermann R., Gölz C., Lorenzen-Schmidt H., Senne T., Stutz J., Trost B., Unold W., and Platt U. (1997), DOAS-Observation of Halogen Radical- catalysed Arctic Boundary Layer Ozone Destruction During the ARCTOC-campaigns 1995 and 1996 in Ny-Alesund, Spitsbergen, Tellus 49B, 533-555.

Van Roozendael, M., Wagner T., Richter A., Pundt I., Arlander D.W., Burrows J.P., Chipperfield M., Fayt C., Johnston P. V., Lambert J.-C., Kreher K., Pfeilsticker K., Platt U., Pommereau J.-P., Sinnhuber B.-M., Tørnkvist K. K., and Wittrock F. (2001), Intercomparison of BrO Measurements from ERS-2 GOME, Ground-Based and Balloon Platforms, Proc. COSPAR meeting.

Vogt R., Crutzen P.J., Sander R. (1996), A mechanism for halogen release from sea-salt aerosol in the remote marine boundary layer, Nature 383, 327-330.

Wagner T., Platt U. (1998), Observation of Tropospheric BrO from the GOME Satellite, Nature 395, 486-490.

Wagner T., Leue C., Wenig M., Pfeilsticker K. and Platt U. (2001), Spatial and temporal distribution of enhanced boundary layer BrO concentrations measured by the GOME instrument aboard ERS-2, J. Geophys. Res. 106, 24225-24235.

Wayne R.P., Poulet G., Biggs P., Burrows J.P., Cox R.A., Crutzen P.J., Haymann G.D., Jenkin M.E., LeBras G., Moortgat G.K., Platt U., and Schindler R.N. (1995), Halogen oxides: radicals, sources and reservoirs in the laboratory and in the atmosphere, Atmosph. Environ. 29, 2675-2884.

Wennberg P.O. (1999), Bromine Explosion, Nature 397, 299-301.

Wingenter O.W., Kubo M.K., Blake N.J., Smith T.W., Blake D.R., and Rowland F.S. (1996), Hydrocarbon and halocarbon measurements as photochemical and dynamical indicators of atmospheric hydroxyl, atomic chlorine, and vertical mixing obtained during Lagrangian flights, J. Geophys. Res. 101, 4331-4340.

Kinetic and Mechanistic Studies in the Laboratory

Recent Developments in Peroxy Radical Chemistry

Robert Lesclaux

LPCM - Université Bordeaux I - 33405 Talence Cedex – France
E-mail : lesclaux@cribx1.u-bordeaux.fr

Introduction

Peroxy radicals are key intermediate species in oxidation processes of hydrocarbons and other organic compounds containing oxygen, nitrogen or halogens in the atmosphere. The increasing need of accurate kinetic and mechanistic data for modelling the atmospheric chemical processes has motivated a large amount of work during the last fifteen years, resulting in significant improvement of our knowledge of the reactivity of this class of radicals. Six reviews dealing specifically of this topic have been published during the last decade (Lesclaux, 1997; Lightfoot *et al.*, 1992; Nielsen *et al.*, 1997; Tyndall *et al.*, 2001; Wallington *et al.*, 1992; Wallington *et al.*, 1997).

In general, peroxy radicals exhibit fairly low reactivity, particularly towards molecular (closed shell) species and, as a result, combination reactions with themselves or with other radical (open shell) species are their most common reactions. The kinetics and mechanisms of such combination reactions of interest in atmospheric chemistry are examined in this overview.

Formation of Peroxy Radicals in the Atmosphere

Peroxy radicals RO_2 are formed in the atmosphere according to the following principal reactions.

- *Hydrogen abstraction by OH radicals from saturated hydrocarbons*

$$RH + OH \rightarrow R + H_2O$$
$$R + O_2 + M \rightarrow RO_2 + M$$

- *Addition to unsaturated compounds*

$$OH + {>}C{=}C{<} + M \rightarrow {>}C(OH){-}C{<} + M$$
$${>}C(OH){-}C{<} + O_2 + M \rightarrow {>}C(OH){-}C(OO){<} + M$$

- *Decomposition of alkoxy radicals*

$$R^1R^2R^3CO + M \xrightarrow{O_2} R^1R^2C{=}O + R^3O_2 + M$$

- *HO_2 addition to aldehydes*

$$HO_2 + R\text{-}CHO \rightleftharpoons RCH(OH)OO$$

This latter reaction has no importance in the atmosphere (equilibrium shifted towards the left) but must be taken into account in laboratory studies.

Peroxy radical reactions

Peroxy radicals are not very reactive species ; under atmospheric conditions, they only react with other radical species :

- Nitrogen oxides : NO and NO_2 *(in polluted atmospheres)*
- Other peroxy radicals *(in clean atmospheres)*, the most abundant being HO_2, CH_3O_2 and $CH_3C(O)O_2$.

In the laboratory, self reactions $RO_2 + RO_2$ and cross reactions $RO_2 + R'O_2$ always play an important role in reaction mechanisms and in kinetic analyses. Thus, the kinetics and mechanisms of these reactions must be investigated.

I. Barnes (ed.), Global Atmospheric Change and its Impact on Regional Air Quality, 79–84.

Laboratory Investigations of Peroxy Radical Reactions

For kinetic studies, molecular modulation spectroscopy, flash - laser / flash photolysis have been used, radical concentrations being most of the time monitored using UV absorption spectrometry, as RO_2 radicals absorb fairly strongly (σ = (2-10) x 10^{-18} cm^2 $molecule^{-1}$) in the 200 – 300 wavelength range. Occasionally, mass spectrometry has also been used.

Mechanistic studies have essentially been performed in reaction chambers, using end-product analyses.

Reactions with nitrogen oxides

Reactions of RO_2 radicals with NO and NO_2 are among the most important reactions in the atmosphere. They are either propagating (NO) or terminating (NO_2) steps in reaction chains forming ozone and they can form peroxynitrates, such as PANs, which can transport NO_x away from sources.

- *Reactions with NO* :

$$RO_2 + NO \rightarrow RO + NO_2 \quad (a)$$
$$+M \rightarrow RONO_2 + M \quad (b)$$

RO is the corresponding alkoxy radical which propagates chain reactions and $RONO_2$ is a nitrate which can be formed in significant amounts for large RO_2 radicals.

Table 1: *Examples of branching ratios for nitrate formation* $\beta = k_b/(k_a + k_b)$

Radical	β
1-propyl-O_2	0.019
2-propyl-O_2	0.049
1-butyl-O_2	0.041
t-butyl-O_2	0.180
1-hexyl-O_2	0.121
1-heptyl-O_2	0.195
1-octyl-O_2	0.360
3-octyl-O_2	0.37

Concerning the kinetics of RO_2 + NO reactions, rate constants ($k_a + k_b$) are fall within a fairly narrow range, $\approx$ 0.4 to 2 $\times$ 10^{-11} cm^3 $molecule^{-1}$ s^{-1}

- *Reactions with NO_2* :

$$RO_2 + NO_2 + M \rightleftharpoons RO_2NO_2 + M$$

Formation of a peroxynitrate is the only channel of this reaction. Concerning the kinetics, the rate constants are all similar, 6 - 10 $\times$ 10^{-12} cm^3 $molecule^{-1}$ s^{-1} (298 K, 1atm), independent of the RO_2 structure. Particularly important are the reactions forming PANs:

$$RC(O)O_2 + NO_2 + M \rightleftharpoons RC(O)O_2NO_2 + M$$

Due to their stability, PANs are efficient reservoirs of NO_x, and thus, can carry NO_x species away from sources. This reaction is also an efficient terminating step in chain reactions forming ozone.

A recent evaluation (Tyndall *et al.*, 2001) gives for the rate constants of the $CH_3C(O)O_2$ radical reactions with NO and NO_2 : (2.0 ± 0.3) and (1.0 ± 0.2) x 10^{-11} cm^3 $molecule^{-1}$ s^{-1}, respectively, at 298K and 1 atm pressure of air.

Self reactions of peroxy radicals

RO_2 self reactions generally play a minor role in the atmosphere, except in weakly polluted atmospheres for the most abundant radicals, mainly $\cdot CH_3O_2$ and $CH_3C(O)O_2$. However, they are always important in laboratory studies of peroxy radical reactions and the rate constants must be known for assessing the rate constants for cross reactions with the most abundant peroxy radicals. Self reactions generally exhibit two channels:

$$RO_2 + RO_2 \rightarrow 2\,RO + O_2 \quad \text{(a)}$$
$$\rightarrow ROH + RO_{(-H)} + O_2 \quad \text{(b)}$$
$$\text{with } \alpha = k_a / (k_a + k_b)$$

The mechanism and kinetics of this type of reactions are difficult to characterise as α can vary from nearly 0 to 1, according to the temperature and the structure of the RO_2 radical and, rate constants vary over more than five orders of magnitude (from 3×10^{-17} to $\approx 10^{-11}$ cm^3 $molecule^{-1}$ s^{-1}) and alkoxy radicals RO generally react by forming new peroxy radicals. Thus separate studies of reaction mechanisms are necessary for investigating the kinetics. It is essentially the class of radicals, tertiary, secondary or primary, which is the most sensitive parameters for determining self reaction rate constants.

- *Self reactions of tertiary peroxy radicals*

$$R_1R_2R_3CO_2 + R_1R_2R_3CO_2 \rightarrow 2\,R_1R_2R_3CO + O_2$$

Only channel (a) is possible for this class of reactions. The few rate constants available to date are given in Table 2.

Table 2 – *Rate constants for self reactions of tertiary peroxy radicals.*

Radical	k / cm^3 $molecule^{-1}$ s^{-1} 298 K	(E/R)/K
$(CH_3)_3CO_2$	3×10^{-17}	4200
$HO(CH_3)_2CC(CH_3)_2O_2$	5×10^{-15}	1220
$Br(CH_3)_2CC(CH_3)_2O_2$	$\approx 2 \times 10^{-14}$	
1,2-dimethyl-2-OH-cyclohexyl-O_2	6×10^{-14}	

It is seen that the reactions are slow and that rate constants can vary by several orders of magnitude upon substitution by various groups. It is difficult to recommend structure activity relationships (SAR) for this type of radical.

- *Self reactions of secondary peroxy radicals*

$$R_1R_2CHO_2 + R_1R_2CHO_2 \rightarrow 2\,R_1R_2CHO + O_2$$
$$\rightarrow R_1R_2CHOH + R_1R_2CO + O_2$$
$$\alpha = 0.3 \text{ to } 0.6$$

Table 3 – *Rate constants for self reactions of secondary peroxy radicals.*

Radical	k / cm^3 molecule^{-1} s^{-1} 298 K	(E/R)/K
$(CH_3)_2CHO_2$	1.1×10^{-15}	2200
sec-$C_5H_{11}O_2$	3.3×10^{-14}	
sec-$C_{10}H_{23}O_2$	9.4×10^{-14}	
sec-$C_{12}H_{25}O_2$	1.4×10^{-13}	
cyclo-$C_5H_9O_2$	4.5×10^{-14}	550
cyclo-$C_6H_{11}O_2$	4.2×10^{-14}	200
$HO(CH_3)CHCH(CH_3)O_2$	6.6×10^{-13}	-1300
$Br(CH_3)CHCH(CH_3)O_2$	7.3×10^{-13}	-1200
cyclo-$C_6H_{10}(\beta\text{-}OH)O_2$	1.2×10^{-12}	

Here the rate constants are significantly higher than for tertiary radicals and increase with the number of carbon atoms (for alkylperoxy radicals) and upon substitution. Fairly reliable SAR can be provided from these data.

- *Self reactions of primary peroxy radicals*

$$RCH_2O_2 + RCH_2O_2 \rightarrow 2\,RCH_2O + O_2$$
$$\rightarrow RCH_2OH + RCHO + O_2$$

$\alpha = 0.2$ to 0.8

For alkylperoxy radicals, the rate constant increases with the number of carbon atoms (if CH_3O_2 is taken apart)(Table 4) and thus, it is fairly easy to derive an SAR.

Table 4 – *Rate constants for self reactions of primary alkylperoxy radicals.*

Radical	k(298K) 10^{-13} cm^3 molecule^{-1} s^{-1}
CH_3O_2	3.5
$C_2H_5O_2$	0.72
n-$C_3H_7O_2$	3.0
n-$C_5H_{11}O_2$	3.9
neo-$C_5H_{11}O_2$	12.0
allyl-O_2	7.0

☛ For substituted primary peroxy radicals (with OH, halogens, C=O,...), the rate constants do not change much and most values can be accounted for by the following mean value:

$$k(298K) \approx (4 \pm 2) \times 10^{-12}\ cm^3\ molecule^{-1}\ s^{-1}$$

☛ Concerning acylperoxy radicals, $RC(O)O_2$, rate constants for self reaction,

$$RC(O)O_2 + RC(O)O_2 \rightarrow 2\,RC(O)O + O_2$$
$$\hookrightarrow R + CO_2$$

are all the same within uncertainties : $k = (1.3 \pm 0.1) \times 10^{-11}$ cm^3 molecule^{-1} s^{-1}, with the exception of fluorine substituted radicals which exhibit slightly smaller values. Obviously, the $C(O)O_2$ group determines the rate constant, as it does for cross reactions (see below).

Cross reactions

The cross reactions of RO_2 radicals important in the atmosphere are the reactions with the most abundant radicals, HO_2, CH_3O_2 and $CH_3C(O)O_2$. Reliable structure activity relationships are now available for those three important classes of reactions.

Cross reactions of RO_2 with HO_2

$$RO_2 + HO_2 \rightarrow ROOH + O_2$$

Other reaction channels have never been clearly characterised and thus, this reaction is an important terminating step of chain reactions forming ozone. Rate constants measured to date are in the range (0.5 – 2.2) x 10^{-11} cm^3 $molecule^{-1}$ s^{-1}, with a rather clear correlation with the number of carbon atoms. Recent measurements in our laboratory, using a simplified method, compared to previous ones, have given the results listed in Table 5.

Table 5 : *Rate constants for RO_2 + HO_2 reactions at 298 K*

RO_2 radical	k_1 [a]	Literature value [a]
Radicals obtained by hydrogen abstraction		
CH_3O_2	5.1	5.2 [REF]
$C_2H_5O_2$	7.8	7.8 [REF]
cyclohexyl-O_2	17.2	17 [REF]
neopentyl-O_2	13.8	15 [REF]
Radicals obtained by OH addition to unsaturated compounds		
HO-(C_2H_4)-O_2	11.7	13 [REF]
HO-(1,2-$C_2H_2Cl_2$)-O_2	11.5	10 [REF]
HO-(TME)-O_2	15.1	≈20 [REF]
HO-(cyclohexene)-O_2	22.4	
HO-(α-pinene)-O_2	21	
HO-(γ-terpinene)-O_2	19.7	
HO-(d-limonene)-O_2	22.1	

[a]) units of cm^3 $molecule^{-1}$ s^{-1}

All the data are reasonably well accounted for by the following expression, n being the number of carbon atoms : $k = 2.3 \times 10^{-11} (1-\exp(-0.245n))$ cm^3 $molecule^{-1}$ s^{-1} [REF]. One exception is the reaction $CH_3C(O)O_2 + HO_2$: $k = 1.4 \times 10^{-11}$ cm^3 $molecule^{-1}$ s^{-1}. Note that the rate constants for self reactions of RO_2 radicals originating from terpenes are all near the upper limit, (2-2.2) x 10^{-11} cm^3 $molecule^{-1}$ s^{-1}.

- *Cross reactions of RO_2 with CH_3O_2*

Cross reactions of RO_2 radicals with CH_3O_2 are fairly well described by the geometrical mean of the self reaction rate constants of the two radicals :

$$k(\text{cross}) = 2\sqrt{k(RO_2\text{-self}) \times k(CH_3O_2\text{-self})}$$

Again, one exception is the reaction $CH_3C(O)O_2 + CH_3O_2$: $k = 1.0 \times 10^{-11}$ cm^3 $molecule^{-1}$ s^{-1}.

- *Cross reactions of RO_2 with $CH_3C(O)O_2$*

As emphasised above, all reactions of the $CH_3C(O)O_2$ radical have their kinetic properties determined by the $C(O)O_2$ group and thus, all the rate constants investigated have been found equal to 1.0 x 10^{-11} cm^3 $molecule^{-1}$ s^{-1}, within uncertainties. It is likely that all acylperoxy $RC(O)O_2$ radicals react with RO_2 radicals with the same rate constant. This has been verified recently (Tomas *et al.*, 2000) for the $(CH_3)_2CHC(O)O_2$ and $CH_3)_3CC(O)O_2$ radicals.

References

Lesclaux R., Combination of peroxy radicals in the gas phase, *Peroxy radicals*, Z. Alfassi, Ed., Wiley (1997), 82-112.

Lightfoot P.D., R.A. Cox, J.N. Crowley, M. Destriau, G.D. Hayman, M.E. Jenkin, G.K Moortgat and F. Zabel, Organic peroxy radicals : kinetics, spectroscopy and tropospheric chemistry, *Atmospheric Environment*, 26A (1992), 1805-1961.

Nielsen O.J., T.J. Wallington, Ultraviolet absorption spectra of peroxy radicals in the gas phase, *Peroxy radicals*, Z. Alfassi, Ed., Wiley (1997), 69-80.

Tomas A., E Villenave and R. Lesclaux, Kinetics of the $(CH_3)_2CHCO$ and $(CH_3)_3CCO$ Radical Decomposition: Temperature and Pressure Dependencies, *Phys. Chem. Chem. Phys.*, 2 (2000), 1165-1174.

Tyndall G.S., R. A. Cox, C. Granier, R. Lesclaux, G. K. Moortgat, M. J. Pilling, A. R. Ravishankara and T.J. Wallington, Atmospheric chemistry of small organic peroxy radicals, *J. Geophys. Res.* 106 (2001), 12157-12183.

Wallington T.J., P. Dagaut, M.J. Kurylo, Ultraviolet absorption cross sections and reaction kinetics and mechanisms for peroxy radicals in the gas phase,*Chem. Rev.*,92 (1992), 667-710.

Wallington T.J., O.J. Nielsen, J. Sehested, Reactions of organic peroxy radicals in the gas phase, *Peroxy radicals*, Z. Alfassi, Ed., Wiley (1997), 113-172.

Atmospheric Chemistry of Oxygenated VOCs

Georges Le Bras, Abdelwahid Mellouki, Isabelle Magneron, Valérie Bossoutrot
and Gérard Laverdet

Laboratoire de Combustion et Systèmes Réactifs, LCSR-CNRS, 45071 Orléans Cédex 2 (France)

Introduction

Oxygenated volatile organic compounds (VOCs) present in the atmosphere cover a large variety of compound families of both anthropogenic and biogenic origins. Alcohols, ethers, esters, glycol ethers and other polyfunctional compounds have mainly primary sources. Aldehydes, ketones, acids and peroxides have mainly secondary sources, although ketones, for instance have also significant primary sources. The anthropogenic component of the oxygenated VOCs emissions is expected to increase as a result of the increasing use of such compounds as solvents and fuel additives.

The atmospheric kinetics and mechanisms of oxygenated VOCs have recently been the subject of laboratory research in order to assess the impact of these compounds on, e.g., tropospheric ozone formation. We report here kinetic and mechanistic data for gas phase OH reactions of oxygenated VOCs, which is the major atmospheric oxidation process for such compounds. We discuss their impact on ozone formation and also on acetone formation which is currently suggested to influence the oxidative power of the upper troposphere (e.g. *Jaeglé et al.*, 2001).

Kinetics of OH reaction with oxygenated VOCs

The rate constants the reaction of OH radicals with oxygenated VOCs range from 1×10^{-13} to a few 10^{-11} cm^3 $molecule^{-1}$ s^{-1} at room temperature, corresponding to atmospheric lifetimes from 1 month (e.g. acetone) to less than one day (e.g. dipropyl ether). The reactivity of oxygenated VOCs is different to that of corresponding hydrocarbons, for small compounds but becomes similar when the compounds become larger.

Rate constants for the reaction of OH with a large number of alcohols, ethers, diethers, hydroxyethers, esters (formates, acetates, methyl esters), ketones have been measured in our laboratory using mostly the laser photolysis-LIF method (e.g. *El Boudali et al.*, 1996; *Le Calvé et al.*, 1997; *E. Porter et al.*, 1997; *Le Calvé et al.*, 1997). The obtained data have been used to develop structure-activity relationships such as the SAR method of Atkinson (*Kwok et Atkinson*, 1995). In this method, calculation of the H-atom abstraction for C-H bonds is based on the estimation of $-CH_3$, $-CH_2-$ and $-CH<$ group rate constants, assuming that the group rate constants depend on the identity of the substituents attached to the group. For instance, the SAR method has been applied to a large number of esters and the calculated and experimental rate constants agreed within ±20 % as shown in Figure 1 (*Le Calvé et al.*, 1997).

More generally, the structure-activity relationships established for the existing extensive data base for OH reaction rate constants of oxygenated to VOCs (k_{OH}) can be used with reasonable reliability to predict :

- unknown k_{OH} for oxygenated VOCs (when experimental data are difficult to obtain, for instance, for low volatile compounds);
- relative reactivity of the different sites in these compounds, hence product distribution for their atmospheric oxidation.

I. Barnes (ed.), Global Atmospheric Change and its Impact on Regional Air Quality, 85–90.

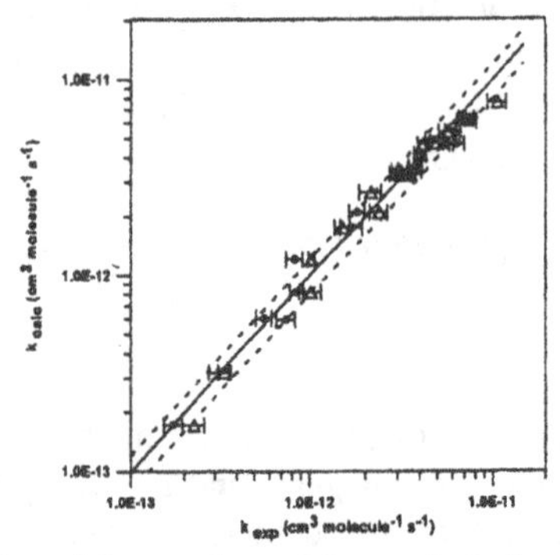

Figure 1 : Comparison of the experimental, k_{exp}, and calculated, k_{calc}, rate constants at 298 K for the reaction of OH radicals with 23 esters.

Regarding the atmospheric applications, the established OH kinetic data base for oxygenated VOCs has been used to estimate the atmospheric lifetimes of the compounds. Alcohols and ethers have relatively short lifetimes (hours to one day) whereas ketones and esters have significantly longer lifetimes (days to weeks). Alcohols and ethers will then have some impact on local photooxidant formation whereas ketones and esters will have relatively little impact.

Mechanims of OH-initiated oxidation of oxygenated VOCs

Oxidation mechanisms are investigated in photoreactors. The ones we use are : i) Teflon bags (200 L) irradiated by lamps with FTIR and GC analysis of the species after sampling; ii) the European photoreactor EUPHORE irradiated by sun and equipped with in situ FTIR and DOAS analysis and a large variety of analytical equipment sampling analysis. We present here a typical study of two compounds, hexylene glycol ($(CH_3)_2C(OH)CH_2CH(OH)CH_3$) and diacetone alcohol ($(CH_3)_2C(OH)CH_2C(O)CH_3$), which are used as solvents and intermediates for organic synthesis.

Hexylene glycol

Figure 2a shows typical concentration-time profiles of hexylene glycol and reaction products of its OH-radical initiated oxidation in the presence of NO_x in the EUPHORE photoreactor. The major carbon products are diacetone alcohol and acetone. The linear relation obtained by plotting the product concentrations as a function of hexylene consumed indicates that diacetone alcohol and acetone are primary products with yields of 50 % and 20 %, respectively (Figure 2b).

These results allowed to identify the mechanism of the OH-initiated oxidation of hexylene glycol. Diacetone alcohol and acetone were produced through the following reactions :

$$OH + (CH_3)_2C(OH)CH_2CH(OH)CH_3 \rightarrow (CH_3)_2C(OH)CH_2C^{\bullet}(OH)CH_3 + H_2O$$

$$(CH_3)_2C(OH)CH_2C^{\bullet}(OH)CH_3 + O_2 \rightarrow \underline{(CH_3)_2C(OH)CH_2C(O)CH_3} + HO_2$$

$$OH + (CH_3)_2C(OH)CH_2CH(OH)CH_3\ (+ O_2, NO) \rightarrow (CH_3)_2C(OH)CH(O^{\bullet})CH(OH)CH_3\ (+ H_2O, NO_2)$$

$$(CH_3)_2C(OH)CH(O^{\bullet})CH(OH)CH_3\ (+ M) \rightarrow \underline{CH_3C(O)CH_3} + CH_3CH(OH)CHO$$

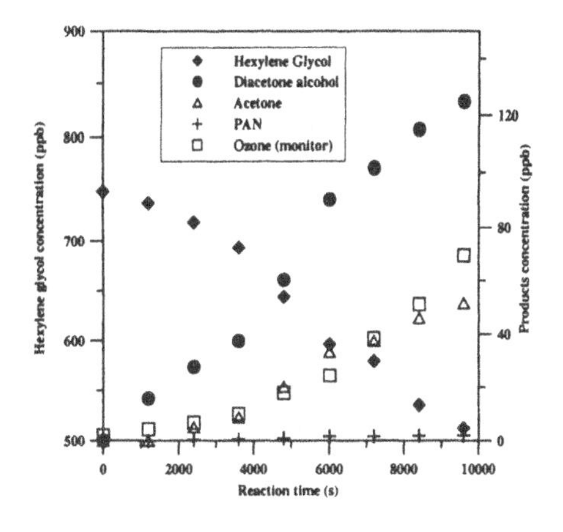

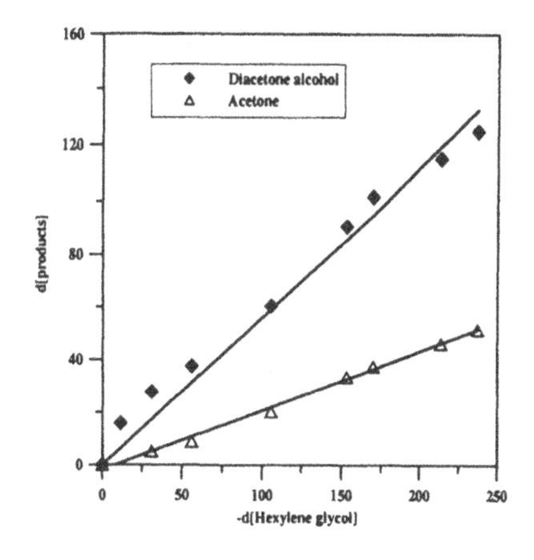

Figure 2 : Reaction OH + hexylene glycol in air in the presence of NO_x : Concentration-time profiles of hexylene glycol and products (2a); formation yields of products (2b) (EUPHORE experiments).

Diacetone alcohol

Similar experiments in EUPHORE on the OH-initiated oxidation of diacetone alcohol in air in the presence of NO_x led to identifation of acetone, formaldehyde and PAN as major and primary products with yields of 90 %, 40 % and 10 %, respectively. For acetone, the following major production route has been identified :

$$OH + (CH_3)_2C(OH)CH_2C(O)CH_3 \ (+ O_2, NO) \rightarrow (CH_3)_2C(OH)CH(O^{\bullet})C(O)CH_3 \ (+ H_2O, NO_2)$$

$$(CH_3)_2C(OH)CH(O^{\bullet})C(O)CH_3 \ (+ M) \rightarrow (CH_3)_2C^{\bullet}OH + CH_3C(O)CHO$$

$$(CH_3)_2C^{\bullet}OH + O_2 \rightarrow \underline{CH_3C(O)CH_3} + HO_2$$

PAN would be produced from the following alternative channel for the decomposition of the above oxy radical :

$$(CH_3)_2C(OH)CH(O^{\bullet})C(O)CH_3 \ (+ M) \rightarrow (CH_3)_2C(OH)CHO + CH_3CO^{\bullet}$$

$$CH_3CO^{\bullet} + O_2 \ (+ NO_2, M) \rightarrow \underline{CH_3C(O)O_2NO_2}$$

For formaldehyde, several formation routes have been considered.

The mechanistic information obtained together with the OH rate constant measurements of the primary VOCs ($k_{OH} = 1.4 \times 10^{-11}$ cm^3 $molecule^{-1}$ s^{-1} for hexylene glycol and 3.6×10^{-12} cm^3 $molecule^{-1}$ s^{-1} for diacetone alcohol at 298 K (*Magneron*, 2001)) provide parameters needed to define and assess the atmospheric impact of these VOCs. These parameters are the atmospheric persistence of the VOCs and yields and persistence of the oxidation products. These parameters are summarised below for hexylene glycol and diacetone alcohol.

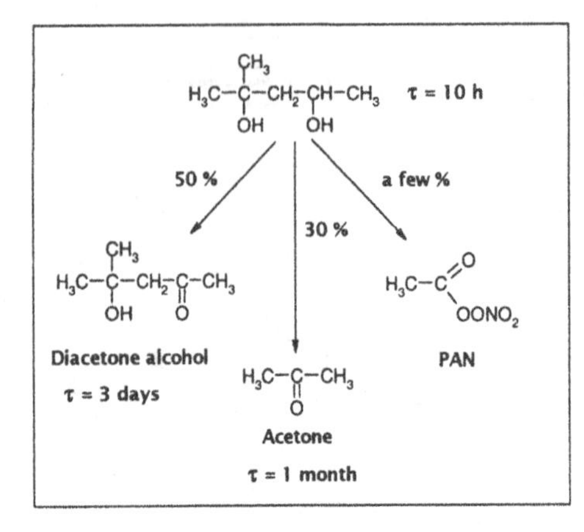

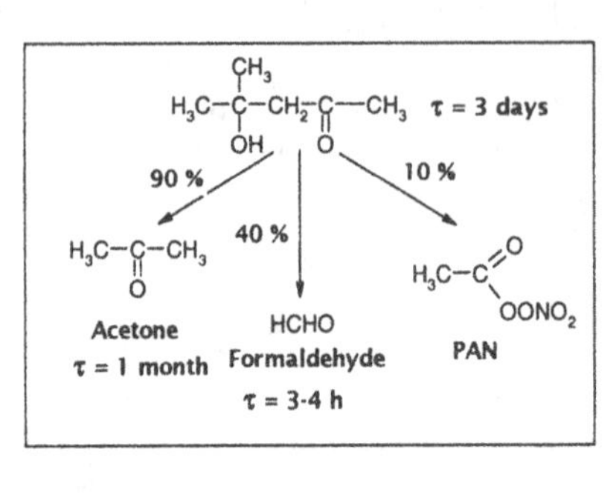

Figure 3: Major products of the OH-initiated oxidation of hexylene glycol and diacetone alcohol.

Atmospheric impact of oxygenated VOCs

The oxygenated VOCs as other types of VOCs oxidise to form local or regional ozone in the presence of NO_x in a catalytic process (Figure 4). The ozone forming potential of the VOC will depend on the nature of the primary carbonyl products. Short-lived aldehydes will enhance the ozone formation by generating HO_x radicals, whereas long-lived species such as esters (produced from ethers) or acetone will decrease the ozone formation. On the other hand, the acetone formation may give to the VOC some impact on the oxidising power of the upper troposphere as already mentioned. The ozone and acetone formation potentials of oxygenated VOCs are discussed below.

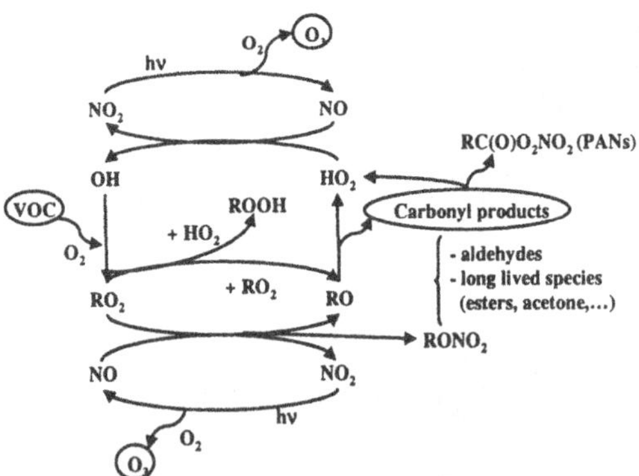

Figure 4: Oxidation mechanism of VOCs.

Ozone forming potential

The ozone forming potential of VOCs can be expressed by the POCP indices which represent the ozone formation calculated using a photochemical trajectory model (*Derwent et al.*, 1996). POCP are given below for selected oxygenated VOCs including hexylene glycol and diacetone alcohol. For these two compounds, POCP have been estimated following the method of Jenkin (*Jenkin*, 1998). The POCP for oxygenated VOCs are generally low compared to those of hydrocarbons (the reference compound, ethylene, has a POCP of 100).

Consequently, the use of oxygenated VOCs as substitutes for hydrocarbons as fuels or solvents should lead to reductions in tropospheric ozone.

Alcohols	POCP	Esters	POCP
Methanol	14	Methyl formate	2.7
Ethanol	39.9	Methyl acetate	5.9
1-butanol	62	n-propyl acetate	28.2
		t-butyl acetate	5.3
Ethers			
Dimethyl ether	18.9	Diethers/Diesters	
Diethyl ether	44.5	Dimethoxy methane	16.4
Methyl-t-butyl ether	17.5	Dimethyl carbonate	2.5
Glycol ethers		Hexylene glycol	47(estimated)
2-methoxy ethanol	30.7	Diacetone alcohol	16(estimated)
2-butoxy ethanol	48.3		

Acetone formation

Acetone, a long-lived species ($\tau \approx 1$ month), which can be convectively transported to the upper troposphere where it can provide a significant source of HO_x radicals which may influence the ozone budget (e.g. *Jaeglé et al.*, 2001). An important parameter needed to assess this impact is the global source strength of acetone. The source inventories of acetone include primary and secondary sources. The secondary ones, considered so far, result form the oxidation of anthropogenic and biogenic hydrocarbons, and they would represent more than half of the total source strengths, which are estimated to be 56 Tg yr^{-1}, however, with a rather large uncertainty range (37-80)Tg yr^{-1} (*Singh et al.*, 2000).
A scientific question that we raise is how significant is the contribution of oxygenated VOCs to the secondary source of acetone in the atmosphere.
Considering the long atmospheric lifetime of acetone, the global strength of its secondary sources, S_G(acetone), can be expressed as follows : $S_G(\text{acetone}) = \Sigma\, Y_i(\text{acetone}) \times S_G(VOC_i)$, where Y_i(acetone) and $S_G(VOC_i)$ are, respectively, the acetone formation yield and the global emission rates of the VOC_i. The acetone yield is determined in the laboratory from the investigation of the oxidation mechanism of VOCs under atmospheric conditions.
From the present mechanistic knowledge, VOC precursors of acetone must contain $(CH_3)_2CH$-, $(CH_3)_2C<$ or $(CH_3)_2C=$ groups. For instance, the atmospheric relevant oxygenated VOCs, isopropanol ($(CH_3)_2CHOH$), isopropyl ethers ($(CH_3)_2CHOR$), isopropyl esters ($(CH_3)_2CHOC(O)R$), isopropanal ($(CH_3)_2CHCHO$), tertiobutyl alcohol ($(CH_3)_3COH$), belong to this category. Acetone yields have already been measured for a substantial number of oxygenated VOCs including hexylene glycol and diacetone alcohol for which data are presented in this paper :

	% Y(acetone)	Reference
$(CH_3)_2CHOH$	0.93	Atkinson et al., 1995
$(CH_3)_3COH$	0.93	Japar et al., 1990
$(CH_3)_2C(OH)CH=CH_2$	0.5-0.6	Ferronato et al., 1998
$(CH_3)_2C(OH)CH_2CH(CH_3)_2$	0.92	Atkinson et al., 1995
$(CH_3)_2CHOCH(CH_3)_2$	0.90	Le Calvé et al., 1999
$(CH_3)_2CHOC(O)CH_3$	0.25	Tuazon et al., 1998

$(CH_3)_2CHCH_2OC(O)CH_3$	0.80	Picquet et al., 2000
$(CH_3)_3COC(O)CH_3$	0.20	Japar et al., 1997
$(CH_3)_2CHC(O)CH_3$	0.55	Le Calvé et al., 1998
$(CH_3)_2CHCH_2C(O)CH_3$	0.7-0.8	Thévenet et al., 2000
$(CH_3)_2CHC(O)H$	1	Atkinson et al., 1994
$(CH_3)_2C(OH)CH_2CH(OH)CH_3$	0.20	Magneron, 2001
$(CH_3)_2C(OH)CH_2C(O)CH_3$	0.90	Magneron, 2001

These laboratory data show that some classes of oxygenated VOCs produce high yields of acetone. However, emission rate data for these VOCs on the global scale are lacking so that it is not presently possible to assess the contribution of this acetone source to the global source. Such a source could be potentially significant when it is considered that an increase in the use of oxygenated VOCs as fuel and solvent alternatives is being considered.

Conclusions

An extensive data base is now available for rate constant of reactions of OH with oxygenated VOCs, such that unknown rate constants can be reasonably predicted from structure-activity relationships. Substantial amounts of mechanistic data have been obtained, but more work is needed. Concerning the atmospheric impact, oxygenated VOCs have relatively low ozone forming potential, but their contribution to acetone global sources remains to be assessed in relation with the potential influence of acetone on the oxidising power of the upper troposphere. Finally, to have a complete picture of the atmospheric chemistry of oxygenated VOCs, their multiphase chemistry needs to be further investigated.

References

Derwent R. G., M. E. Jenkin and S. M. Saunders; Photochemical ozone creation potential for a large number of reactive hydrocarbons under european conditions, *Atmos. Environ.* **30** (1996) 181-199.

El Boudali A., S. Le Calvé, G. Le Bras and A. Mellouki ; Kinetic studies of OH reactions with a series of acetates, *J. Phys. Chem.* **100** (1996) 12364-12368.

Jaeglé L., D. J. Jacob, W. H. Brune and P. O. Wennberg ; Chemistry of HOx radicals in the upper troposphere, *Atmos. Environ.* **35** (2001) 469-489.

Jenkin M. E.; Photochemical ozone and PAN creation potentials : Rationalisation and methods of estimation, Report on the contract EPG 1/3/70 (1998)

Kwok E. S. C. and R. Atkinson; Estimation of hydroxyl radical reaction rate constants for gas-phase organic compounds using a struture-activity relationship : an update, *Atmos. Environ.* **29** (1995) 1685-1695.

Le Calvé S., G. Le Bras and A. Mellouki ; Temperature dependence for the rate coefficients of the reactions of the OH radical with a series of formates, *J. Phys. Chem. A* **101** (1997) 5489-5493.

Le Calvé S., G. Le Bras and A. Mellouki ; Kinetic studies of OH reactions with a series of methyl esters, *J. Phys. Chem. A* **101** (1997) 9137-9141.

Magneron I; Cinétiques et mécanismes de photooxydation atmosphérique de composés oxygénés (aldéhydes insaturés, hydroxyaldéhyde, diol, hydroxycétone et alcool aromatique), PhD thesis, Orléans (2001).

Porter E., J. Wenger, J. Treacy, H. Sidebottom, A. Mellouki, S. Teton and G. Le Bras; Kinetic studies of the reaction of hydroxyl radicals with diethers and hydroxyethers, *J. Phys. Chem.* **101** (1997) 5770-5775.

Singh H., Y. Chen, A. Tabazadeh, Y. Fukui, I. Bey, R. Yantosca, D. Jacob, F. Arnold, K. Wohlfrom, E. Atlas, F. Flocke, D. Blake, N. Blake, B. Heikes, J. Snow, R. Talbot, G. Gregory, G. Sachse, S. Vay et Y. Konds; Distribution and fate of selected oxygenated organic species in the troposphere and lower stratosphere over the atlantic, *J. Geophys. Res.* **105** (2000) 3795-3805.

Acknowledgments

We thank UCD Dublin (Dr H. Sidebottom and co-workers) and CEAM (Dr Klaus Wirtz and co-workers) for cooperation. We gratefully acknowledge financial support from the European Union (Environment programme), CNRS (French programme of atmospheric chemistry), the Ministry of Environment (Air quality programme) and the TotalFinaElf company.

Experimental and Theoretical Study of the Atmospheric Degradation of Aldehydes

Barbara D'Anna and Claus J. Nielsen

Department of Chemistry, University of Oslo, Blindern, N-0315 Oslo, Norway

Introduction

Aldehydes are ubiquitous key components in the chemistry of the troposphere. They are common primary pollutants from biogenic emissions and in residues of incomplete combustion (Ciccioli *et al.*, 1993). Relevant natural sources are vegetation, forest fires and microbiological processes (Kotzias *et al.*, 1997). Aldehydes are also nearly mandatory intermediates in the photo-oxidation processes of most organic compounds in the troposphere (Kerr and Sheppard, 1981; Carlier *et al.*, 1986). Formaldehyde (HCHO) and acetaldehyde (CH_3CHO) are among the most abundant carbonyls in the atmosphere. Ambient levels are in the order of a few tens of pptv in clean background conditions (Zhou et al., 1996; Ayers et al., 1997) but may reach tens of ppbv in polluted urban areas as a consequence of the elevated anthropogenic emissions of aldehydes and their precursors from automobile traffic, industrial and domestic heating, and industrial activity (Carlier *et al.*, 1986; Yokouchi *et al.*, 1990). The atmospheric loss processes include photolysis, day-time reaction with OH radicals and with Cl and Br atoms in the marine boundary layer, and reaction with NO_3 radicals during the night-time. The photolytic cleavage of aldehydes constitute an important source of free radicals, particularly in the moderately and strongly polluted areas (Carlier *et al.*, 1986; Yokouchi *et al.*, 1990). Aldehydes are toxic compounds themselves, and some of their photo-oxidation products, the peroxyacylnitrates, are phytotoxic and strong eye-irritant compounds (Carlier *et al.*, 1986; Carter *et al.*, 1981). Further, peroxyacylnitrates, such as peroxyacetylnitrate (PAN), are long-lived species, which can act as a NO_2 reservoir in the troposphere.

Experimental studies

Two recent research projects, financed through the CEC Environment and Climate program, have adressed the atmospheric chemistry of aldehydes (RADICAL, 2000; CATOME, 2000), and a number of results from these studies have appeared in the literature (Beukes *et al.*, 2000; Ullerstam *et al.*, 2000; D'Anna *et al.*, 2001a, 2001b, 2002). Other results have been published by the Atkinson group (Papagni *et al.*, 2000).

For all the aliphatic aldehydes studied so far the Cl and Br reactions show small, positive activation energies (Atkinson *et al.*, 1997), the OH reactions show negative activation energies (Atkinson, 1994), while the NO_3 reactions show activation energies of more than 10 kJ mol^{-1} (Atkinson, 1994; Ullerstam *et al.*, 2000; D'Anna *et al.*, 2001a). The larger aliphatic aldehydes show reactivities

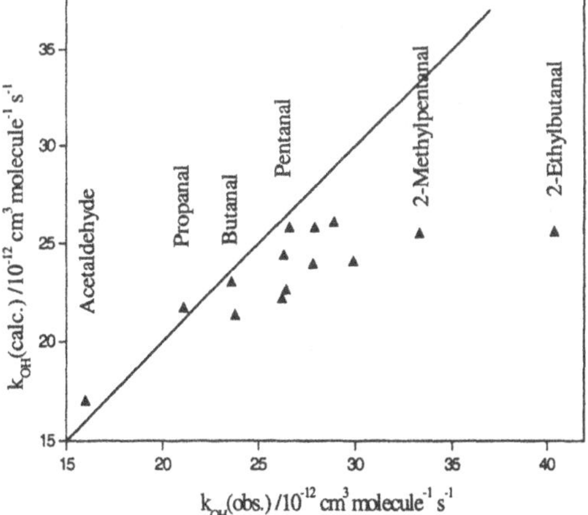

Figure 1.. Calculated *vs.* observed reaction rate coefficients at 298 K for the OH reaction with aliphatic aldehydes. The calculated reaction rate coefficients are based on the structure-reactivity by Kwok and Atkinson (1995).

I. Barnes (ed.), Global Atmospheric Change and its Impact on Regional Air Quality, 91–96.

towards OH and NO_3 radicals deviating from commonly used structure activity relations. This is illustrated in Figure 1 for the OH reactions. One may also note that the structure activity relationship for the NO_3 reactions with organics (Atkinson, 1991) fails completely for the aldehydes (D'Anna *et al.*, 2001b).

In a so-called linear free energy plot showing $\log(k_{NO_3})$ *vs.* $\log(k_{OH})$, Figure 2, the reaction rate coefficients fall close to the correlation line for addition reactions in spite the fact that the reactions are abstraction reactions, see below (D'Anna and Nielsen, 1997; D'Anna *et al.*, 2001b).

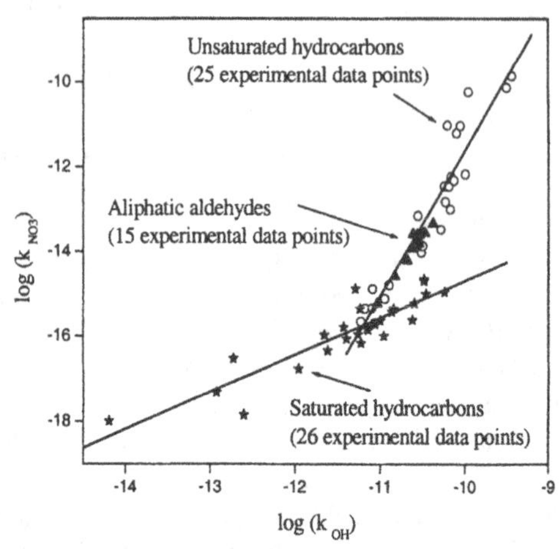

Figure 2. Log of the rate coefficients at 298 K for the reaction of NO_3 *vs.* the log of the rate coefficients for reaction of OH with the same aliphatic aldehyde.

Aliphatic aldehydes may in principle react with radicals either by H-abstraction or by addition of the radical to the π-electron system. This is illustrated below using acetaldehyde as the example:

$$\begin{aligned} X + CH_3CHO &\rightarrow CH_3CXO + H && \text{(a)} \\ &\rightarrow CH_3 + XCHO && \text{(b)} \\ &\rightarrow CH_3CO + HX && \text{(c)} \\ &\rightarrow CH_3 + HX + CO && \text{(d)} \\ &\rightarrow CH_2CHO + HX && \text{(e)} \end{aligned}$$

Channels (a) and (b) may be termed as addition reactions, while the other three are H-abstraction reactions. According to thermodynamic data (Atkinson *et al.*, 1997, and references therein) and to quantum chemical results (Beukes *et al.*, 2000; D'Anna *et al.*, 2002) all channels should be open to reaction for F and OH, channels (c)-(e) should be open to reaction for Cl and NO_3, while Br may only react through channel (c).

There are three product studies of the Cl reaction with acetaldehyde (Niki *et al.*, 1985; Wallington *et al.*, 1988; Beukes *et al.*, 2000) and all conclude that less than 5 % of the reaction proceeds through H_{met}-abstraction. There are no similar studies of the Cl reaction with the larger aldehydes.

Results from studies of the regioselectivity of the NO_3 reaction with aldehydes have been presented (D'Anna *et al.*, 2001a, 2001b). The results for acetaldehyde and butanal both imply that more than 95% of the NO_3 reaction proceeds through H_{ald}-abstraction. There are no systematic studies of the OH reaction paths for the larger aldehydes. However, results for acetaldehyde set upper limits of 2.5 and 2% through the addition channels (a) and (b), respectively, and 10% in total to channel (d) and to the H_{met}-abstraction reaction (D'Anna *et al.*, 2002).

Additional insight into the reaction mechanisms may be obtained from studies of the isotopic signature of reaction. Kinetic isotope effects have been reported for Cl reactions with formaldehyde and acetaldehyde (Niki *et al.*, 1978; Beukes *et al.*, 2000), and unusually large kinetic isotope effects are reported in the similar bromine reactions (Beukes *et al.*, 2000). Isotopic data are available for the OH reactions with formaldehyde (Morris and Niki, 1971; Niki *et al.*, 1984; D'Anna *et al.*, 2002) and acetaldehyde (Taylor *et al.*, 1996; D'Anna *et al.*, 2002). Kinetic isotope effects have also been reported in the NO_3 radical reactions with formaldehyde and acetaldehyde (D'Anna *et al.*, 2001; D'Anna *et al.*, 2002). The experimental kinetic isotope effects observed in the formaldehyde reactions are summarised in Table 1.

Table 1. Experimental kinetic isotope effects for the reactions of Cl, Br, NO_3 and OH radicals with formaldehyde at 298 K.

k_A/k_B	HCHO/DCDO	$H^{13}CHO/DCDO$	HCHO/ $H^{13}CHO$	HCHO/ DCHO
Cl	1.302 ± 0.014 [a]	1.217 ± 0.025 [a]		
	1.3 ± 0.2 [b]			
Br	7.5 ± 0.4 [a]	6.8 ± 0.4 [a]		
NO_3	2.97 ± 0.14 [c]		0.97 ± 0.02 [c]	
OH	1.62 ± 0.08 [c]	1.64 ± 0.12 [c]	0.97 ± 0.11 [c]	~1 [d]

[a] Beukes *et al.*, 2000. [b] Niki *et al.*, 1978. [c] D'Anna *et al.*, 2002. [d] Morris and Niki, 1971.

In summary, there is overwhelming experimental evidence of the Cl, Br, NO_3 and OH reactions with aliphatic aldehydes being primarily H_{ald}-abstraction reactions and not addition reactions. It is therefore somewhat puzzling why their reactivity towards OH and NO_3 radicals apparently fall on the addition line in a linear free energy plot, Fig. 2. It should also be noted that common to the all the experimental isotopic signatures is the fact that it is neither possible to reproduce the trends nor the actual magnitudes satisfactory by conventional transition state theory (Beukes *et al.*, 2000; D'Anna *et al.*, 2002).

Theoretical studies

The reactions of F, Cl, Br, NO_3 and OH radicals with formaldehyde and acetaldehyde were recently studied by quantum chemical methods on the MP2 and CCSD(T) levels of theory (Beukes *et al.*, 2000; D'Anna *et al.*, 2002). The calculations indicate the existence of weak adducts in which the radicals in all cases are bonded to the aldehydic oxygen. Transition states of the reactions X + HCHO $\rightarrow$ products, and X + CH_3CHO $\rightarrow$ products have been located, and energy level diagrams for reactants, intermediates and products in the reactions have been presented and discussed in relation to the observed product distributions. For the F, Cl, Br and NO_3 reactions the rate determining transition states are situated above the entrance channel. However, for the OH reactions the transition states to H_{ald}-abstraction are *below* the entrance channels. The relevant energy levels for formaldehyde/OH are illustrated in Figure 3.

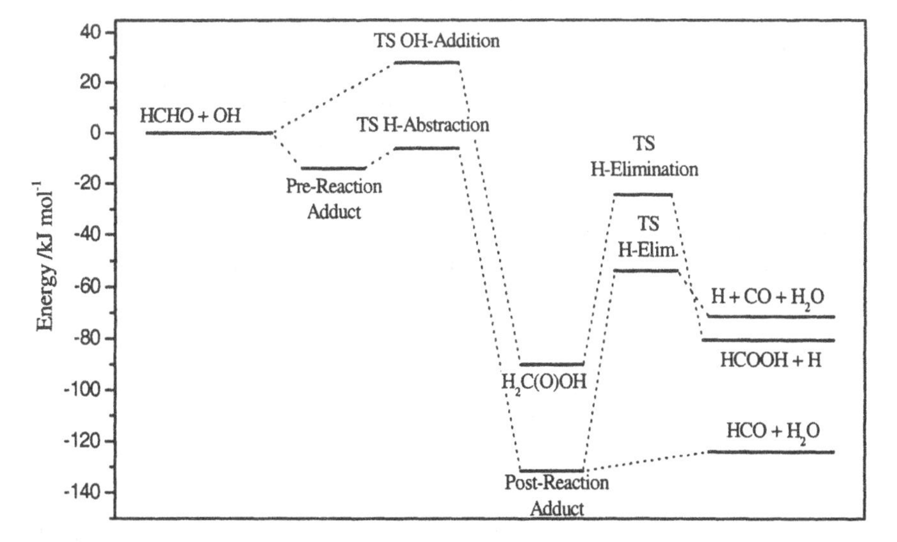

Figure 3. Energy levels, $E_{CCSD(T)} + E_{ZPE}$, of the OH + HCHO reaction system calculated at the CCSD(T)/aug-cc-pVDZ//MP2/aug-cc-pVDZ level.

As can be seen from Figure 3 the barrier to H_{ald}-abstraction is calculated to be negative relative to the reactants energy, $E_{CCSD(T)} + E_{ZPE} = -5.8$ kJ mol^{-1}(D'Anna *et al.*, 2002), and that the OH radicals form both pre- and post-reaction adducts with the substrate. The energy of the pre-reaction adduct, including the zero point energy, is ca. −11 kJ mol^{-1} relative to that of the reactants. As to the relative importance of two channels in the H_{ald}-abstraction reaction, channel (c) conform with the minimum energy path leading from the transition state via an adduct to the products, HCO + H_2O, without any barrier, whereas channel (d) involves a barrier of *ca.* 70 kJ mol^{-1} to the elimination of the hydrogen atom.

Figure 4 shows the OH–HCHO structures of the pre- and post-reaction adducts together with that of the transition state to H_{ald}-abstraction. The figure also includes the more interesting bond distances and the imaginary frequency corresponding to the negative eigenvalue of the Hessian matrix.

The barrier to OH addition to the carbonyl carbon was calculated as ca. 28 kJ mol^{-1} (D'Anna *et al.*, 2002) and the structure of the transition state is included in Figure 4. Previous studies report barriers ranging from 34 to 43 kJ mol^{-1} depending on the method of calculation (Soto and Page, 1990; Alvarez-Idaboy *et al.*, 2001). In any case, the barrier to the OH addition reaction with formaldehyde is so high that pathway (a) to the formation of HCOOH has little importance under atmospheric conditions.

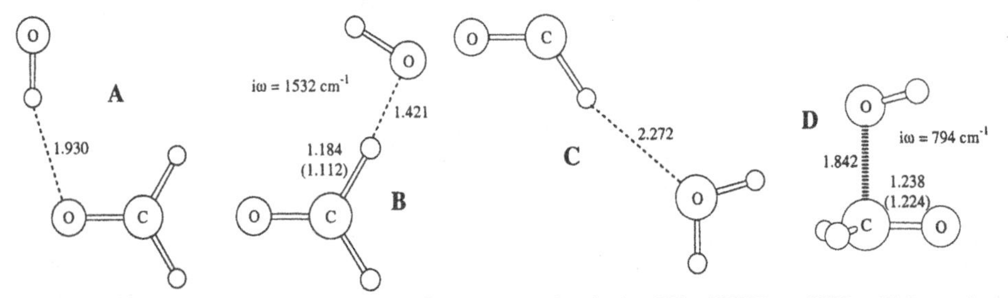

Figure 4. MP2/aug-cc-pVDZ structures of stationary points in the OH + HCHO → HCO + H_2O reaction: (A) the pre-reaction adduct, (B) the transition state, (C) the post-reaction adduct; (D) the transition state of the OH addition reaction, OH + HCHO → HCOOH + H_2O. Distances in Å, values in parenthesis refer to the free molecule.

Very similar results are obtained for the acetaldehyde–OH system (D'Anna *et al.*, 2002). In the H_{ald}-abstraction pathway the OH radicals form pre- and post-reaction adducts with the substrate. The barrier to H_{ald}-abstraction is below the entrance channel, ca. -14 kJ mol^{-1}. The minimum energy path from the transition state to H_{ald}-abstraction is via a water adduct to the reactants, CH_3CO + H_2O, without any barrier. The alternative channel (d) involves a barrier of *ca.* 70 kJ mol^{-1} to the formation of CH_3 + CO + H_2O. Although the transition state towards C-C bond breakage in channel (d) is ca. 50 kJ mol^{-1} below the energy of the reactants, it is more likely that the excess energy of reaction is channeled into the loose bond to water than into C-C bond breakage. That is, from the theoretical results alone channel (c) is expected to dominate the reaction. The minimum energy path to H_{methyl}-abstraction is also via pre- and post-reaction adducts and the barrier to H_{methyl}-abstraction is calculated as only 12 kJ mol^{-1}. In any case, the barrier to H_{methyl}-abstraction is so high that channel (e) is of little importance at atmospheric conditions. The barrier to OH addition to the carbonyl carbon, the first step of channels (a) and (b), is calculated as ca. 31 kJ mol^{-1}. The addition channel is therefore of no importance under atmospheric conditions.

In summary, the theoretical calculations indicate that H_{ald}-abstraction is by far the dominating reaction of OH with acetaldehyde. In all likelihood, the same is true for the larger aldehydes as well.

Rates of the OH reactions with formaldehyde and acetaldehyde have been derived from general expressions based on RRKM theory (D'Anna *et al.*, 2002) following the method by

Jodkowski *et al.* (1998). In the case of acetaldehyde, the OH radical forms two H-bonded adducts, one with the oxygen oriented towards H_{ald} (A_{ald}), and one with oxygen oriented towards H_{methyl} (A_{methyl}), separated by a very low energy barrier ($TS_{met-ald}$). The following model was assumed:

$$OH + CH_3CHO \Leftrightarrow A_{methyl} \Leftrightarrow TS_{met-ald} \Leftrightarrow A_{ald} \Leftrightarrow TS_{Hald-abstraction} \Leftrightarrow \text{Products}$$

Taking the structures and energies obtained by the quantum chemical calculations it was not only possible to reproduce the overall temperature dependence of the reactions, but also to calculate the kinetic isotopic signatures (D'Anna *et al.*, 2002). Figure 5 shows a comparison between the experimental data and the calculated reaction rate coefficient.

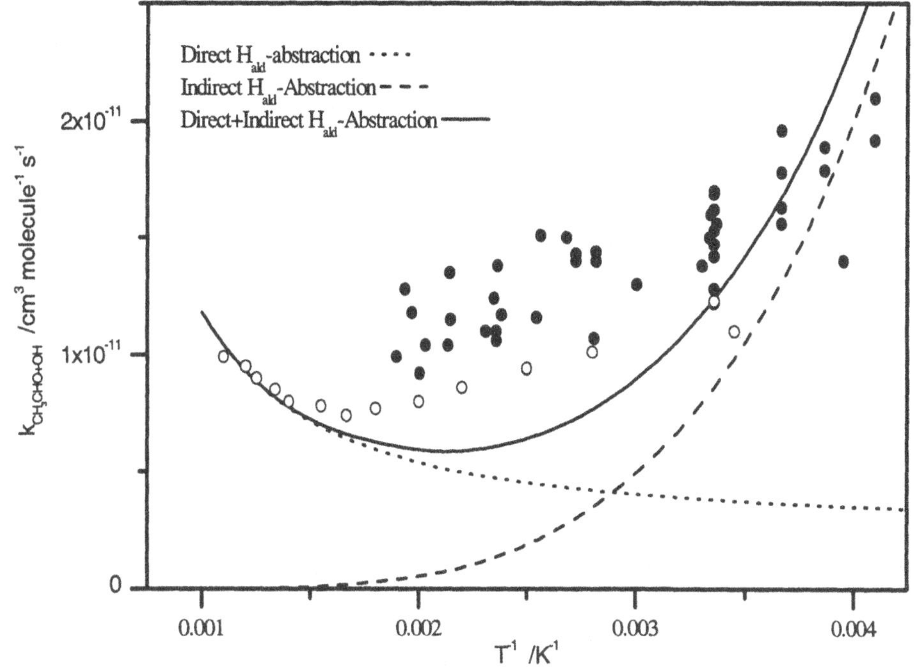

Figure 5. Observed and calculated rate coefficients of the reaction $OH + CH_3CHO \rightarrow$ products. Data taken from Atkinson (1994) (•) and from Taylor *et al.* (1996) (o). Dotted curve: Rate coefficient for direct H_{ald}-abstraction. Dashed curve: Rate coefficient for (indirect) H_{ald}-abstraction via initial adduct formation. Full curve: Sum of the rate coefficients for direct and indirect H_{ald}-abstraction.

Acknowledgement

This work is part of the project "Carbonyls in Tropospheric Oxidation Mechanisms" (CATOME) and has received support from the CEC Environment and Climate program through contract ENV4-CT97-0416.

References

Alvarez-Idaboy, J. R., N. Mora-Diez, R. J. Boyd and A. Vivier-Bunge; On the importance of prereactive complexes in molecule-radical reactions: Hydrogen abstraction from aldehydes by OH, *J. Am. Chem. Soc.* **123** (2001) 2018-2024.

Atkinson, R.; Kinetics and mechanisms of the gas-phase reactions of the NO_3 radical with organic compounds, *J. Phys. Chem. Ref. Data* 20 (1991) 459-507.

Atkinson, R.; Gas-phase tropospheric chemistry of organic compounds, *J. Phys. Chem. Ref. Data* **Monograph 2** (1994).

Atkinson, R., D. L. Baulch, R. A. Cox, R. F. Hampson Jr., J. A. Kerr, M. J. Rossi and J. Troe; Evaluated kinetic, photochemical and heterogeneous data for atmospheric chemistry. 5. IUPAC Subcommittee on Gas Kinetic Data Evaluation for Atmospheric Chemistry, *J. Phys. Chem. Ref. Data* **26** (1997) 521- 1011.

Ayers, G. P., R. W. Gillet, H. Granek, C. de Serves and R. A Cox; Formaldehyde production in clean marine air, *Geophys. Res. Lett.* **24** (1997) 401-404.

Beukes, J. A., B. D'Anna, V. Bakken and C. J. Nielsen; Experimental and theoretical study of the F, Cl and Br reactions with formaldehyde and acetaldehyde, *Phys. Chem. Chem. Phys.* **2** (2000) 4049-4060.

Carlier, P., H. Hannachi and G. Mouvier; The chemistry of carbonyl compounds in the atmosphere - a review, *Atmos. Environ.* **20** (1986) 2079-2099.

Carter, W. P. L., A. M. Winer and J. N. Pitts; Effect of peroxyacetyl nitrate on the initiation of photochemical smog, *Environ. Sci. Technol.* **15** (1981) 831-834.

CATOME "Carbonyls in Tropospheric Oxidation Mechanisms", CEC Environment and Climate program contract ENV4-CT97-0416, Coordinated by C. Dye (2000).

Ciccioli, P., E. Brancaleoni, M. Frattoni, A. Cecinato and A. Brachetti; Ubiquitous occurrence of semivolatile carbonyl-compounds in tropospheric samples and their possible sources, *Atmos. Environ.* **27A** (1993) 1891- 1901.

D'Anna, B. and C. J. Nielsen; Kinetic study of the vapour-phase reaction between aliphatic aldehydes and the nitrate radical, *J. Chem. Soc. Faraday Trans.* **93** (1997) 3479-3483.

D'Anna, B., S. Langer, E. Ljungström, C. J. Nielsen and M. Ullerstam; Rate coefficients and Arrhenius parameters for the reaction of the NO_3 radical with acetaldehyde and acetaldehyde-1d, *Phys. Chem. Chem. Phys.* **3** (2001a) 1631-1637.

D'Anna, B., Ø. Andresen, Z. Gefen and C. J. Nielsen; Kinetic study of OH and NO_3 radical reactions with 14 aliphatic aldehydes, *Phys. Chem. Chem. Phys.* **3** (2001b) 3057-3063.

D'Anna, B. V. Bakken, J. A. Beukes, J. T. Jodkowski and C. J. Nielsen; Experimental and theoretical study of gas phase NO_3 and OH radical reactions with formaldehyde, acetaldehyde and their isotopomers, *Phys. Chem. Chem. Phys.* **4** (2002) *submitted.*

Kerr, J. A. and D. W. Sheppard; Kinetics of the reactions of hydroxyl radicals with aldehydes studied under atmospheric conditions, *Environ. Sci. Technol.* **8** (1981) 960-963.

Kotzias, D., C. Konidari and C. Spartà; Volatile carbonyl compounds of biogenic origin - emission and concentration in the atmosphere, *in Biogenic Volatile Organic Compouns in the Atmosphere – Summary of present knowledge* (Eds. G. Helas, S. Slanina and R. Steinbrecher), SPB Academic Publishers, Amsterdam, 1997, 67-78.

Morris, E. D. Jr. and H. Niki; Mass spectrometric study of the reaction of hydroxyl radical with formaldehyde, *J. Chem. Phys.* **55** (1971) 1991-1992.

Niki, H., P. D. Maker, L. P. Breitenbach and C. M. Savage; FTIR studies of the kinetics and mechanism for the reaction of chlorine atom with formaldehyde, *Chem. Phys. Lett.* **57** (1978) 596-599.

Niki, H., P. D. Maker, C. M. Savage and L. P. Breitenbach; An Fourier transform infrared study of the kinetics and mechanism for the reaction of hydroxyl radical with formaldehyde, *J. Phys. Chem.* **88** (1984) 5342-5344.

Niki, H., P. D. Maker, C. M. Savage and L. P. Breitenbach; FTIR study of the kinetics and mechanism for chlorine-atom-initiated reactions of acetaldehyde, *J. Phys. Chem.* **89** (1985) 588-591.

Papagni, C., J. Arey and R. Atkinson; Rate constants for the gas-phase reactions of a series of C-3 - C-6 aldehydes with OH and NO_3 radicals, *Int. J. Chem. Kin.* **32** (2000) 79- 84.

RADICAL "Evaluation of Radical Sources in Atmospheric Chemistry through Chamber and Laboratory Studies", CEC Environment and Climate program contract ENV4-CT97-0419, Coordinated by G. Moortgat (2000).

Soto, M. R. and M. Page; Features of the potential energy surface for reactions of hydroxyl with formaldehyde, *J. Phys. Chem.* **94** (1990) 3242-3246.

Taylor, P. H., M. S. Rahman, M. Arif, B. Dellinger and P. Marshall; Kinetics and mechanistic studies of the reaction of hydroxyl radicals with acetaldehyde over an extended temperature range, *26th International Symposium on Combustion* (1996) 497-504.

Ullerstam, M., S. Langer and E.Ljungström, Gas phase rate coefficients and activation energies for the reaction of butanal and 2-methyl-propane with nitrate radicals, *Int. J. Chem. Kit.* **32** (2000) 294- 303.

Wallington, T. J., L. M. Skewes, W. O. Siegel, C. H. Wu and S. M. Japar; Gas phase reaction of chlorine atoms with a series of oxygenated organic species at 295 K, *Int. J. Chem. Kin.* **20** (1988) 867-875.

Yokouchi, Y., H. Mukai, K. Nakajima and Y. Ambe; Semivolatile aldehydes as predominant organic gases in remote areas, *Atmospheric Environment* **24A** (1990) 439-442.

Zhou, X. L., Y-N. Lee, L. Newman, X. H. Chen and K. Mopper; Tropospheric formaldehyde concentration at the Mauna Loa observatory during the Mauna Loa observatory photochemistry experiment 2, *J. Geophys. Res.* **101** (1996) 14711- 14719.

The Photochemical Degradation of Chloral and Oxalyl Chloride

Lorraine Nolan , Tanya Kelly, John Wenger, Jack Treacy, Klaus Wirtz
and Howard Sidebottom
Department of Chemistry, University College Dublin, Belfield, Dublin 4, Ireland

Introduction

The major source of chloral, CCl_3CHO in the atmosphere is from the degradation of 1,1,1-trichloroethane, methyl chloroform, (Nelson *et al.*, 1990, Platz *et al.*, 1995). In comparison, direct emissions of chloral from its use as a feedstock chemical are negligible. While methyl chloroform was formerly widely used as a solvent, it is now strictly regulated as an ozone-depleting substance under the Montreal Protocol. As a consequence, estimates of annual global emissions have fallen by approximately 95% over the last 10 years (Midgeley and McCulloch, 1999). These calculations are borne out by recent observations of atmospheric background concentrations of methyl chloroform (Montzka *et al*, 1999). Methyl chloroform degrades mainly by reaction with the hydroxyl radical in the troposphere to give chloral (~85%) with minor contributions from hydrolysis in the oceans (~5%) and destruction in the stratosphere (~10%) (Cox, 1998). Chloral is expected to have a relatively short atmospheric lifetime due to its loss *via* reaction with the OH radical, UV photolysis and wet deposition (Nelson *et a.l*, 1990, Platz *et al.*, 1995, Rattigan *et al.*, 1993, Barry *et al.*, 1994, Sidebottom and Franklin, 1996, Rattigan *et al.*, 1998, Talukdar *et al.*, 2001). It has generally been assumed that the major tropospheric loss process for chloral is by photolysis. This suggestion is based on the UV absorption cross sections of chloral and an assumed quantum yield for photodissociation of unity (Rattigan *et al.*, 1993, Sidebottom and Franklin, 1996, Rattigan *et al.*, 1998). This work is concerned with a determination of the quantum yields for photodissociation of CCl_3CHO and for formation of the resulting products under atmospheric conditions.

Experimental

Photolysis experiments for chloral and oxalyl chloride in the presence of a Cl atom trap were carried out in Dublin. The experiments were performed at 298 ± 2 K and atmospheric pressure in a Teflon cylindrical reaction vessel (volume ~ 50 litre) surrounded by a bank of fluorescent lamps. The loss of reactants and formation of products was monitored by FTIR spectroscopy using an evacuable multi-path cell (pathlength 10m). Analyses of the reaction mixtures were also carried out by capillary gas chromatography using flame ionisation detection.

The photolysis of chloral and oxalyl chloride in the presence of cyclohexane were also investigated at the European Photoreactor. The experiments were performed in a hemispherical reaction chamber made of FEP Teflon foil with a volume of approximately 195,000 litres. A White mirror system installed inside the chamber and aligned with an optical pathlength of 526.8m was used for *in situ* measurements by FTIR spectroscopy. The concentration of carbon monoxide was measured using a commercial CO monitor. Samples of the reaction mixtures were also analysed by gas chromatography using flame ionisation and photoionization detectors.

Results and Discussion

Photolyses of chloral carried out in the laboratory in the presence of excess cyclohexane as a Cl atom trap and at $\lambda > 300$nm gave CCl_2O and CO as major products. The yields of these products were 0.87 ± 0.06 and 1.1 ± 0.1 respectively and were based on the loss of chloral. The concentration - time profile for reactants and products for the photolysis of

I. Barnes (ed.), Global Atmospheric Change and its Impact on Regional Air Quality, 97–101.

chloral carried out at the EUPHORE facility, Figure 1., gave the following product yields CO (1.0 ± 0.05), CCl_2O (0.84 ± 0.05), c-C_6H_{12} -(1.2 ± 0.05), cyclohexanone (0.47 ± 0.3) and cyclohexanol (0.11 ± 0.01), Figure 2.

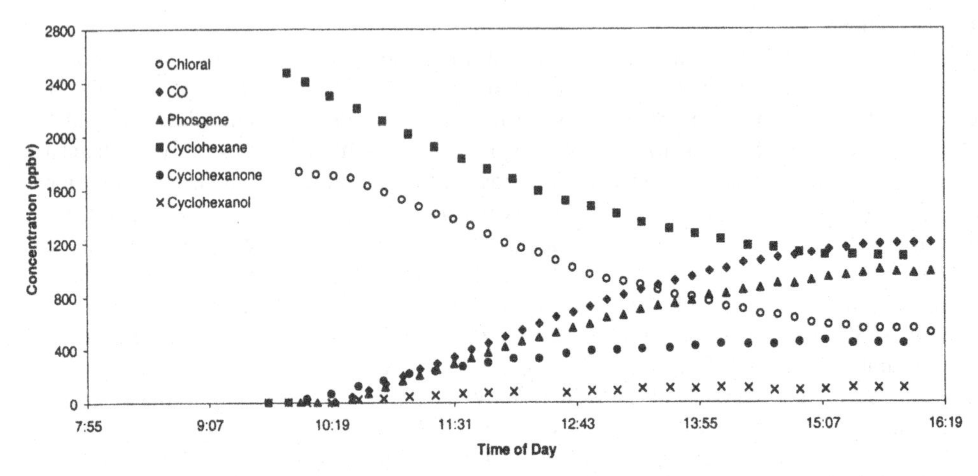

Figure 1. Concentration-time profiles of reactants and products during the sunlight photolysis of chloral (1700 ppbv) in the presence of cyclohexane (2400 ppbv) at EUPHORE

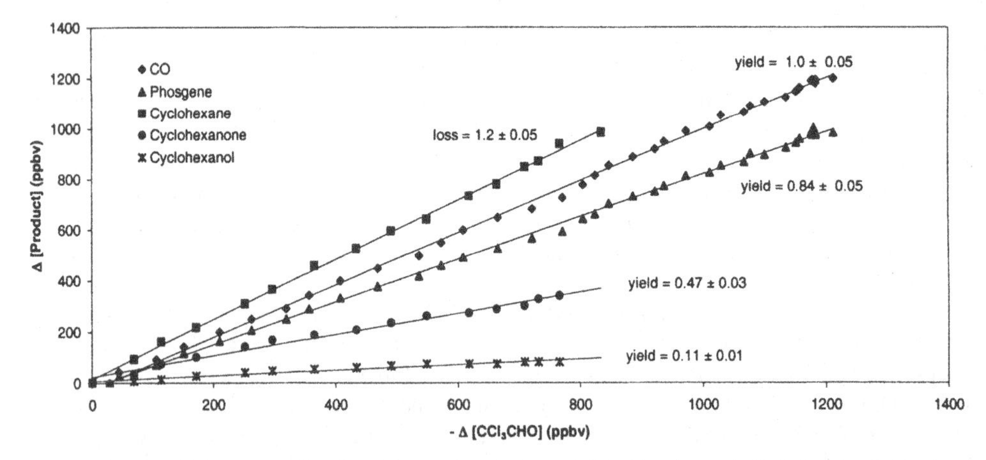

Figure 2. Plots of the yields of products against the loss of chloral for the sunlight photolysis of chloral (1700 ppbv) in the presence of cyclohexane (2400 ppbv) at EUPHORE

The maximum theoretical loss rates of chloral in experiments carried out at EUPHORE were calculated from the solar flux intensity measurements of the spectroradiometer, reported absorption cross-section data (Rattigan *et al.*, 1993, 1998) and assuming a quantum efficiency of unity across the actinic absorption range of chloral. A comparison of the observed photolysis rate with the calculated maximum theoretical loss rate for each experiment allows the effective quantum efficiency to be determined. The results yield a value of 1.00 ± 0.05 for the photodissociation quantum yield for chloral, which validates the assumption of Rattigan *et al.* (1993, 1998) in calculating a chloral tropospheric lifetime in the range of 4-5 hours.

There are four possible reaction pathways for the photolysis of chloral in the actinic region of the troposphere, > 290nm.

$CCl_3CHO + h\nu$	$\rightarrow$	$CCl_3 + CHO$	(1)
	$\rightarrow$	$CCl_3CO + H$	(2)
	$\rightarrow$	$Cl + CCl_2CHO$	(3)
	$\rightarrow$	$CCl_3H + CO$	(4)

The primary quantum yields for O, H and Cl atoms from the photolysis of CCl_3CHO by detecting them via atomic resonance fluorescence following pulsed excimer laser photolysis at 308 nm have recently been reported, (Talukdar *et al.*, 2001). The quantum yield of Cl atoms was 1.3 ± 0.3, while the yields of O and H atoms were negligible. The authors concluded from their results that chlorine atoms were produced in the primary photolytic process with a quantum yield close to unity. They proposed that the photodissociation of CCl_3CHO occurs mainly via reaction (3) to give a Cl atom and a CCl_2CHO radical. The yield of CCl_2O of 0.84 ± 0.05 found in this work indicates that if reaction (3) is the major photodissociation pathway then CCl_2CHO does not react quantitatively to form CCl_2O as previously proposed, (Talukdar *et al.*, 2001). Assuming that all of the Cl atoms are effectively scavenged by reaction with cyclohexane, then comparison of the loss of chloral and cyclohexane provides a value of 1.2 ± 0.05 for the Cl atom yield from the photolysis of chloral. The observed yields of cyclohexanone and cyclohexanol are consistent with previous studies on the Cl atom initiated oxidation of cyclohexane (Rowley *et al.*, 1991). The results can be rationalised in terms of the following mechanism:

$CCl_2CHO + O_2 + M$	$\rightarrow$	$C(O)HCCl_2O_2 + M$	(5)
$2C(O)HCCl_2O_2$	$\rightarrow$	$2C(O)HCCl_2O + O_2$	(6)
$C(O)HCCl_2O$	$\rightarrow$	$HCO + CCl_2O$	(7)
	$\rightarrow$	$C(O)HC(O)Cl + Cl$	(8)
$HCO + O_2$	$\rightarrow$	$CO + HO_2$	(9)

Thus, decomposition of $C(O)HCCl_2O$ occurs by either C-C or C-Cl bond fission. Loss of a Cl atom from the $C(O)HCCl_2O$ radical yields two atoms of Cl per molecule of chloral photolysed and hence the yield of Cl will be higher than unity as observed in this work. The results of the present study indicate that the yield of $COCl_2$ (0.84) is lower than that for CO (1.0). According to the proposed mechanism the yield of CCl_2O and CO should be equal, however the product C(O)HC(O)Cl would be expected to be rapidly photolysed to produce additional CO and Cl atoms as observed experimentally.

A sample of C(O)HC(O)Cl was not available, however, the photolysis of the structurally similar compound oxalyl chloride was carried out in the laboratory and at EUPHORE. The major products from the photolysis of oxalyl chloride at $\lambda > 300$nm in laboratory studies were CO (1.6 ± 0.1) and CCl_2O (0.39 ± 0.05). The photolytic lifetime of C(O)ClC(O)Cl under sunlight conditions was around 60 minutes, Figure 3. The major products were CO (1.8 ± 0.1) and CCl_2O (0.45 ± 0.05), Figure 4. The results are consistent with the following reaction scheme

$C(O)ClC(O)Cl + h\nu$	$\rightarrow$	2COCl	(10)
	$\rightarrow$	$CCl_2O + CO$	(11)
ClCO	$\rightarrow$	$Cl + CO$	(12)

The relative short lifetime of C(O)ClC(O)Cl and distribution of products indicates that the analogous α-dicarbonyl C(O)HC(O)Cl, proposed as a product in the photolysis of chloral, may degrade to a certain extent during the time frame of the present experiment to give additional amounts of CO and Cl atoms.

Figure 3. Concentration-time profiles of reactant and products during the sunlight photolysis of oxalyl chloride (650 ppbv) at EUPHORE

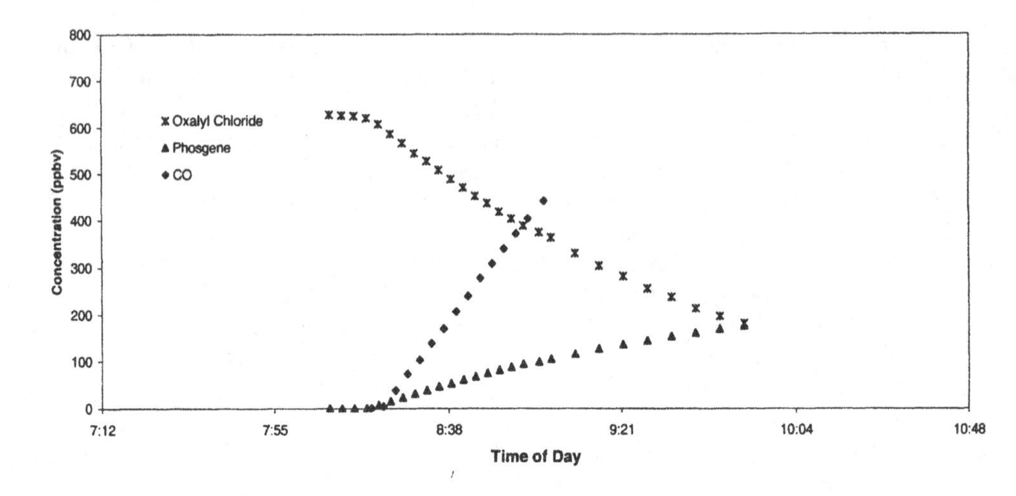

Figure 4. Plots of the yields of products against the loss of oxalyl chloride for the sunlight photolysis of oxalyl chloride (650 ppbv) at EUPHORE.

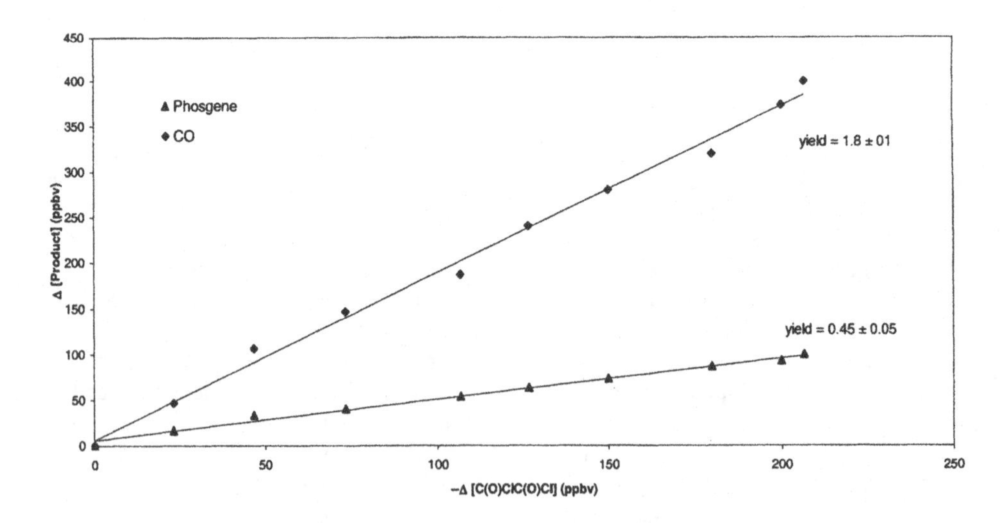

Conclusion and Atmospheric Implications

The results of the present study show that the quantum yield for the photodissociation of chloral is close to unity and that the major product is phosgene. The tropospheric photolysis lifetime is around 4-5 hours while that for both reaction with the OH radical or uptake by clouds is of the order 10-15 days. Hence it is clear that the major loss process for chloral is photolysis to produce CCl_2O. It has been shown from a modelling study that tropospheric removal of CCl_2O is predominantly *via* wet deposition with a lifetime of around 70 days, (Kindler *et al.,* 1995). Degradation of CCl_3CH_3 following reaction with OH in the troposphere leads to the formation of CCl_3CHO which rapidly photolyses to give predominantly CCl_2O. Thus, oxidation of CCl_3CH_3 in the troposphere generates mainly CCl_2O, which may provide a flux of Cl into the stratosphere.

References

Barry J., D.J. Scollard, J.J. Treacy, H.W. Sidebottom, G. Le Bras, G. Poulet, S. Teton, A. Chichinin, C.E. Canosa-Mas, D.J. Kinnison, R.P. Wayne, O.J. Nielsen; Kinetic data for the reaction of hydroxyl radicals with 1,1,1-trichloroacetaldehyde at 298 ± 2K, *Chem. Phys. Lett.* **221** (1994) 353-358.

Cox R.A.; Atmospheric degradation of halocarbon substitutes. In: *Global Ozone Research and Monitoring Project-Report No. 37, Scientific Assessment of Ozone Depletion:1994,* World Meteorological Organization, Geneva. (1995) 12.1-12.23.

Kindler T.P., W.L. Chameides, P.H. Wine, D.M. Cunnold, F.N. Alyea and J.A. Franklin; The fate of atmospheric phosgene and the stratospheric chlorine loadings of its parent compounds: CCl_4, C_2Cl_4, C_2HCl_3, CH_3CCl_3 and $CHCl_3$, *J. Geophys. Res.* **100 (D1)** (1995) 1235-1251.

Midgeley P.M, and A. McCulloch; Production, sales and emissions of halocarbons from industrial sources. In: *The Handbook of Environmental Chemistry, Vol. 4, Part E,* Reactive Halogen Compounds in the Atmosphere. Eds: P. Fabian and O.N. Singh, Springer-Verlag, Berlin-Heidelberg. (1999) 155-190.

Montzka S.M., J.H. Butler, J.W. Elkins, T.M. Thompson, A.D. Clarke and L.T. Lock; Present and future trends in the atmospheric burden of ozone-depleting halogens, *Nature* **398** (1999) 690-694.

Nelson L., I. Shanahan, H.W. Sidebottom, J. Treacy and O.J. Nielsen; Kinetics and mechanism for the oxidation of 1,1,1-trichloroethane, *Int. J. Chem. Kinet.* **22** (1990) 577-590.

Platz J., O.J. Nielsen, J. Sehested and T.J. Wallington; Atmospheric chemistry of 1,1,1-trichloroethane: UV absorption spectra and self-reaction kinetics of CCl_3CH_2 and $CCl_3CH_2O_2$ radicals, kinetics of the reactions of the $CCl_3CH_2O_2$ radical with NO and NO_2 and the fate of the alkoxy radical CCl_3CH_2O, *J. Phys. Chem.* **99 (17)** (1995) 6570-6579.

Rattigan O.V., O. Wild, R.L. Jones and R.A. Cox; Temperature-dependent absorption cross-sections of CF_3COCl, CF_3COF, CH_3COF, CCl_3CHO and CF_3COOH, *J. Photochem. Photobiol. A: Chem.* **73** (1993) 1-9.

Rattigan O.V., O. Wild and R.A. Cox; UV absorption cross-sections and atmospheric photolysis lifetimes of halogenated aldehydes, *J. Photochem. Photobiol. A: Chem.* **112** (1998) 1-7.

Rowley D.M., P.D. Lightfoot, R. Lesclaux and T.J. Wallington; UV absorption spectrum and self-reaction of cyclohexlperoxy radicals, *J. Chem. Soc. Faraday Trans.* **87** (1991) 3221-3225.

Sidebottom H., and J. Franklin; The atmospheric fate and impact of hydrofluorocarbons and chlorinated solvents, *Pure and Appl. Chem.* **68** (1996) 1757-1769.

Talukdar R.K., A. Mellouki, J.B. Burkholder, M.K. Gilles, G. Le Bras and A.R. Ravishankara; Quantification of the tropospheric removal of chloral (CCl_3CHO): Rate coefficient for the reaction with OH, UV-absorption cross sections and quantum yields, *J. Phys. Chem.* **105** (2001) 5188-5196.

Fluorinated Radical Reactions of Atmospheric Importance

Walter Hack, Markus Hold, Karlheinz Hoyermann,
Igor I. Morozov, and Evgenii S.Vasiliev

Semenov Institute of Chemical Physics RAS, Kosygin str.4, 117334, Moscow, Russia

Introduction

The three main pathways for the destruction of organic molecules in the atmosphere are radical reactions mainly with OH, atom reactions mainly with O (^{1}D) and UV photolysis (Scientific, 1995). In these processes organic radicals are formed which react mainly with O_2 molecules and form peroxy radicals. The same mechanism is valid for partially fluorinated hydrocarbons and partially fluorinated ethers. These hydrofluoroethers are proposed as an alternative for chlorofluorocarbons. The rate constants for the reaction of OH with fluorinated ethers are in the range $2.10^8 \leq k$ (cm^3mol^{-1} s^{-1}) $\leq 4.4.10^{10}$ under atmospheric conditions (Zhang *et al.*, 1992; Orkin *et al.*, 1999; Tokuhashi *et al.*, 1999). The possible reaction schemes of the radicals are shown in Good *et al.* (1998). As alternative radical sources Cl atom and NO_3 radical reactions are mentioned in Hack *et al.* (1999) and Kambanis *et al.* (1998). Atmospheric measurements of Cl atom concentrations together with the measured rate constants (Kambanis *et al.*, 1998), however, show that these reactions are of no significant importance as a degradation channel. NO_3, which is important in the atmospheric night chemistry, has been shown in the paper by Hack *et al.* (2001) to be too slow to be of importance.

In the second step of the scheme given in Good *et al.* (1999) the partially fluorinated ether radical mainly reacts with O_2 molecules and forms peroxy radicals. This reaction is fast due to the high rate constant, as shown in this work, and due to the high O_2 concentration in the atmosphere. The fluorinated peroxy radicals than react in an analogous way as hydrocarbon peroxy radicals. Additionally, their final fate should be the oxidative degradation into environmentally "innocent" species. Before a global turn into some new type of refrigerants, a process demanding expensive and time-consuming changes, their physical and chemical properties should be studied in detail. The chemical studies should involve their reactions with all atmospherically important species: NO radicals, molecular oxygen, and oxygen atoms. Bis(2,2,2-trifluoroethyl) ether ($C_4H_4OF_6$) and bis(fluoromethyl) ether ($C_2H_4OF_2$) are some of the most characteristic representatives of partially fluorinated ethers. The oxidation mechanism in the presence of atmospheric radicals includes formation of the radicals $CF_3CHOCH_2CF_3$ ($C_4H_3OF_6$), CF_2OCHF_2 (C_2HOF_4), and $CHFOCH_2F$ ($C_2H_3OF_2$), which are further oxidized to the final products through a multistep process.

Despite the intensive efforts in laboratory studies there are many questions in the mechanism of fluorinated ether radical oxidation including primary intermediates and final products. Up to now no direct studies have been done on the kinetics of the radical $CF_3CHOCH_2CF_3$ with O_2 and the $CFHOCH_2F$ radical with O_2 and NO. The atmospheric oxidation of fluorinated ethers is initiated by reactions with OH radicals. The fate of the primary radicals is of great interest. The reactions with O_2 molecules,

$CF_3CHOCH_2CF_3 + O_2 + M \rightarrow$ products (1)

$CHFOCH_2F + O_2 + M \rightarrow$ products (2)

are of special interest, although they are found to be slow, due to the high concentration of O_2 in the atmosphere. The reactions of the radical $CHFOCH_2F$ (written as $C_2H_3OF_2$) with NO radicals and O atoms (3) and (4)

$CHFOCH_2F + NO \rightarrow$ products (3)

I. Barnes (ed.), Global Atmospheric Change and its Impact on Regional Air Quality, 103–108.

$CHFOCH_2F + O \longrightarrow$ products (4)

are also important for atmospheric chemistry. The following reactions are useful as reference reactions.

$CH_3 + O_2 \rightarrow (CH_3O_2)^* \rightarrow CH_3O_2$ (5)

$CH_3 + NO \rightarrow$ products (6)

$CH_2OCH_3 + O \rightarrow$ products (7)

Experimental Section

Experimental arrangements of the type / fast flow reactor/ molecular beam sampling/ mass spectrometer/ have been used with three different methods of mass spectrometric detection (quadrupole, magnetic deflection and time-of-flight-mass spectrometer).

The arrangement with the modulated molecular beam quadrupole mass spectrometer MC 7303 (electron impact ion source) was equipped with a molecular beam sampling system via a nozzle (0.3-0.8 mm diameter). A modulated molecular beam mass spectrometer was used to detect radicals and stable particles continuously sampled at the downstream end of the flow tube via a three chambers beam inlet system. The energy of ionizing electrons was varied 29 - 70 eV. The molecular beam was modulated at 33 Hz via a running type chopper. The detection limit of e.g. fluorinated ether (m/z=182) was about $2 \cdot 10^{-14}$ mol cm^{-3} for an integration time of 300 s. The reactor was a Pyrex flow tube 20 cm length (reaction distance ≤ 15 cm) and (1.9 cm id.) containing a central injector of 0.9 cm id. The flow tube was thermostated by a fluid passed through the outer jacket in the temperature range $253 \le T/K \le 373$. The reactor tube was coated with HF to reduce the heterogeneous loss of atoms and radicals. The fluorine atoms were generated by microwave discharge (2450 MHz) of F_2. Ceramic (Al_2O_3) inserts were placed into the quartz discharge tube to reduce F atom reactions. A MKS Baratron measured the pressure in the flow tube. A MKS mass flow controller (1100 Series, type 1160 B) maintained the calibrated flow rates. Variable reaction times were obtained by moving the sliding injector from two up to 30cm. Typical flow rates were in the range $6 \le v$ (m s^{-1}) ≤ 12. The experiments were carried out at a total pressure in the range $0.6 \le p(mbar) \le 1.5$ with He as a carrier gas. A MKS Bararton measured the pressure in the flow tube. The experimental arrangement with the time-of-flight mass spectrometer (Hack *et al.*, 2001) was used for multiphoton ionization studies. It consists of a conventional discharge flow reactor, a molecular beam sampling system and a combined ion source for electron impact or resonance enhanced multi-photon-ionization. A pulsed dye laser (Lambda Physic FL 2002) pumped by a XL eximer-laser (Lambda Physic LPX 200) is used for the photoionization.

A major problem using the mass spectrometric technique is the identification of the free radical mass peaks in the presence of fragments of the same mass originating from other components in the system. Therefore the REMPI technique was used for the kinetics study of the rate constant of the reaction (4) $O + C_2H_3OF_2 \rightarrow$ products.

The concentrations of the molecular fluorine and oxygen were determined by measuring the decrease of pressure in a calibrated volume. The O atom concentration was calculated using the amount of dissociation of O_2 by the discharge as measured by the mass spectrometer.

Chemicals: The chemicals were of commercial grade: Ar (Messer-Griesheim, 99.998%), and O_2 (Messer-Griesheim, 99.995%). Bis(fluoromethyl) ether (Fluorochem 97%) a clear colorless liquid, has a melting point of about 307 K and a boiling point of about 343 K. The vapor pressure has been measured as 601 mbar at 293 K. We have noted a pink color when $C_2H_4OF_2$ was added to glass vessels and this was followed by a deposit of a white film.

The substances were used without further purification. The source of F_2 was a commercial mixture constituted of 5% of F_2 (98%) in He (99.9949%). As a carrier gas, He with a commercial purity (99.9949%). We determined the mass spectra of some fluorinated ethers listed above.

Results and Discussion

The radicals $C_4H_3OF_6$, $C_2H_3OF_2$, CH_2OCH_3 (C_2H_5O), CHF_2 and CH_3 (Hack *et al.*, 2001) were generated from the fast fluorine atom reactions:

$F + CF_3CH_2OCH_2CF_3 \rightarrow CF_3CHOCH_2CF_3 + HF$ (8)

$F + CH_2FOCH_2F \rightarrow \dot{C}HFOCH_2F + HF$ (9)

$F + CH_3OCH_3 \rightarrow CH_2OCH_3 + HF$ (10)

$F + CH_2F_2 \rightarrow CHF_2 + HF$ (11)

$F + CH_4 \rightarrow CH_3 + HF$ (12)

The reaction of $C_4H_3OF_6$ radicals with molecular oxygen

$CF_3CHOCH_2CF_3 + O_2 + M \rightarrow CF_3CO_2HOCH_2CF_2$ (1)

was studied at a pressure of P = 1.33 mbar (mainly He) at room temperature. The experimental study of the reaction (1) was performed by the relative method with the reference reaction $CH_3 + O_2 \rightarrow (CH_3O_2)^* \rightarrow CH_3O_2$ (5). The changes of the concentrations of the $C_4H_3OF_6$ and CH_3 radicals were determined by their ions m/z = 112 and m/z = 15, respectively. The determination was performed by simultaneously producing $C_4H_3OF_6$ - and CH_3-radicals via reaction (8) and (12) in a $C_4H_4OF_6/CH_4$ mixture and adding increasing amounts of molecular oxygen. Low F atom concentrations were chosen such that radical-radical reactions were suppressed. In the absence of secondary reaction, the ratio of the rate constants can be evaluated by the relation

$\ln \{ [C_4H_3OF_6]_{-O2} / [C_4H_3OF_6]_{+O2} \} == k_1 / k_5 \ln \{ [CH_3]_{-O2} / [CH_3]_{+O2} \}$ (I)

where the indices ($\pm O_2$) represent the radical concentrations in the presence and of absence molecular oxygen. The details of the experimental results are given in (Hack *et al.*, 2001). The least-squares analysis kinetic data of this reaction yields a rate constant ratio k_1 / k_5 = (0.47±1 0.14). Using $k_5 = 4.65\text{-}10^9$ cm^3mol^{-1} s^{-1} (Atkinson *et al.*, 1997), the $k_1 = (2.2\pm0.7)\cdot10^9$ cm^3mol^{-1} s^{-1} is obtained.

The reaction $CHFOCH_2F + O_2 \rightarrow$ products (2) was studied in the temperature range of $248 \leq T/K \leq 353$. The reaction rate and the mechanism were determined. The $C_2H_3OF_2$ radicals were produced in the fast reaction (9). The rate coefficient k_2 was determined relative the reference reaction $O_2 + CH_3 \rightarrow CH_3O_2$ (5). The experiments to measure the rate constants of the reaction (2) were analogous to the experiments described, for reaction (1) (see above). The pressure in the flow tube was P= 0.93 mbar and the main He flow was = 18.5 cm^3/s, and methane - 0.019 cm^3/s. To follow the change of the concentration of the $C_2H_3OF_2$ radicals, the signal at m/z = 81 was used. The signal at m/z = 81 {I (m/z = 81)} was corrected with respect to the contribution of the molecule on mass m/z = 81. The I (m/z = 81) / I (m/z = 82) ratio in the CH_2FOCH_2F fragment mass spectrum was measured in each run.

The temperature dependence of the ratio of the rate constants (k_2/k_5) are determined. For the rate of reaction (5) the rate constant $k_5(T) = 1.85.10^{16} \exp(4.65\ \text{kJ/mol/(RT)})$ [He] cm^6mol^{-2} s^{-1} was used (Atkinson *et al.*, 1997). The Arrhenius plot for the reaction (2) gives a linear relation-ship in the temperature range examined as shown:

$k_2(T) = (1.5\pm1.1)\cdot10^{15} \exp((13.5\pm1.8)\ (\text{kJ/mol})/(RT))$ [He] $cm^6mol^{-2}s^{-1}$.

Reaction mechanism

The C^*HFOCH_2F radical can add an oxygen molecule; $C^*HFOCH_2F + O_2 \rightarrow CHFO(O_2)CH_2F$. A weak product signal was observed at m/z = 113, corresponding to the mass of the primary $CH_2FOCHFOO$ peroxi radical. The signal m/z = 113 could also be a fragment of the recombination product $(CH_2FOCHFOO)_2$ but it did not disappear even at very low initial radical concentrations. From the observation of a signal at m/z = 48, which can be attributed to CHFO, it was concluded that CHFO is one product of reaction (2). From the kinetic behavior of CHFO it was concluded that CHFO was a final product. The reaction mechanism can be illustrated in the following way:

$CH_2F{-}O{-}CHF$ (m/z = 81) + O_2 ⇌ $CH_2F{-}O{-}CHF{-}O{-}O\cdot$ (m/z = 113) → $CH_2F{-}O{-}CHF{-}O{-}O{-}H$

− OH

→ $CH_2F{-}O{-}CHF{-}O$ → $HC(O)F$ (m/z = 48) + $HC(O)F$ (m/z = 48)

The rate constant for the radical-radical reaction $CHFOCH_2F$ + NO → products was measured relative to the rate constant of the reaction CH_3 + NO → products. These measurements were performed by simultaneously producing $C_2H_3OF_2$ and CH_3 in reaction (9) and (12) and adding increasing amounts of NO. The function $\ln\{[C_2H_3OF_2]_{-NO}/[C_2H_3OF_2]_{+NO}\}$ versus $\ln\{[CH_3]_{-NO}/[CH_3]_{+NO}\}$ is measured in experiments. The rate constant k_3 over temperature range $248 \leq T/K \leq 353$, are determined. For the rate of reaction (6) the rate constant $k_6(T) = 7.24 \cdot 10^{16} \exp(1.75\ kJ/mol/(RT))\ [He]\ cm^6 mol^{-2} s^{-1}$ was inserted (Atkinson *et al.*, 1997): The Arrhenius plot give a linear relationship in the temperature range examined, With the absolute value of k_6, this leads to

$$k_3 = (7.4 \pm 1.5) \cdot 10^{16} \exp(2.5\ kJ/mol/(RT))\ [He]\ cm^6 mol^{-2} s^{-1}.$$

To study the reaction of oxygen atoms with $CHFOCH_2F$ radicals REMPI time of flight technique was used to measure the concentration profiles of the $C_2H_3OF_2$ radicals (Hack *et al.*, 2001). The REMPI spectrum of the radical $C_2H_3OF_2$ (m/z = 81) was determined.

The rate constant of the reaction of the $C_2H_3OF_2$ radicals with O atoms, reaction (4), was determined relative to O + CH_2OCH_3 → products. Provided that the reactant $C_2H_3OF_2$ and the radical C_2H_5O are consumed only by reaction with atoms, it can be shown that:

$$\ln\{[C_2H_3OF_2]_{-o}/[C_2H_3OF_2]_{+o}\} = k_4/k_7 \ln\{[C_2H_5O]_{-o}/[C_2H_5O]_{+o}\}$$

where $[C_2H_3OF_2]_{-o}$ and $[C_2H_5O]_{-o}$ denote the initial concentration of $C_2H_3OF_2$ radicals and C_2H_5O radicals in the absence of O atoms and $[C_2H_3OF_2]_{+o}$ and $[C_2H_5O]_{+o}$ are the concentrations of the corresponding radicals in the presence of O atoms. The measurements of the ratio of the rate constants k_4/k_7 was performed by the parent ions m/z = 81 as well as the daughter ions m/z = 33 for the $C_2H_3OF_2$ radical and m/z = 45 for the C_2H_5O radical. Typical experimental conditions were: pressure p = 1.4 mbar; mean velocity $\langle v \rangle$ = 18 m/s; reaction times t_R = 2-3 ms. The initial radical concentration was in the range $1 \leq [C_2H_3OF_2]10^{-10}$ (mol cm^{-3}) ≤ 5.

The ratio k_4/k_7 was studied over temperature range $245 \leq T/K \leq 301$. The temperature range was limited by a reaction of $C_2H_4OF_2$ with glass and metal surface. Thus the temperature range was too small to derive a precise value for the activation energy. The ratio $k_4/k_7 = (0.63 \pm 0.2)$ was determined. Using $k_7 = (1.54 \pm 0.03)\ 10^{14}\ (T/298/K)^{0.15}\ cm^3 mol^{-1}\ s^{-1}$ (Hoyermann *et al.*, 1996) the absolute value of the rate constant is determined to be $k_4 = (9.5 \pm 1)\ 10^{13}\ cm^3 mol^{-1}\ s^{-1}$ independent on temperature in the range $245 \leq T/K \leq 301$.

The results for the reactions of oxidation processes of the fluorinated ethers studied are listed in (Hack *et al.*, 2001). The obtained values of rate constants are determined for the first time. The error limits of our results are standard deviations derived from the linear least-squares fit of the plot of first-order rate constants versus reactant concentration of absolute measurements, and the systematic errors are not considered. The systematic errors in our experiments are estimated to be 10 %. The rate constants measured by the two different methods agree with each other within the estimated uncertainties.

The rate constant for the reaction of the $C_4H_3OF_6$ radical with O_2 is about one order of magnitude lower than that of the $C_2H_3OF_2$ radical with O_2. This fact may be due to the excess energy of the reaction, which prevents formation of the O-H bond in the dissociating peroxy radical.

Conclusion and Atmospheric Implications

On the basis of the kinetic experimental data obtained in this study and the literature values (De More *et al.*, 1997), we propose a scheme for the degradation of the partially fluorinated ethers in the atmosphere. This scheme is applicable to both the troposphere and stratosphere. A key question is: could the degradation products of the fluorinated ether generate species that can destroy ozone in the stratosphere? If long-lived species are produced, can they be transported into stratosphere from the troposphere? If ozone-destroying radicals other than chlorine are released from degradation products, erroneous ozone depletion potential will result (Scientific., 1995).

The major fate of fluorinated ethers in the troposphere is the reaction with OH radicals. The reaction rates of Cl atoms are always higher than those of OH radicals (Kambanis *et al.*, 1998). The concentration of the OH radical however is several times higher than the concentration of the Cl atoms (average tropospheric concentrations of OH is in the order of $[OH] = 2.10^{-20}$ mol cm^{-3}). In the reaction with OH the $C_4H_3OF_6$ radicals are formed which will react with O_2 to form the peroxy radical $C_4H_3O_3F_6$. The presence of H atoms in these fluorinated ethers makes each of the studied species susceptible to OH attack in troposphere. Their lifetime in the troposphere τ and, hence, the extent of their transport into the stratosphere depends critically on the value of the rate coefficient of their reaction with OH radicals. (Good *et al.*, 1998) determined the lifetimes of several fluorinated ethers to be in the range of $5.7 \leq \tau$ (years) ≤ 165. The atmospheric lifetime of $CF_3CH_2OCH_2CF_3$ was determined to be $\tau = 2.9$ years (Kambanis *et al.*, 1998). These results suggest that a penetration of these ethers into the stratosphere is possible.

The $C_4H_3O_3F_6$ radical further reacts rapidly with NO to produce $CF_3CHOOCH_2CF_3$ + NO_2. The $CF_3CO^*HOCH_2CF_3$ radical is unstable and decomposes to CF_3 and $O{=}CHOCH_2CF_3$ very rapidly. The CF_3 radicals react with O_2 to produce peroxy radicals CF_3O_2. The reaction with NO and other peroxy radicals remove the peroxy radicals. The atmospheric fate of the alkoxy radical CF_3O is shown in Figure 1. The fate of the major and relatively stable intermediates $O{=}CHOCH_2CF_3$, $CF_3CHOOHOCH_2CF_3$ and $CF_3CHOONO_2OCH_2CF_3$ differ in the troposphere or in the stratosphere. This scheme of atmospheric degradation of fluorinated ethers was confirmed in (Nohara *et al.*, 1998) by the detection of final products CF_2O, CO_2, and HF in the system of degradation processes of all perfluoroalkyl methyl ethers. CHFO was found as the main product of the $C_2H_3OF_2 + O_2$ reaction in the present work. The amount of CO_2 and CF_2O as the final products was supported by the carbon mass balance (Nohara *et al.*, 1998). [Good *et al.*, 1999] showed that a major product formed from the oxidation of CF_3OCH_3, CHF_2OCHF_2, and CHF_2OCF_3 is carbonyl fluoride.

Acknowledgements. This work received financial support from VW Foundation and European Commission. (Proposal ENVK2 -1999 -00099). Financial support by the Fonds der chemischen Industrie is acknowledged.

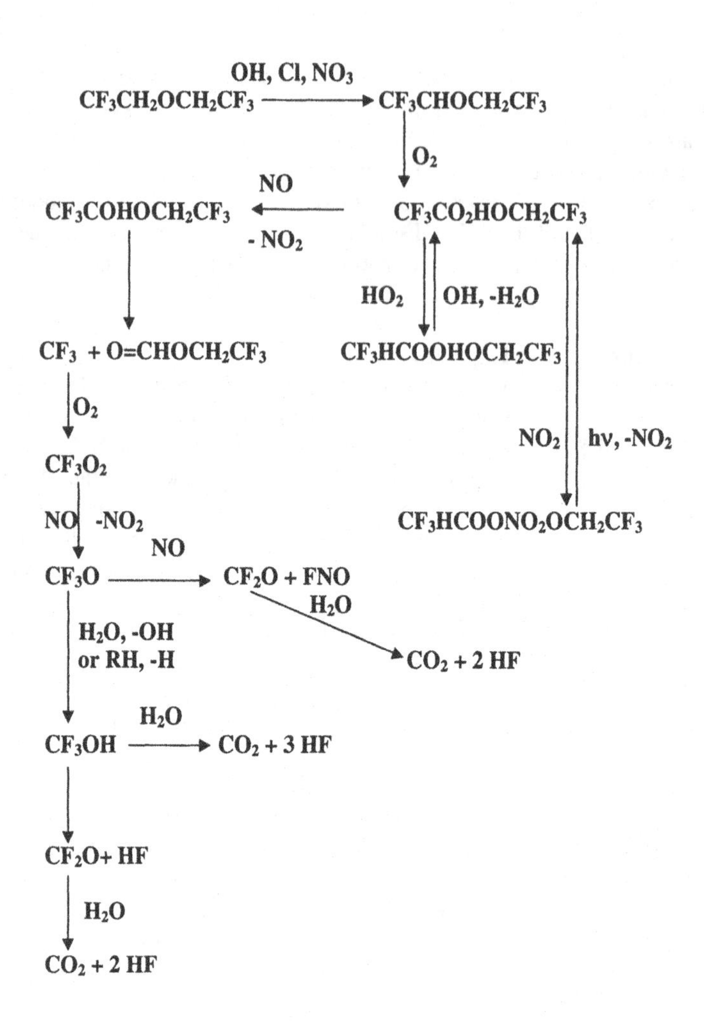

Figure 1. Scheme of atmospheric degradation of $CF_3CH_2OCH_2CF_3$.

References

Atkinson, R., D.L. Baulch, R.A. Cox, R.F. Jr. Hampson, J. A. Kerr, M.J. Rossi, J.Troe, J. Phys. Chem. Ref. Data 1997,26,521.

Good, D.A., M. Kamboures, R. Santiano, and J .S. Francisco, J. Phys. Chem. A 103, 9230 (1999)

Good, D.A., J.S. Francisco, A.K. Jain, D.J. Wuebbles, J. Geophys. Res. 103,28181 (1998) Hack, W., M. Hold, K. Hoyermann, I. Morozov, E. Vasiliev, Proceedings of EUROTRAC Symposium '98, Editors: Patricia M. Borrell and Peter Borrell, WIT press, Southampton and Boston 1999, vol1, p. 99-103.

Hack, W., M. Hold, K. Hoyermann, I. Morozov, E. Vasiliev, PCCP to be published 2001.

Hoyermann, K., F. Nacke, Symp. Int. Combust. Proc. 26, 595 (1996).

Kambanis, K.G., Y .G. Lazarou, and P. Papagiannakopoulos, J. Phys. Chem. A 1998, 102,8620.

De More, W.B., Sander, S.P., Golden, D.M., Hampson, R.F., Kurylo, M.J., Howard C.J., Ravishankara, A.R., Kolb, C.E., Molina M.J . JPL Publication 90-1, 1990.

Nohara, K., Kutsana, S., Ibusuki, T., 6th FECS Conference on Chemistry and the Environment, 5, p. 177, 1998.

Orkin, V.L., Villenave, E., Huie, R.E., Kurylo, M.J.: J. Phys. Chem. A 1999, 103,9770.

Scientific Assessment of Ozone Depletion: Report No.37 , W .M.O.: Geneva 1995.

Tokuhashi, K., Takahashi, A., Kaise, M., Kondo, S., and Sekiga, A., Int. J. Chem. Kinet. 31, 846 (1999)

Zhang, Z., Saini, R.D., Kurylo, M.J., Huie, R.E.: J. Phys. Chem. 1992,96,9301.

Heterogeneous Reactions Affecting Chlorine Activation in the Troposphere

M.Yu.Gershenzon, V.M.Grigorieva, S.D.Il'in, R.G.Remorov, D.V.Shestakov, V.V.Zelenov, E.A.Aparina and Yu.M.Gershenzon

N.Semenov Institute of Chemical Physics RAS, Moscow 119991, Russian Federation

Introduction

Recent indirect and direct observations have shown that chlorine and bromine are active initiators of oxidation in the marine and coastal regions. Bromine effectively depletes ozone at Arctic sunrise and reacts with some hydrocarbons in the marine boundary layer (MBL). Chlorine reacts with many natural and anthropogenic pollutants. Morning Cl concentrations were estimated to reach in some cases 10^5 molecule/cm^3 (Spicer *et al.*, 1998; Singh *et al.*, 1996). This can markedly increase the oxidation potential of the coastal areas where polluted air from the continent is purified. Moreover, the back-transport of chlorine-enriched air causes changes in the oxidation potential of the continental troposphere.

Sea salt is the largest source of halogens in the lower troposphere. Waves produce small droplets of seawater in the MBL. When the droplets are transported to higher altitudes or inland where the relative humidity is lower, they can crystallize to form solid particles. Transport of the sea salt particles has been observed as far as 900 km away from the ocean (Shaw, 1991). There are other cases when salts were found at high concentrations in particles. For example, the products of oil wells burning in Kuwait contained significant amounts of salt (Lownthal *et al.*, 1993). Aqueous salts are typical of the remote MBL, while solid salts covered with adsorbed water may be more frequent in the coastal area. Thus the chemistry of these regions should differ. In addition the remote MBL chemistry proceeds under NO_x poor conditions whereas liberation of halogens in polluted coastal regions from solid particles can occur in heterogeneous reactions with nitrogen-containing compounds such as NO_2, NO_3, N_2O_5, and $ClONO_2$.

Bromine and tropospheric ozone depletion in Polar Regions

Over the past 15 years ozone destruction (from 30-40 pptv to less than 2 pptv) has been observed at polar sunrise in several Arctic sites. It was found that during these ozone loss episodes BrO increases up to several 10 pptv (Hausmann and Platt, 1994), while ClO may reach only a few pptv (Perner *et al.*, 1999). Later on, a few studies have confirmed these observations in the Arctic (Tuckermann *et al.*, 1997; Martinez *et al.*, 1999; Hönninger and Platt, 2001). Earlier BrO was also observed in Antarctica (Kreher *et al.*, 1996).

Satellite observations show that regions with elevated BrO cover over several million square kilometers including the Dead Sea and the Caspian Sea (Wagner and Platt, 1998; Richter *et al.*, 1998; Wagner *et al.*, 2001). The photolytic precursors of atomic bromine and chlorine, Br_2 and BrCl were found to present at 5-10 pptv and 35 pptv respectively (Foster *et al.*, 2001). Note that the Cl^-/Br^- ratio in snowfall ranges from 50 to 200, while the $BrCl/Br_2 <10$. Molina and co-workers showed that sea salt aerosols and aerosols deposited on the polar ice pack are most likely to be liquid under polar sunrise conditions (Kopp *et al.*, 2000). The ClO/BrO and $BrCl/Br_2$ ratios as well as the correlation between ozone destruction and BrO increase show that ozone depletion in spring is caused mainly by aqueous bromine chemistry in spite of the excess of chlorine ions present in salt ice.

I. Barnes (ed.), Global Atmospheric Change and its Impact on Regional Air Quality, 109–113.

Chlorine and bromine in the mid-latitude MBL and urban coastal area

Sea salt particles generated by breaking ocean waves should have a chemical composition similar to that of seawater. However, recent analysis of the marine aerosol composition carried out in the Southern Ocean during an extended period of time showed that the Cl^- deficiency in sea droplets is only few percent whereas the Br^- deficiency is high (ranging from ~ -10% to -80% and more (Ayers *et al.*, 1999)). These observations and some other indirect evidence shows that the Br/Cl ratio in the mid-latitude MBL air is enhanced in comparison with the 1:630 Br^-/Cl^- ratio in mid-latitude seawater. The mechanism responsible for the enhanced aqueous bromine chemistry is based on the acid catalyzed (Keene *et al.*, 1998) gas-liquid cycle (Sander and Crutzen, 1996), in which HOBr plays a key role.

$$\underline{HOBr} \xrightarrow{Br^- + H^+} Br_2 \xrightarrow{h\nu} 2Br \xrightarrow{2O_3} 2BrO \xrightarrow{2HO_2} \underline{\underline{2HOBr}}$$

Thus, the field observations supported by model studies based on laboratory measurements confirm the enhancement of bromine chemistry in aqueous solutions of sea salt both in the polar and mid-latitude NO_x-poor remote MBL.

Nevertheless, there is an important observation (Spicer *et al.*, 1998) in which "unexpectedly" high concentrations (up to 150 pptv) of molecular chlorine at night in the North American coastal site (Long Island) were observed. Unlike the other observations the Br_2 concentration was estimated in this experiment to be 25 times lower than that of Cl_2 in the same observation (Foster *et al.*, 2001). Spicer *et al.*, 1998, couldn't find a source for the nighttime Cl_2 located at a ground. They estimated that to simulate the observed night Cl_2 it is necessary to introduce a source of 330 pptv Cl_2 working during the whole day.

A new possible photolytic source of molecular chlorine in the MBL has been recently proposed. The mechanism is most probably relevant to formation of $[OHCl]^-$ interface complex followed by its self-reaction (Oum *et al.*, 1998)

$$2\,[OH...Cl]^- \rightarrow 2\,OH^- + Cl_2$$

Other photolytical aqueous-phase mechanisms producing halogen proposed earlier were also not able to explain so high Cl_2 coastal concentration. It is important that a non-photolytic nighttime Cl_2 source is necessary to support the extra high nighttime Cl_2 concentration.

This paper does not aim to explain the experimental observation of Spacer *et al.*, 1998. First of all because the authors consider that the meteorological situation at the point of their observation (2.8 km north of Atlantic coast) was close to the MBL conditions. For example in their local model a poor NO_x = 5 pptv was accepted. We intend here to find the enriched NO_x coastal conditions at which high nighttime Cl_2 may be formed and to identify the mechanism of the process.

We consider the nighttime heterogeneous reactions of NO_3 and $ClONO_2$ with solid NaCl proceeding in the urban coastal zone in presence of NO_x as a possible source of high Cl_2 concentrations. For this simple qualitative study we did not include aqueous phase reactions. In this preliminary study non-methane hydrocarbons chemistry was also excluded. CH_4 concentration was variable between 10^{13} and $4*10^{13}$ cm^{-3}. A flux of stratospheric O_3 to the troposphere was taken to be equal $5*10^{10}$ molecule/cm^2s and O_3 dry deposition is $4*10^{-2}$ cm/s. The vertical NO flux of $4*10^9$ molecule/cm^2s was introduced to imitate the inland NO_x transport.

Reaction NO_3 + solid NaCl

Indirect evidence that Cl atoms are the primary products of the heterogeneous reaction between nitrate radical and solid NaCl was proposed at the first time in (Seisel *et al.*, 1997). They proposed the reaction

$$NO_3 + NaCl \rightarrow NaNO_3 + Cl \quad (R1)$$

as a non-photolytic source of Cl. More detailed studies were carried out in a wide range of initial NO_3 concentrations at T = 20 - 100°C later (Gershenzon *et al.*, 1999). The mechanism with adsorbed atomic chlorine $(Cl)_{ad}$ as a long lived product was suggested for dry (Gershenzon *et al.*, 1999; Zelenov *et al.*, 2002) and humidified (Zelenov *et al.*, 2002a; Zelenov and Aparina, this issue) NaCl.

$$NO_3 + NaCl \rightarrow (Cl)_{ad} + NaNO_3 \quad (R2)$$

Both gaseous and adsorbed Cl can produce ClO in reactions

$$Cl + O_3 \rightarrow ClO + O_2 \quad (R3)$$

$$(Cl)_{ad} + O_3 \rightarrow ClO + O_2 \quad (R4)$$

The ClO reacts with NO_2

$$ClO + NO_2 \rightarrow ClONO_2 \quad (R5)$$

and the following reaction

$$ClONO_2 + NaCl \rightarrow Cl_2 + NaNO_3 \quad (R6)$$

produces Cl_2.

Seisel *et al.*, 1997, Gratpanche and Sawerysyn, 1999, and Gershenzon *et al.*, 1999, measured the initial probability of NO_3 reactive uptake on dry NaCl. Their results are similar (γ_{NO3}~(2-5)*10^{-2}). Gershenzon *et al.*, 1999, show that this value doesn't depend on $[NO_3]$ at $[NO_3]<10^{11}$ cm^{-3}. Quite recently it was shown that the probability of the reactive uptake increases with humidity (Gershenzon *et al.*, 1998; Zelenov and Aparina, this book, and Zelenov *et al.*, 2002a). It may reach unity under real conditions in the coastal zone. Caloz *et al.*, 1998, measured very high value of reactive uptake for reaction (R6), γ=0.26.

Autocatalytic chlorine increase

After the sunrise the following chain branched process may proceed

$$\underline{Cl_2} \xrightarrow{h\nu} 2Cl \xrightarrow{2O_3} 2ClO \xrightarrow{2NO_2} 2ClONO_2 \xrightarrow{2NaCl} \underline{\underline{2Cl_2}}.$$

The termination step takes place in atomic Cl reaction with hydrocarbon

$$Cl + CH_4 \rightarrow HCl + CH_3 \quad (R7).$$

The rate constants for heterogeneous reactions (R2) and (R6) were accepted to be limited by diffusion, k = 4 π r D N. Here N is the number density of particles with an average radius r≈10^{-4} cm, D is diffusion coefficient, D_{NO3} = 0.1 cm^2 atm s^{-1} (Rudich *et al.*, 1996), $D_{ClONO2} = D_{NO3}\left(\frac{m_{NO3}}{m_{ClONO2}}\right)^{1/2} \approx 0.064$ cm^2 atm s^{-1}, m is the molecular mass. For preliminary estimation we used $k_2 \approx k_4 \approx 10^{-4}$ N

The Cl_2 and Cl concentrations are given in Figure 1. One may see a very high level of molecular chlorine at night and atomic chlorine during the daytime. Both are strongly dependent on k_2, k_4 and k_7.

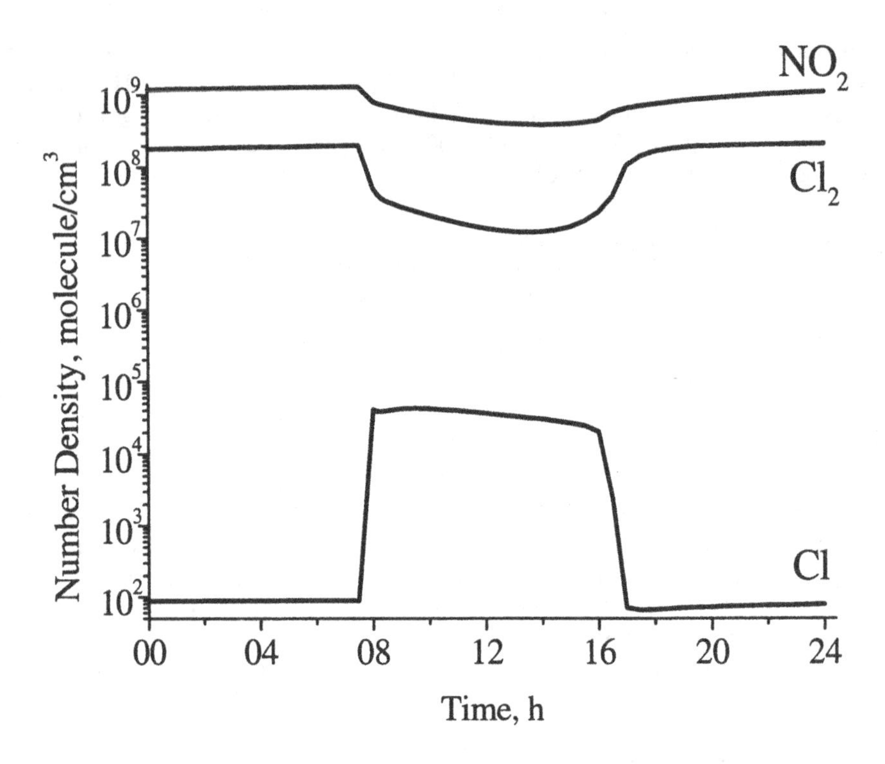

Figure 1. Modeling of Cl_2 (Cl) abundance in polluted coastal area. Established number densities on the 5th day, in December, 42°N. Accepted fluxes: NO_2 from the ground 5×10^9 molecule$\times$cm^{-2} s^{-1}, O_3 from free troposphere 5×10^{10} molecule$\times$cm^{-2} s^{-1}, rate of dry O_3 deposition 4×10^{-2} cm s^{-1}, $[CH_4]_0=1.2\times10^{13}$ molecule cm^{-3}, aerosol density N=3 particle$\times$cm^{-3}, with radius 10^{-4} cm.

Acknowledgement

This study was supported by RFBR (grants N 00-03-32999 and N 00-05-65062) and VW foundation (grant I/72931). Useful discussions with Dr. M. Rossi and Professor R. Zellner are highly appreciated. V.M. Grigorieva thanks Drs. Roeth and Wagner for help and leadership during her work in the Essen-Uni.

References

Ayers, G.P., R.W. Gillett, J.M Cainey and A.L. Dick; Chloride and bromide loss from sea-salt particles in southern ocean air, *J. Atmos. Chem.* **33** (1999) 299-319.

Caloz, F., F.F. Fenter and M.J. Rossi; Heterogeneous kinetics of the uptake of ClONO2 on NaCl and KBr, *J. Phys. Chem.* **100** (1996) 7494-7501.

Foster, K.L., R.A. Plastridge, J.W. Bottenheim, P.B. Shepson and B.J. Finlayson-Pitts; The role of Br_2 and BrCl in surface ozone destruction at polar sunrise, *Science* **291** (2001) 471-474.

Gershenzon, M.Yu., Jr.; Heterogeneous uptake of NO_3 radicals on NaCl surface: Kinetics, mechanism and role in the atmospheric chemistry, *M.S. thesis*, Moscow Institute of Physics and Technology, Moscow (1998) 55, in Russian.

Gershenzon, M.Yu.,Jr., S.D. Il'in, N.G. Fedotov and Yu.M. Gershenzon; The mechanism of reactive NO_3 uptake on dry NaX (X = Cl, Br), *J. Atm. Chem.* **34** (1999) 119-135.

Gratpanche, F. and J.-P. Sawerysyn; Uptake coefficients of NO_3 radicals on solid surfaces of sea-salts, *J. Chim. Phys.* **96** (1999) 213-231.

Hausmann, M. and U. Platt; Spectroscopic measurement of bromine oxide and ozone in the high Arctic during Polar Sunrise Experiment 1992, *J. Geoph. Res.*, **99** (1994) 25399-25413.

Hönninger, G. and U. Platt; Observations of BrO and its vertical distribution during surface ozone depletion at Alert, *Atm. Env.* (2001), in press.

Keene, W.C.; R. Sander, A.P. Pszenny, R. Vogt, P.J. Crutzen and J.N. Galloway; Aerosol pH in the marine boundary layer: a review and model evaluation, *J. Aerosol Sci.* **29** (1998) 339-356.

Kopp, T., A. Kapilasharami, L.T. Molina and M.J. Molina; Phase transitions of sea-salt/water mixtures at low temperatures: Implications for ozone chemistry in the polar marine boundary layer, *J. Geoph. Res.* **105** (2000) 26393-26402.

Kreher, K., P.V. Johnston, S.W. Wood and U. Platt; Ground-based measurements of tropospheric and stratospheric BrO at Arrival Heights (78°S), Antarctica, *Gephys. Res. Lett.* **24** (1996) 3021-3024.

Lownthal, D.H.; R.D. Borys, C.F. Rogers, J.C. Chow, R.K. Stevens, J.P. Pinto and J.M. Ondov; A fine-particle sodium tracer for long-range transport of the Kuwaiti oil-fire smoke, *Geophys. Res. Lett.* **20** (1993), 691-693.

Martinez, M., T. Arnold and D. Perner; The role of bromine and chlorine chemistry for arctic ozone depletion events in Ny-Alesund and comparison with model calculations, *Ann. Geophys.* **17** (1999) 941-956.

Oum, K.W., M.J. Lakin, D.O. de Haan, T. Brauers and B.J. Finlayson-Pitts; Formation of molecular chlorine from the photolysis of ozone and aqueous sea-salt particles, *Science* **279** (1998) 74-77.

Perner, D., T. Arnold, J. Crowley, T. Klupfel, M. Martinez and R. Seuwen; The measurement of active chlorine in the atmosphere by chemical amplification, *J. Atmos. Chem.* **34** (1999) 9-20.

Richter, A., F. Wittrock, M. Eisinger and J.P. Burrows; GOME observations of tropospheric BrO in northern hemispheric spring and summer 1997, *Geophys. Res. Lett.* **25** (1998) 2683-2686.

Rudich, Y., R.K. Talukdar and A.R. Ravishankara; Reactive uptake of NO_3 on pure water and ionic solutions, *J. Geophys. Res.* **101** (1996) 21023-21031.

Sander, R. and P.J. Crutzen; Model study indicating halogen activation and ozone destruction in polluted air masses transported to the sea, *J. Geophys. Res.* **101** (1996) 9121-9138.

Seisel, S., F. Caloz, F. Fenter, H. van den Bergh and M.J. Rossi; The heterogeneous reaction of NO_3 with NaCl and KBr. A nonphotolytic source of halogen atoms, *Geoph. Res. Lett.* **24** (1997) 2757-2760.

Shaw, G.E.; Aerosol chemical components in Alaska air mass. 2. Sea salt and marine product, *J. Geoph. Res.* **96** (1991) 22369-22372.

Singh, H.B., G.L. Gregory, B. Anderson, E. Browell, G.W. Sachse, D.D. Davis, J. Crawford, J.D. Bradshaw, R. Talbot, D.R. Blake, D. Thornton, R. Newell and J. Merrill; Low ozone in the marine boundary layer of the tropical Pacific Ocean: Photochemical loss, chlorine atoms and entrainment, *J. Geophys. Res.* **101** (1996) 1907-1917.

Spicer, C.W.; E.G. Chapman, B.J. Finlayson-Pitts, R.A. Plastridge, J.M. Hubbe, J.D. Fast and C.M. Berkowitz; Unexpectedly high concentrations of molecular chlorine in coastal air, *Nature* **394** (1998) 353-356.

Tuckermann, M., R. Ackermann, C. Gölz, H. Lorenzen-Schmidt, T. Senne, J. Stutz, B. Trost, W. Unold and U. Platt; DOAS-observation of halogen radical-catalyzed Arctic boundary layer ozone destruction during the ARCTOC-campaigns 1995 and 1996 in Ny-Alesund, Spitsbergen, *Tellus* **49B** (1997) 533-535.

Wagner, T. and U. Platt; Satellite mapping of enhanced BrO concentration in the troposphere, *Nature* **395** (1998) 486-490.

Zelenov, V.V., E.V. Aparina, M.Yu. Gershenzon, Jr., S.D. Il'in and Yu.M. Gershenzon; Kinetic mechanisms of atmospheric gases uptake on sea salt surfaces. 3.Reactive NO_3 uptake on humidified sea salts NaX (X=Cl, Br) under steady state conditions, *Khimicheskaya Fizika* (2002a), in press, in Russian.

Zelenov, V.V., E.V. Aparina, Yu.M. Gershenzon, M.Yu. Gershenzon, Jr. and S.D. Il'in; Kinetic mechanisms of atmospheric gases uptake on sea salt surfaces. 2.Reactive NO_3 uptake on dry salts NaX (X=Cl, Br), *Khimicheskaya Fizika* (2002), in press, in Russian.

Photochemical Processes in the Atmosphere

Geert K. Moortgat

Max-Planck-Institut für Chemie, Atmospheric Chemistry Department, P.O.Box 3060, D-55020 Mainz, Germany

Introduction

Atmospheric trace gases are primarily removed by reactions with radicals and therefore radicals play an important role in the chemistry of the atmosphere. Radicals in the atmosphere are produced by the photodissociation of molecules, or by the reaction of a primary radical with another molecule. In Table 1 a short overview of several important atmospheric radical sources is given. From this table it can also be seen that many of the follow-up reactions produce the source molecules, for example the NO_2 produces $O(^3P)$ which in turn can produce O_3, which may lead to OH radicals. The OH radical reacts with many different compounds, cleaning the atmosphere of various trace gases (Ehhalt, 1999).

The photochemical reactions that can take place depend on many factors the most important being: the energy of the photons, the temperature, the absorption cross-section of the molecules and the efficiency by which they eventually dissociate and into which compounds if there is more than one possibility. The energy and number of the photons that are available in the lower atmosphere depend on the absorption of the light in higher layers of the atmosphere. Therefore, the distinction between the lower and the higher atmosphere must be made. In the higher atmosphere all light below about 300 nm is absorbed, mainly by the Hartley-band of the ozone molecule and the electronic transitions in oxygen and nitrogen. This limits the energy available to dissociate molecules in the lower atmosphere. In Figure 1 the absorption spectra of the molecules cited in Table 1 and the actinic flux is given.

For many photolabile molecules, the lifetime in the atmosphere is limited by both photodissociation and reactions with radicals, primarily OH. These two mechanisms compete with each other, and accurate measurements are needed to determine the lifetime of these molecules due to each of the processes, especially since photodissociation often produces radicals, which may influence the measurements. Since it is not possible to measure the lifetime for each compound, some theorical (e.g. spectroscopic) background is needed to predict lifetimes based on the general structure of the specific molecule.

The photodissociation of ozone and the production of radicals through the photolysis of carbonyl compounds in the lower atmosphere will be discussed in more detail.

Table 1. Important radical sources by photolytic action.

$O_3 + h\nu \rightarrow O(^1D) + O_2(a\,^1\Delta_g)$	$HCHO + h\nu \rightarrow HCO + H$
$\rightarrow O(^1D) + O_2(X\,^3\Sigma_g^-)$	$\rightarrow H_2 + CO$
$\rightarrow O(^3P) + O_2(X\,^3\Sigma_g^-)$	$HCO + O_2 \rightarrow HO_2 + CO$
$O(^1D) + H_2O \rightarrow 2\,OH$	$H + O_2 + M \rightarrow HO_2 + M$
$NO_2 + h\nu \rightarrow NO + O(^3P)$	$HO_2 + NO \rightarrow OH + NO_2$
$O(^3P) + O_2 + M \rightarrow O_3 + M$	$HONO + h\nu \rightarrow OH + NO$
	$H_2O_2 + h\nu \rightarrow 2\,OH$

I. Barnes (ed.), Global Atmospheric Change and its Impact on Regional Air Quality, 115–120.

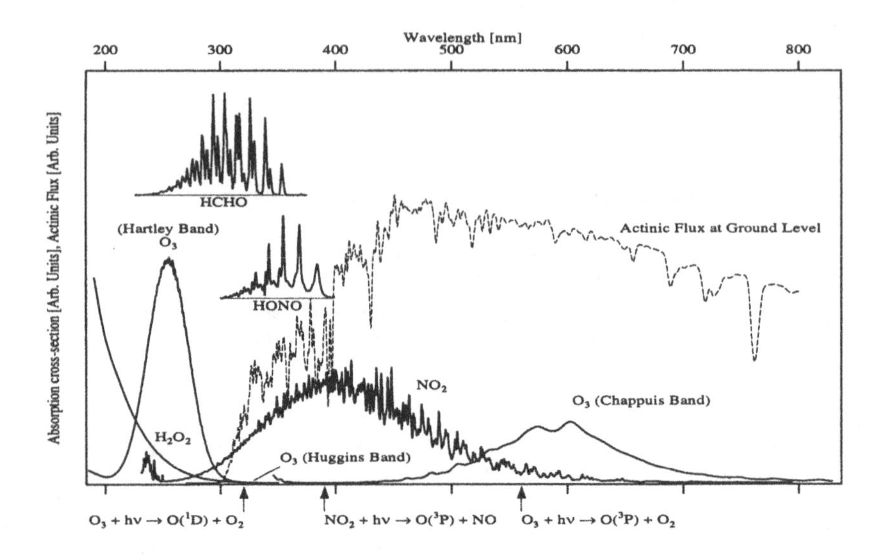

Figure 1. The absorption spectrum of various chemical species and the actinic flux at ground-level. The actinic flux is calculated using MODTRAN4 (Anderson *et al.*, 1995), for an US Standard Atmosphere 1976, with a typical springtime/summer loading of aerosols. The visibility is 23 kilometre. The solar zenith angle is 23°, the surface albedo is 0.05. Note that the vertical scales are *not* the same. The vertical scale for O_3 is split, the Chappuis band and the onset of the Huggins band use a different scale from the Hartley band. All scales are linear, not logarithmic (Moortgat *et al.* , 2001)

Photolysis of Ozone: Yield of $O(^1D)$

Among the many important roles played by ozone in the atmosphere is the role it plays in the generation of OH radicals, which are responsible for initiating the oxidation of a wide variety of atmospheric trace constituents (e.g. organic compounds, reduced sulphur species, etc), thereby removing them from the atmosphere (Ehhalt, 2000). While photolysis of ozone in the peak of the Hartley (200-300 nm) bands is important for the stratospheric chemistry of ozone, absorption of ozone in the Huggins bands at $\lambda>300$ nm dominates the photochemical activity in the troposphere and the lower stratosphere. In this region, the OH production occurs dominantly from the formation of the excited $O(^1D)$ species in the UV photolysis of ozone, followed by the rapid reaction of $O(^1D)$ with H_2O vapour: $O(^1D) + H_2O \rightarrow 2OH$. Thus the $O(^1D)$ production rate will depend on the photolysis frequency $J_{(O1D)}$ and the ozone concentration. $J_{(O1D)}$ is given by the product of the actinic flux F, the ozone absorption cross section σ, and the $O(^1D)$ quantum yield $\varphi_{O(1D)}$, all of which are wavelength dependent:

$$J_{O(1D)} = \int F(\lambda) \,.\, \sigma(\lambda) \,.\, \varphi_{O(1D)}(\lambda) \,.\, d\lambda$$

Above 290 nm, the $O(^1D)$ production rate is a critical component for model calculations: the solar actinic flux increases by more than four orders of magnitude, the temperature-dependent

O_3 absorption cross sections in the structured Huggins bands drop from 10^{-19} to 10^{-23} cm^2 molecule^{-1}, and the temperature-dependent $O(^1D)$ quantum yield decreases sharply with increasing wavelengths. Thus even small changes in the quantum yield in this region affect the $O(^1D)$ production rate.

The photochemistry of ozone is very complex, as the relatively weak bonds in ozone allow different states of the O (^3P and ^1D) and O_2 ($^3\Sigma_g^-$, $^1\Delta_g$ and $^1\Sigma_g^+$) photoproducts to be accessed. Above 290 nm, there are five thermodynamically allowed photolysis channels, shown with their wavelength threshold (at 0 K):

$$O_3 + h\nu \rightarrow O(^1D) + O_2\,(a^1\Delta_g) \qquad \lambda \leq 310\ \text{nm} \qquad (1)$$
$$\rightarrow O(^1D) + O_2\,(^3\Sigma_g^-) \qquad \lambda \leq 411\ \text{nm} \qquad (2)$$
$$\rightarrow O(^3P) + O_2\,(b^1\Sigma_g^+) \qquad \lambda \leq 463\ \text{nm} \qquad (3)$$
$$\rightarrow O(^3P) + O_2\,(a^1\Delta_g) \qquad \lambda \leq 611\ \text{nm} \qquad (4)$$
$$\rightarrow O(^3P) + O_2\,(^3\Sigma_g^-) \qquad \lambda \leq 1180\ \text{nm} \qquad (5)$$

Ozone photolysis dominates throughout the intense Hartley absorption band via the spin-allowed processes (1), with a reported $O(^1D)$ quantum yield in the range 0.9 to 0.95. The $O(^1D)$ quantum yield drops to near zero below the thermodynamic threshold at 310 nm, since the other (O^1D) producing channel (2) is spin-forbidden. However the exact shape of the $O(^1D)$ quantum yield curve at $\lambda > 310$ nm has been the subject of many experimental (Moortgat and Warneck, 1975; Brock and Watson, 1980; Trolier and Wiesenfeld, 1988; Armerding *et al.*, 1995; Ball and Hancock, 1995; Takehashi *et al.*, 1996, 1997, 1998; Ball *et al.*, 1997; Silvente *et al.*, 1997; Talukdar *et al.*, 1997, 1998; Bauer *et al.*, 2000) and theoretical (Michelson *et al.*, 1994) studies over the last two decades, leading to controversial results. In particular the existence of a longer wavelength "$O(^1D)$ quantum yield tail" has now been established over the temperature range 312-227 K (Talukdar *et al.*, 1998) and is now recommended for atmospheric calculations (Sander *et al.*, 2000). In addition, the principle investigators of several experimental groups combined their efforts to provide the modeling community with the best possible data on the quantum yield for the $O(^1D)$ production (Matsumi *et al.*, 2001).

Essentially, the experimentally measured $O(^1D)$ quantum yield tail is consistent with the participation of the temperature-dependent spin-allowed process (1) of vibrationally and rotationally excited ozone in the range 305 to 325 nm, and the temperature-independent constant spin-forbidden process (2) contributing with a constant value of about 0.06. The latter process was found to extend exclusively to 375 nm at 295 K, consistent with absorption to a single excited state (Bauer *et al.*, 2000). In these measurements (Talukdar *et al.*, 1997, 1998; Bauer *et al.*, 2000) the production and detection of OH radicals as spectroscopic marker for $O(^1D)$ was used, and revealed to be directly correlated with the ozone absorption cross sections. This recent study suggested that the $O(^1D)$ yield may be constant out to the thermodynamic threshold at 411 nm, however studies were inconclusive due to unreliable O_3 absorption cross sections above 340 nm. (Malicet *et al.*, 1995; Brion *et al.*, 1998; Voigt *et al.*, 2001). The importance of the longer "wavelength tail" in the $O(^1D)$ formation has been confirmed by comparing *in situ* $J_{(O1D)}$ measurements using spectroradiometric and/or chemical actinometers with calculated photolysis frequencies (Ehhalt, 2000; Müller *et al.*, 1995; Talukdar *et al.*, 1998). The new recommended temperature-dependent $O(^1D)$ quantum yields (including the tail) yield substantially larger $O(^1D)$ (and consequently larger OH) production rates especially at higher solar zenith angles, low temperatures and large overhead ozone columns (Bauer *et al.*, 2000), particularly at high latitudes in the troposphere from late autumn to early spring.

Photolysis of Carbonyl Compounds

Other significant species, producing HO_x (OH, HO_2) species upon photolysis in the atmosphere, include HONO, H_2O_2, CH_3OOH and carbonyl compounds (mainly aldehydes and ketones). The latter are emitted as primary pollutants (combustion, vegetation) or are produced as reaction intermediates from NO_x-mediated photooxidation of volatile organic compounds (VOC) emitted into the atmosphere. It is well established that the main degradation processes of carbonyl compounds are controlled by photolysis and/or reaction with OH radicals. The photolysis of these intermediate species is one of the major uncertainties in the VOC oxidation chain in gas-phase atmospheric chemistry. Photochemical data for the calculation of the photodissociation rates can be found in NASA and IUPAC publications on photochemical data (DeMore *et al.*, 1997; Atkinson *et al.*, 1997).

The photolysis pathways of the most important aldehyde, HCHO, is shown in Table 1, and photolytic parameters have been reported (Moortgat and Warneck, 1979; Meller and Moortgat, 2000). However, the photolysis of most carbonyl compounds remains one of the major uncertainties in the VOC oxidation chain in gas-phase atmospheric chemistry. Only quantum yields for the lower aldehydes have been measured, and in many cases unity quantum yield is assumed in models.

Because carbonyl compounds produce free radicals as photolysis products, it is important to assess and quantify reliable free radical yields from the photolytic action. This topic was addressed in an EU-funded project "RADICAL" on the photolysis, induced by natural UV/VIS solar radiation, of selected carbonyl compounds involved in the atmospheric oxidation of key hydrocarbons. An estimation of the average or effective quantum yield φ_{eff} is possible, by knowing the absorption spectrum, which can be measured in the laboratory, and by directly measuring *in situ* photolysis rates J_{exp}. If the photolysis rate J_{model} is calculated using simple radiation transfer model, using unity quantum yield, then the average quantum yield is

$$\varphi_{eff} = J_{exp}/J_{model}$$

A selection of compounds was based on their tropospheric importance and their occurence in the boundary layer. Selected were a series of alkanals (C_2 to C_9), bifunctional oxygenated compounds (glyoxal, pyruvic . acid) substituted hydroxycarbonyl (glycolaldehyde) and unsaturated (metylvinylketone, methacrolein, acrolein acid crotonaldehyde) carbonyl compounds; finally photooxidation products of biogenic compounds. Photodissociation rates of individual carbonyl compounds were measured directly during several campaigns in the EUPHORE outdoor smog chamber (Valencia) and these data were compared with model calculations (Moortgat, 2000).

The results are summarised in Table 2. Also, the photolytic lifetimes were compared to the lifetime against the OH radical reaction. Effective quantum yields, φ_{eff}, were determined by comparison of the individual photolysis frequencies with theoretical decomposition rates using the measured actinic fluxes. For most of the carbonyl compounds the determined φ_{eff} is significantly *below unity*.

The effective quantum yield for C_3 to C_9 straight chain aldehydes is 0.25 to 0.30, and is generally about twice as large for α-branched aldehydes. The dicarbonyl glyoxal has a very low φ_{eff} (0.038), such as found for methylglyoxal in a separate study (Koch and Moortgat, 1998). For unsaturated compounds (methylvinylketone, methacrolein, acrolein and *trans*-crotonaldehyde) φ_{eff} is negligible, although those compounds possess absorption spectra reaching the near visible. The photolytic decay from a few carbonyl compounds derived from the photooxidation from monoterpenes is variable, so need to be determined individually, since absorption spectra are not available. In currently worldwide used atmospheric models, quantum yields are often assumed to be unity, which is seldom the case, as is shown in this study. This leads to an overestimation of the calculated photodissociation rates and the associated radicals formed in photolysis processes. These new data will reduce the present uncertainties associated with photolysis processes, so that future assessments can be made with more confidence.

Table 2. Summary of the photodissociation rates measured at the EUPHORE facility near Valencia.

COMPOUND	PHOTOLYTIC RATE	J_{calc} (ϕ =1)	φ_{eff}	τ_{phot}	τ_{OH} [a]
Acetaldehyde	$(2.9 \pm 2.7) \times 10^{-5}$	4.9×10^{-5}	0.06 ± 0.05	4 d	17 h
Propionaldehyde	$(1.1 \pm 0.1) \times 10^{-5}$	3.8×10^{-5}	0.28 ± 0.04	1.2 d	14 h
Butyraldehyde	$(1.0 \pm 0.2) \times 10^{-5}$	5.1×10^{-5}	0.20 ± 0.04	1.2 d	12 h
i-Butyraldehyde	$(3.7 \pm 0.1) \times 10^{-5}$	5.2×10^{-5}	0.71 ± 0.02	7.5 h	11 h
n-Pentanal	$(1.6 \pm 0.2) \times 10^{-5}$	5.4×10^{-5}	0.30 ± 0.02	17 h	10 h
2-Methylbutyraldehyde	$(3.8 \pm 0.1) \times 10^{-5}$	5.2×10^{-5}	0.72 ± 0.03	7.3 h	12 h[b]
3-Methylbutyraldehyde	$(1.25 \pm 0.1) \times 10^{-5}$	4.7×10^{-5}	0.27 ± 0.01	22 h	12 h[b]
Pivaldehyde	$(1.45 \pm 0.1) \times 10^{-5}$	2.6×10^{-5}	0.56 ± 0.05	19 h	10 h
n-Hexanal	$(1.65 \pm 0.3) \times 10^{-5}$	5.9×10^{-5}	0.28 ± 0.05	17 h	15 h
n-Nonanal	$(1.15 \pm 0.2) \times 10^{-5}$	4.9×10^{-5}	0.23 ± 0.03	1 d	10 h
Glyoxal	$(1.05 \pm 0.3) \times 10^{-4}$	2.7×10^{-3}	0.038 ± 0.01	2.6 h	25 h
Glycolaldehyde	$(1.14 \pm 0.25) \times 10^{-5}$	8.6×10^{-6}	1.32 ± 0.30	1 d	28 h
Pyruvic acid	$(1.0 \pm 0.15) \times 10^{-4}$	2.3×10^{-4}	0.43 ± 0.07	2.8 h	3.2 m
Methylvinylketone	$< 10^{-6}$	5.2×10^{-4}	< 0.004	> 6 d	15 h
Methacrolein	$< 10^{-6}$	5.2×10^{-4}	< 0.004	> 6 d	8 h
Acrolein	$< 10^{-6}$	4.3×10^{-4}	< 0.004	> 6 d	13 h
Trans-Crotonaldehyde	$(1.2 \pm 0.15) \times 10^{-5}$	4.0×10^{-4}	0.03 ± 0.01	1 d	7.4 h
Pinonaldehyde	$(1.15 \pm 0.10) \times 10^{-5}$	8.0×10^{-5}	0.14 ± 0.01	1d	3.1 h
Nopinone	3.1×10^{-5}			> 4 d	16 h
Limonaketone	3.1×10^{-5}			> 4 d	2.0 h

a) assuming [OH] = 1×10^{6} molecule cm^{-3}, b) OH rate constants from references [*Atkinson*, 1994; *Moortgat*, 2000].

References

Anderson, G.P., F.X. Kneizys, J.H. Chetwynd, J. Wang, M.L. Hoke, L.S. Rothman, L.M. Kimball and R.A. McClatchey, FASCODE/MODTRAN/LOWTRAN: Past/present/future, *18th Annual Review Conference on Atmospheric Transmission Models* (1995).

Armerding, W., F.J. Comes and B. Schülke, O(^{1}D) quantum yield of ozone photolysis in the UV from O_3 nm to ist threshold and at 355 nm, *J. Chem. Phys.* **99** (1995) 3137-3143.

Atkinson, R., Gas-phase tropospheric chemistry of organic compounds, *J. Phys. Chem. Ref. Data*, (1994), Monograph 2.

Atkinson, R., D.L. Baulch, R.A. Cox, R.F. Hampson, Jr., J.A. Kerr, M.J. Rossi and J. Troe, Evaluated kinetic, photochemical and heterogeneous data for atmospheric chemistry: Supplement V, *J. Phys. Chem. Ref. Data*, **26** (1997) 521-1010.

Ball, S.M. and G. Hancock, The relative quantum yields of $O_2(a^1\Delta_g)$ from the photolysis of ozone at 227 K,*Geophys. Res. Lett.* **22** (1995) 1213-1216.

Ball, S.M., G. Hancock, S.E. Martin and J.C. Pinot de Moira, A direct mesurement of O(^{1}D) quantum yields from the photodissociation of ozone between 300 and 328 nm, *Chem. Phys. Lett.* **264** (1997) 531-538.

Bauer, D., L. D'Ottone and A.J. Hynes, O^1D quantum yields from O_3 photolysis in the near UV region between 305 and 375 nm, *Phys. Chem. Chem. Phys.* **2** (2000) 1421-1424.

Brion, J., A. Chakir, J. Charbonnier, D. Daumont, C. Parisse and J. Malicet, Absorption spectra measurements for the ozone molecule in the 350–830 nm region, *J. Atmos. Chem.* **30** (1998) 291-299.

Brock, J.C. and R.T. Watson, Ozone photolysis: laser flash photolysis of ozone: $O(^1D)$ quantum yields in the fall-off region 297-325 nm, *Chem. Phys.* **71** (1980) 477-484.

DeMore, W. P., S. P. Sander, D. M. Golden, R. F. Hampson, M. J. Kurylo, C. J. Howard, A. R. Ravishankara, C. E. Kolb, and M. J. Molina, Chemical kinetics and photochemical data for use in stratospheric modeling, Evaluation No. 12, *JPL Publication 97-4* (1997).

Ehhalt, D.H., Photooxidation of trace gases in the troposphere, *Phys. Chem. Chem. Phys.* **1** (1999) 5401-5408.

Koch, S. and G.K. Moortgat, Photochemistry of methylglyoxal in the vapor phase, *J. Phys. Chem.* **102** (1998) 9142-9153.

Malicet, J., D. Daumont, J. Charbonnier, C. Parisse, A. Chakir and J. Brion, Ozone UV spectroscopy. II. Absorption cross-sections and temperature dependence, *J. Atmos. Chem.* **21** (1995) 263-273.

Matsumi, Y., F.J. Comes, G. Hancock, A. Hofzumahaus, A. J. Hynes, M. Kawasaki, and A.R. Ravishankara, Quantum yields for production of $O(^1D)$ in the ultraviolet photolysis of ozone: recommendation based on evaluation of laboratory data, *J. Geophys. Res.* (2001), in press.

Meller, R. and G. K. Moortgat, Temperature dependence of the absorption cross sections of formaldehyde between 223 and 323 K in the wavelength range 225-375 nm, *J. Geophys. Res.* **105** (2000) 7089-7101.

Michelson, H.A., R.J. Salawitch, P.O. Wennberg and J.G. Anderson, Production of $O(^1D)$ from photolysis of O_3, *Geophys. Res. Lett.* **21** (1994) 2227-2230.

Moortgat, G.K., Evaluation of radical sources in atmospheric chemistry through chamber and laboratory studies, Final Report of EU-Project ENV4-CT-0419, RADICAL, February 2000.

Moortgat, G.K., R. Lang, H. Nüss and M. Sneep, Important photochemical processes in the lower atmosphere, Proceedings of the COACh International School, Obernai, France March 2001, (in press).

Moortgat, G.K. and P. Warneck, Relative $O(^1D)$ quantum yields in the near UV photolysis of ozone at 298 K. Z. *Naturforsch.* **30a** (1975) 835-844.

Moortgat, G. K. and P. Warneck, CO and H_2 quantum yields in the photodecomposition of formaldehyde in air, *J. Chem. Phys.*, **70** (1979) 3639-3651.

Müller, M., A. Kraus and A. Hofzumahaus, $O_3 \rightarrow O(^1D)$ photolysis frequencies determined from spectroradiometric measurements of solar actinic UV-radiation: comparison with chemical actinometer measurements, *Geophys. Res. Lett.* **22** (1995) 679-682.

Sander, S.P., R.R. Friedl, W.B. DeMore, D.M. Golden, M.J. Kurylo, R.F. Hampson, R.E. Huie, G.K. Moortgat, A.R. Ravishankara, C.E. Kolb and M.J. Molina, Chemical linetics and photochemical data for use in stratospheric modeling, *JPL Publ.* (2000), 00-3.

Silvente, E., R.C. Richter, M. Zheng, E.S. Salzman and A.J. Hynes, Relative quantum yields for $O(^1D)$ production in the photolysis of ozone between 301 and 366 nm: evidence for the participation of a spin-forbidden channel, *Chem. Phys. Lett.* **264** (1997) 309-315.

Takehashi, K., M. Kishigami, Y. Matsumi, M. Kawasaki and A.J. Orr-Ewing, Observation of the spin-forbidden $O(^1D) + O_2(X\ ^3\Sigma^-_g)$ channel in the 317-327 nm photolysis of ozone, *J. Chem. Phys.* **105** (1996) 5290-5293.

Takehashi, K., Y. Matsumi, M. Kawasaki and K. Takehashi, Photofragment excitation spectrum for $O(^1D)$ from the photodissociation of jet-cooled ozone in the wavelength range 305–329 nm, *J. Chem. Phys.* **106** (1997) 6390-6397.

Takehashi, K., N. Taniguchi, Y. Matsumi, M. Kawasaki and M.N.R. Ashfold, Wavelength and temperature dependence of the absolute $O(^1D)$ yield from the 305-329 nm photodissociation of ozone, *J. Chem. Phys.* **108** (1998) 7161-7172.

Talukdar, R.K., M.K. Gilles, F. Battin-Leclerc, A.R. Ravishankara, J.-M. Fracheboud, J. Orlando and G.S. Tyndall, Photolysis of ozone at 308 and 248 nm: Quantum yield of $O(^1D)$ as function of temperature, *Geophys. Res. Lett.* **24** (1997) 1091-1094.

Talukdar, R.K., C.A. Longfellow, M.K. Gilles and A.R. Ravishankara, Quantum yield of $O(^1D)$ in the photolysis of ozone between 289 and 389 nm as a function of temperature, *Geophys. Res. Lett.* **25** (1998) 143-146.

Trolier, M. and J.R. Wiesenfeld, Relative quantum yield of $O(^1D_2)$ following ozone photolysis between 275 and 325 nm,*Geophys. Res. Lett.* **22** (1988) 7119-7124.

Voigt, S., J. Orphal, K. Bogumil and J. P. Burrows, The temperature dependence (203-293 K) of the absorption cross sections of O_3 in the 230-850 nm region mesured by Fourier-transform spectroscopy, J. *Photochem. Photobiol. A: Chem.* **143** (2001) 1-9.

Atmosphere-Related Aspects of the Aqueous Phase Oxidation of Sulphur Dioxide

Wanda Pasiuk-Bronikowska

Institute of Physical Chemistry, Polish Academy of Sciences, Kasprzaka 44/52, 01-224 Warsaw, Poland
wpb@ichf.edu.pl

Introduction

A significant number of papers on the occurrence of a variety of organic and inorganic chemical species in the atmosphere have been published in the last two decades (e.g. Munger *et al.*, 1984; Delmelle *et al.*, 2001; Poschl *et al.*, 2001; Warneke *et al.*, 2001) indicating the complexity, previously unexpected, of atmospheric chemistry and the still growing need to extend the fundamental knowledge in this area. The situation gained even more drastic expression when the initially neglected input of atmospheric water (cloud droplets, falling raindrops, fog) to the chemistry of the atmosphere was realised. Although clouds makeup only a 5 percent volume fraction in the total troposphere and droplets in a cloud only one per ten million to one per million of its volume (Dentener *et al.*, 1994), the atmosphere aqueous phase should be treated as an accumulation of myriad of minute but efficient reactors where many chemical reactions take place at concentrations of reactants usually much higher than in the atmosphere gas phase (Table 1). The aim of this paper is to discuss some less common, but not less important, directions of facing the phenomena resulting from the chemical complexity of the atmosphere, as seen from the laboratory experimentation perspective.

One of the important atmospheric phenomena is the oxidation of sulphur dioxide to sulphuric acid or sulphate. The oxidation products add significantly to the acidification of the atmosphere and provide nuclei for precipitation. Main sources of this pollutant and the corresponding global emissions are given, respectively, in Figure 1 and Table 1. The aqueous phase oxidation of SO_2 accounts for about 50% of the atmospheric turnover of tetravalent sulphur (Mentel, 1999). Therefore, it is instructive to consider some of the consequences of the growing complexity of the atmosphere using the aqueous phase conversion of SO_2 as an example.

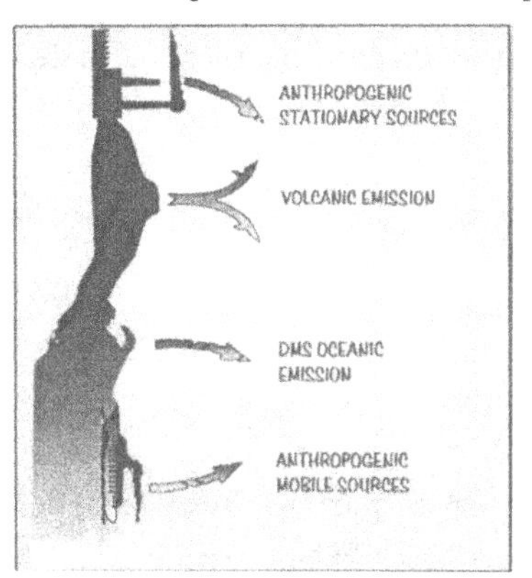

Figure 1. Sources of atmospheric SO_2.

Table 1.
Global emission of sulphur dioxide*

	10^{14} gS/yr	
Sulphur, total	0.98	
Anthropogenic		0.70
Biomass burning		0.025
Oceans		0.16
Volcanoes		0.085
Soil and plants		0.01

*Lagner and Rodhe, 1991

Chemical pathways of aqueous phase oxidation of SO_2

As known, both from field and laboratory measurements, there are several chemical pathways by which the oxidation of SO_2 may occur in atmospheric water. Their relative

I. Barnes (ed.), Global Atmospheric Change and its Impact on Regional Air Quality, 121–127.

efficiency depends in part on the interaction between oxidants of S(IV) and other pollutants capable of competing for the same oxidants. The main reactions leading to S(VI) are listed in Table 2.

Table 2. Aqueous phase oxidation of SO_2 - main chemical pathways

Reaction	*Rate constant* $dm^3/mol\ s$ (298 K)	*Reference*
oxidation of SO_2 by ozone		
$HSO_3^- + O_3 \rightarrow HSO_4^- + O_2$	3.2×10^5	Hoigne *et al.*, 1985
$SO_3^{2-} + O_3 \rightarrow SO_4^{2-} + O_2$	1.0×10^9	Hoigne *et al.*, 1985
oxidation of SO_2 by H_2O_2		
$HSO_3^- + H_2O_2 \rightarrow HSO_4^- + H_2O$	3.9×10^4 (pH 3)	Zellner (ed.), 1999
oxidation of SO_2 by molecular oxygen	chain reaction with the composite rate constant	
$HSO_3^-/SO_3^{2-} + 1/2O_2 \rightarrow HSO_4^-/SO_4^{2-}$	dependent on initiators and inhibitors	

In order to understand the complex reaction as a whole, it is important to determine rate constants for its component steps. In the case of a chain reaction this is not a trivial task.

Answers to problems

Why the oxidation of SO_2 by molecular oxygen is so difficult to be absolutely quantified? Let us now focus our attention on the mechanism of SO_2 oxidation by molecular oxygen (see Table 3). The reaction transients, sulphoxy radical anions $^{\bullet}SO_3^-$, $SO_4^{\bullet -}$ and $SO_5^{\bullet -}$, are very reactive, not only with respect to sulphite but also toward many other inorganic or organic species. Therefore, the rate of SO_2 oxidation by oxygen strongly depends on trace contaminants (water dissolved or solid) which scavenge the radical anions, thus making the overall reaction slower.

Table 3. Reduced chain mechanism of S(IV) autoxidation

Initiation

initiator + $HSO_3^-/SO_3^{2-} \rightarrow (^{\bullet}\mathbf{SO_3^-} + H^+)/^{\bullet}\mathbf{SO_3^-} + ...$ $\quad k_i$

Propagation

$^{\bullet}\mathbf{SO_3^-} + O_2 \rightarrow \mathbf{SO_5^{\bullet -}}$ $\quad k_{p1}$

$\mathbf{SO_5^{\bullet -}} + HSO_3^-/SO_3^{2-} \rightarrow HSO_5^-/SO_5^{2-} + {}^{\bullet}\mathbf{SO_3^-}$ $\quad k_{p2(a,\,b)}$

$\rightarrow HSO_4^-/SO_5^{2-} + \mathbf{SO_4^{\bullet -}}$ $\quad k_{p3(a,\,b)}$

$\mathbf{SO_4^{\bullet -}} + HSO_3^-/SO_3^{2-} \rightarrow HSO_4^-/SO_4^{2-} + {}^{\bullet}\mathbf{SO_3^-}$ $\quad k_{p4(a,\,b)}$

$HSO_5^- + HSO_3^-/SO_3^{2-} \rightarrow 2SO_4^{2-} + 2H^+/H^+$

Termination

$\mathbf{SO_5^{\bullet -}} + \mathbf{SO_5^{\bullet -}} \rightarrow S_2O_8^{2-} + O_2$ $\quad k_t$

This may be one of the reasons why values of rate constants for individual steps have not yet been properly co-ordinated. They may differ even by orders of magnitude (Table 4). Laboratory studies of this reaction require the utmost care to avoid the interaction between sulphoxy radicals and the material of a reactor (cuvette, vessel) or trace impurities of chemicals and water used for the preparation of reacting solutions. A critical analysis of the kinetic data available in literature, obtained by various experimental techniques, should be performed.

What are the possible initiators of SO_2 oxidation in the troposphere? It is clear that the oxidation of S(IV) should be enhanced with the increase of the rate of chain initiation.

Table 4. Rate constants for the oxidation of S(IV) by peroxymonosulphate radicals

$$SO_5^{\bullet-} + HSO_3^- \rightarrow {}^{\bullet}SO_3^- + HSO_5^- \qquad k_{p2a}$$
$$\rightarrow SO_4^{\bullet-} + HSO_4^- \qquad k_{p2b}$$

Rate constant, dm^3/mol s			Reference
k_{p2a}	k_{p2b}	$k_{p2a} + k_{p2b}$	
		$3x10^6$	Huie and Neta, 1984
$2.5x10^4(9.5x10^3)$*	$7.5x10^4(2.8x10^4)$	$3x10^5$ $(1.1x10^5)$	Huie and Neta, 1987
$5.2x10^4(4.3x10^4)$	$2.9x10^5(2.5x10^5)$	$3.4x10^5(2.9x10^5)$	Ziajka *et al.*, 1995, rcld
		$3.6x10^3$	Yermakov *et al.*, 1995
$8.6x10^3(4.7x10^3)$	$3.6x10^2(2.0x10^2)$	$9.0x10^3(5.0x10^3)$	Buxton *et al.*, 1996

$$SO_5^{\bullet-} + SO_3^{2-} \rightarrow {}^{\bullet}SO_3^- + SO_5^{2-} \qquad k_{p3a}$$
$$\rightarrow SO_4^{\bullet-} + SO_4^{2-} \qquad k_{p3b}$$

Rate constant, dm^3/mol s			Reference
k_{p3a}	k_{p3b}	$k_{p3a} + k_{p3b}$	
$3.2x10^6(4.6x10^5)$	$9.8x10^6(1.4x10^6)$	$1.3x10^7(1.9x10^6)$	Huie and Neta, 1987
$2.1x10^5(1.2x10^5)$	$5.5x10^5(3.3x10^5)$	$7.6x10^5(4.6x10^5)$	Buxton *et al.*, 1996

*values in parentheses are recalculated for zero ionic strength

The latter effect may result, for instance, on the additional production of sulphite radicals in parallel initiation steps. The possible modes of initiation may be divided into two main groups.

1. Photoinitiation

- photolysis of hydrogen peroxide (initiation via hydroxyl radicals)

$$H_2O_2 + h\nu \rightarrow 2\ OH$$
$$HSO_3^-/SO_3^{2-} + OH \rightarrow \mathbf{{}^{\bullet}SO_3^-} + H_2O/OH^-$$

- photolysis of nitroxy anions (initiation via hydroxyl radicals)

$$NO_2^- + h\nu + H^+ \rightarrow NO + OH$$
$$NO_3^- + h\nu + H^+ \rightarrow NO_2 + OH$$
$$HSO_3^-/SO_3^{2-} + OH \rightarrow \mathbf{{}^{\bullet}SO_3^-} + H_2O/OH^-$$

-photolysis of transition metal ions, hydroxylated or bonded with sulphoxy anions (initiation via hydroxyl radicals or via sulphate radicals, respectively)

$$[Fe(OH)]^{2+} + h\nu \rightarrow Fe^{2+} + OH$$
$$[Fe(OH)_2]^{2+} + h\nu \rightarrow Fe^{2+} + OH^- + OH$$
$$HSO_3^-/SO_3^{2-} + OH \rightarrow \mathbf{{}^{\bullet}SO_3^-} + H_2O/OH^-$$

or

$$[Fe(SO_4)]^+ + h\nu \rightarrow Fe^{2+} + SO_4^{\bullet-}$$
$$HSO_3^-/SO_3^{2-} + SO_4^{\bullet-} \rightarrow \mathbf{{}^{\bullet}SO_3^-} + HSO_4^-/SO_4^{2-}$$

2. Chemical initiation

- oxidation of S(IV) by transition metal ions and their complexes

$$HSO_3^-/SO_3^{2-} + (ML)^{(z+1-n)+} \rightarrow (\mathbf{{}^{\bullet}SO_3^-} + H^+)/\mathbf{{}^{\bullet}SO_3^-} + (ML)^{(z-n)+}$$

where M = Fe, Mn, Cu, Co, V;

- thermal decomposition of peroxydisulphate anions (initiation via sulphate radicals)

$$S_2O_8^{2-} \rightarrow 2\ SO_4^{\bullet-}$$

$$HSO_3^-/SO_3^{2-} + SO_4^{\bullet-} \rightarrow {}^{\bullet}\mathbf{SO_3}^- + HSO_4^-/SO_4^{2-}$$

- oxidation of phenolic compounds by transition metal ions (initiation via phenoxyl radicals)

$$Ph + M^{(z+1)+} \rightarrow Ph^{\bullet} + M^{z+}$$

$$SO_3^{2-} + Ph^{\bullet} \rightarrow {}^{\bullet}\mathbf{SO_3}^- + Ph$$

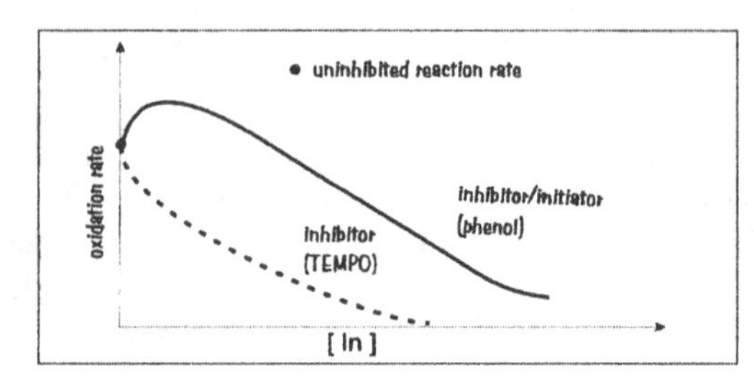

Figure 2. Effect of simple inhibitor (TEMPO) and inhibitor/initiator (phenol) on the S(IV) oxidation rate (neutral solutions).

Recently, indirect proof for the initiation via phenoxyl radicals was obtained from experiments on the oxidation of S(IV) in the presence of phenol (Pasiuk-Bronikowska *et al.*, 2000). Figure 2 shows that phenol, a known scavenger of sulphoxy radicals, enhances the S(IV) oxidation rate over that measured in the absence of phenol. It seems that a list of organic compounds capable of initiating the S(IV) oxidation chain is far from being complete - industrial areas may be rich both in S(IV) and organic initiators. Also, dihalogenide radical ions are known to initiate the S(IV) oxidation by oxygen.

Is the impact of scavengers of sulphoxy radicals important for the atmospheric fate of SO_2? As proven recently, terpenes and their derivatives are efficient scavengers of sulphoxy radicals (Buxton *et al.*, 1999; Ziajka and Pasiuk-Bronikowska,1999; Pasiuk-Bronikowska *et al.*, 1999). The global emission of terpenes attains 4.8×10^{14} gC/yr (Zimmerman *et al.*, 1978), so their interaction with intermediates in S(IV) oxidation by molecular oxygen seems worthy of consideration.

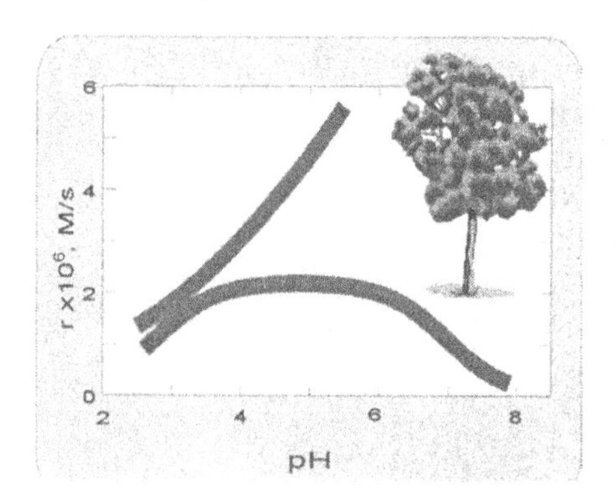

Figure 3. Inhibition by α-pinene: rate of S(VI) formation (r) for the un-inhibited (upper curve) and inhibited reaction (lower curve).

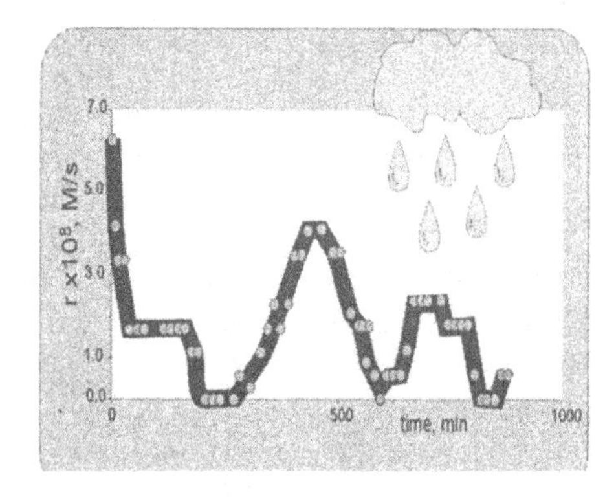

Figure 4. Inhibition by sobrerol: rate of S(VI) formation for the inhibited reaction observed in long experiments.

Laboratory studies of the influence of terpenes on the kinetics of sulphuric acid/sulphate formation by the chain mechanism revealed that sulphoxy-radical scavengers may initiate various phenomena. The rate of SO_2 oxidation may be markedly lowered, the reaction may be temporarily "frozen" or it may have oscillating character (Figures 3 and 4). Another effect is on the site of reaction within a droplet. Possible reaction sites within a droplet are shown in

Figure 5. According to the theory of gas absorption with chemical reaction (Astarita, 1967) the simultaneous diffusion and chemical reaction impose the kinetics of a gas-liquid reaction described by the equation of a form depending on the relative rates of mass transfer and chemical transformation. Under the condition of a constant rate of diffusion, the decreasing rate of chemical reaction has the consequence that the site of reaction shifts from the gas-liquid interface towards the liquid bulk. One should expect such an effect with increasing inhibition (e.g. due to the increasing concentration of an inhibitor) as illustrated in Figure 6.

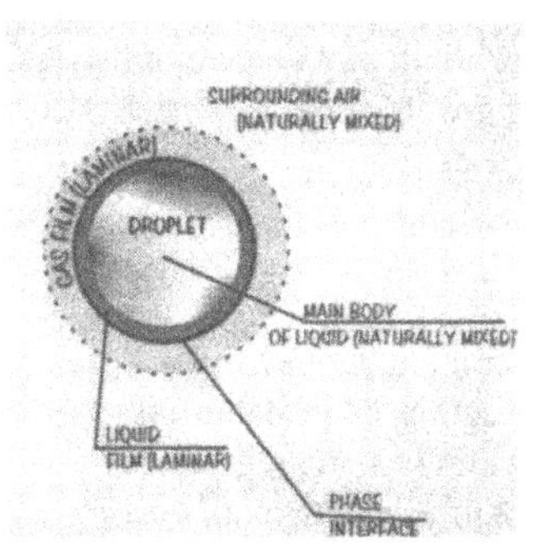

From the number of effects on which considerations of sulphoxy-radical scavenging may be focused the following emerge clearly from hitherto reported laboratory experiments:

- the concentration of sulphoxy radicals decreases causing a drop in the rate of acidity formation;
- the long-range transport of unreacted aqueous SO_2 is possible;
- sulphoxy radical scavengers undergo transformations into secondary pollutants among which organic compounds with sulphonic acid functions are formed;
- depending on the scavenger amount and activity the S(IV) oxidation by molecular oxygen may change from a surface controlled reaction to one controlled by a volume.

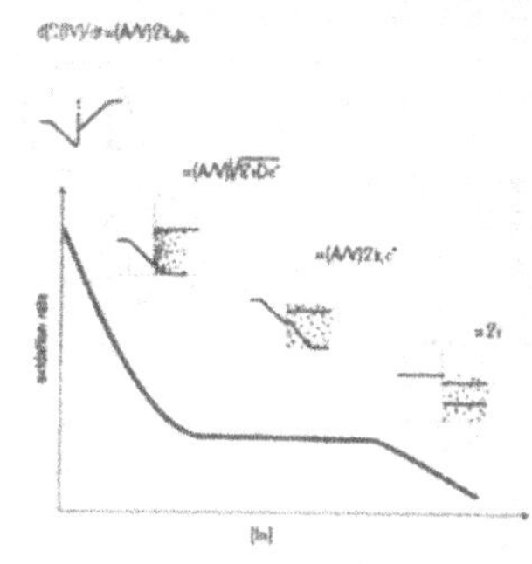

Figure 6. Kinetic effects of the increasing inhibition (concentration profiles are given in circles).

Can oxygen successfully compete with other atmospheric oxidants of S(IV)? It is almost commonly believed that ozone is one of the most effective oxidants of S(IV) in atmospheric waters (Georgii and Warneck, 1999), whereas molecular oxygen is not so effective. However, model calculations reported recently (Herrmann *et al.*, 2000) lead to the result that, for instance, under urban conditions the direct oxidation by O_3 is much less important than the chain pathway involving oxygen. Similarly, the predominance of oxygen was observed in laboratory experiments when oxygen and ozone were absorbed simultaneously into a solution containing S(IV) (Pasiuk-Bronikowska et al., 1998). In these experiments the ratio [S(IV)]/[O_3] was varied from 2.5×10^1 to 2.5×10^2 (in the CAPRAM mechanism (Herrmann *et al.*, 2000) the ratio [S(IV)]/[O_3] from 1.6×10^2 to 3.6×10^2) and the concentration of oxygen was 2.5×10^{-4} M. Figure 7 shows that there is practically is no influence of ozone on the rate of S(VI) formation by a chain reaction involving oxygen. The following conclusions may be derived from the experiments:

- although oxidants such as O_3 or OH attack S(IV) at rate constants of the order of 10^9 $M^{-1}s^{-1}$, their contribution to the oxidation of S(IV) within a droplet may be marginal because of the

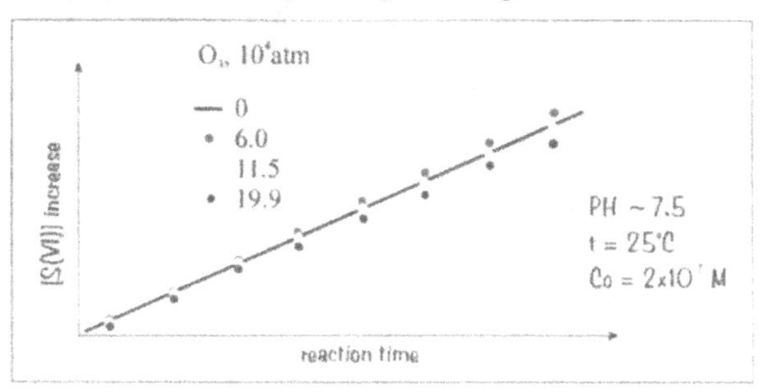

Figure 7. S(VI) formation at simultaneous uptake of O_3 and O_2.

aqueous phase concentration of these oxidants is lower, by as much as even six orders of magnitude, than the concentration of oxygen;

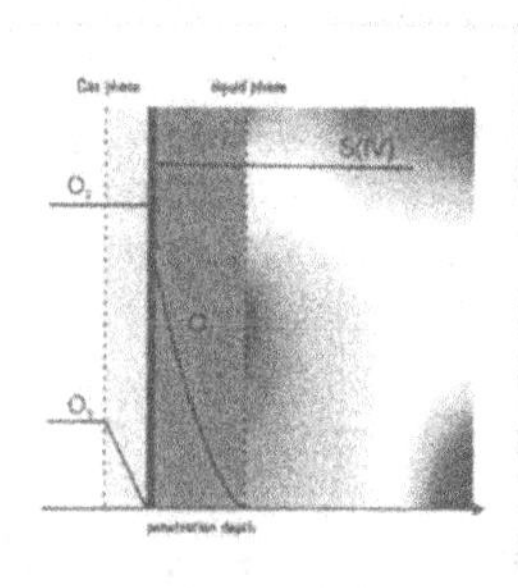

Figure 8. Concentration profiles for simultaneous absorption of O_3 and O_2.

- under certain conditions the reaction site for a trace oxidant (O_3, OH) becomes separated from the site of reaction for oxygen See Figure 8), so that in the first case the oxidation of S(IV) is gas-side diffusion controlled, whereas in the second case the reaction takes place in the fast-reaction regime (in the sense of Astarita, 1967);
- diverse situations are possible, depending on the chain initiators and inhibitors, the most striking being when a trace oxidant is not capable of entering the aqueous phase particle.

Conclusions

The oxidation of S(IV), as seen from the mechanistic-kinetic point of view, is an instructive example of possible effects of chemical composition change in the atmosphere, regardless of the origin of this change (human activity, climate anomalies or natural incidents like volcanic eruptions). In this paper attention was focused on the following aspects of this reaction: the atmosphere as a reservoir of trace constituents capable of accelerating and decelerating the free radical chain oxidation of S(IV), diffusion of reagents as an important factor determining the location of the reaction zone with respect to the surface of a droplet and effects of the oxygen overbalance when simultaneously absorbed with other oxidants into a droplet.

Laboratory studies are the indispensable link between atmospheric chemists and modellers to understand chemical processes in the atmosphere and to provide proper data for their quantification. Otherwise, the atmospheric modelling would not be able to evolve from an art into science.

References

Astarita G., *Mass Transfer with Chemical Reaction*, Elsevier, Amsterdam 1967.

Buxton G. V., S. McGowan, G. A. Salmon, J. E. Williams and N. D. Wood; A study of the spectra and reactivity of oxysulphur-radical anions involved in the chain oxidation of S(IV): a pulse and γ-radiolysis study, *Atmos. Environ.* **30** (1996) 2483-2493.

Delmelle P., J. Stix, P.-A. Bourque, P. J. Baxter, J. Garcia-Alvarez and J. Barquero; Dry deposition and heavy acid loading in the vicinity of Masaya Volcano, a major sulfur and chlorine source in Nicaragua, *Environ. Sci. Technol.* **35** (2001) 1289-1293.

Dentener F. J., P. J. Crutzen and J. Lelieveld; Chemical reactions in clouds: consequences for the global budget of O_3, in: *Physico-Chemical Behaviour of Atmospheric Pollutants*, vol. 1, OOPEC, Luxembourg 1994 (Report EUR 15609/1 EN), pp. 14-17.

Georgii H. W. and P. Warneck; Chemistry of the tropospheric aerosol and of clouds, in: R. Zellner, H. Baumgartel, W. Grunbein and F. Hensel (eds), *Global Aspects of Atmospheric Chemistry*, Springer, New York 1999, pp. 111-179.

Herrmann H., B. Ervens, H.-W. Jacobi, R. Wolke, P. Nowacki and R. Zellner; CAPRAM2.3: A chemical aqueous phase radical mechanism for tropospheric chemistry, *J. Atmos. Chem.* **36** (2000) 231-284.

Hoigne J., H. Bader, W. R. Haag and J. Staehelin; Rate constants of reactions of ozone with organic and inorganic compounds in water III: inorganic compounds and radicals, *Water Res.* **19** (1985) 993-1004.

Huie R. E. and P. Neta; Chemical behavior of SO_3^- and SO_5^- radicals in aqueous solutions, *J. Phys. Chem.* **88** (1984) 5665-5669.

Huie R. E. and P. Neta; Rate constants for some oxidations of S(IV) by radicals in aqueous solutions, *Atmos. Environ.* **21** (1987) 1743-1747.

Lagner J. and H. Rodhe, 1991 - data taken from M. Kanakidou, The role played by anthropogenic emissions in determining global O_3/NO_x/S/hydrocarbon budgets, in: P. M. Borrell, P. Borrell, T. Cvitas and W. Seiler (eds), *Photo-oxidants: Precursors and Products*, SPB Academic Publishing, Den Haag 1993, pp. 269-275.

Munger J. W., D. J. Jacob and M. R. Hoffmann; The occurrence of bisulfite-aldehyde addition products in fog- and cloud-water, *J. Atmos. Chem.* **1** (1984) 335-350.

Mentel, T. F.; Acid deposition, in: R. Zellner, H. Baumgartel, W. Grunbein and F. Hensel (eds), *Global Aspects of Atmospheric Chemistry*, Springer, New York 1999, pp. 295-322.

Pasiuk-Bronikowska W., K. J. Rudzinski, T. Bronikowski and J. Ziajka; Transformations of atmospheric constituents and pollutants induced by S(IV) autoxidation - chemistry and kinetics, in: *EUROTRAC2/CMD Annual Report 1999*, International Scientific Secretariat, Munich 2000, pp. 137-140.

Poschl U., J. Williams, P. Hoor, H. Fischer, P. J. Crutzen, C. Warneke, R. Holzinger, A. Hansel, A. Jordan, W. Lindinger, H. A. Scheeren, W. Peters and J. Lelieveld; High acetone concentrations throughout the 0-12 km altitude range over the tropical rainforest in Surinam, *J. Atmos. Chem.* **38** (2001) 115-132.

Warneke C., R. Holzinger, A. Hansel, A. Jordan, W. Lindinger, U. Poschl, J. Williams, P. Hoor, H. Fischer, P. J. Crutzen, H. A. Scheeren and J. Lelieveld; Isoprene and its oxidation products methyl vinyl ketone, methacrolein and isoprene related peroxides measured online over the tropical rain forest of Surinam in March 1998, *J. Atmos. Chem.* **38** (2001) 167-185.

Yermakov A. N., B. M. Zhitomirsky, G. A. Poskrebyshev and D. M. Sozurakov; Kinetic study of SO_5^- and HO_2 radicals reactivity in aqueous phase bisulphite oxidation, *J. Phys. Chem.* **99** (1995) 3120-3127.

Zellner (ed); *Global Aspects of Atmospheric Chemistry*, Springer, New York 1999, p. 328.

Ziajka J., W. Pasiuk-Bronikowska and P. Warneck; Reaction of peroxodisulphate with hydrogensulphite, in: K. H. Becker (ed), *Proc. Joint Workshop LACTOZ/HALIPP, Air Pollution Res. Rep. 54*, OOPEC, Luxembourg 1995, 445-450.

Zimmerman P. R., R. B. Chatfield, J. Fishman, P. J. Crutzen and P. L. Hanst; Estimates on the production of CO and H_2 from the oxidation of hydrocarbon emissions from vegetation, *Geophys. Res. Lett.* **5** (1978) 679-682.

Soot Particles from Different Combustion Sources: Composition, Surface groups, Oxidation under Atmospheric Conditions

Michail N. Laritchev & Jean C. Petit*

INEP CP RAS, Moscow, Leninsky Prospect 38, 2, Russia 117 829
E-mail: mlarichev@chph.ras.ru
*CNRS/LCSR, 45071 Orleans, Cedex 2, France

Introduction

Soot particles are a product of incomplete combustion of organic fuel and are considered as an important component of atmospheric heterogeneous chemistry (Penner and Novakov, 1996). They have a complex composition and consist of at least of two main fractions: one is a graphitic part (elemental carbon) on which the functional (surface) groups containing heteroatoms (such as O, H, S, N) are chemically bonded and the second fraction consists of organic compounds, mainly PAH (Polycyclic Aromatic Hydrocarbons) and their derivatives containing heteroatoms, which are generally considered to be physically adsorbed on the surface of the graphitic fraction. These compounds can be removed by organic solvent extraction. Graphitic and organic fractions of soot are also multi-component systems (Akhter et al., 1985). The composition and structure of soot particles are determined by the composition of both fuel and oxidant, and also by parameters of the combustion process.

The reactivity of the soot surface is determined mainly by the individual reactivity of the component fractions of the soot but the reciprocal effect can be remarkable. The soot reactivity depends on the presence of reactive centres (for example, uncompensated valences of carbon atoms or heteroatoms) and reactive functional groups containing heteroatoms on the soot surface. Reactive centres are the result of the following two processes: growth of the soot particles and the reaction of the soot surface with gaseous components (fuel, oxidant and products of combustion) in the combustion source. These processes occur over a wide range of temperatures (e.g. between temperatures of combustion and temperatures of the exhaust) and they determine the surface structure and the initial reactivity of the soot at the moment of soot exhaust. The difference in the reactivity of the soot particles formed by different sources of combustion is the result of differences in their composition and structure.

Immediately after leaving the exhaust the surface reactions of soot particles with active components of the atmosphere at temperatures dropping from the exhaust temperature to ambient temperature and at ambient temperature are very important. Oxygen and water vapour have the highest concentrations among the active components of atmospheric air. The soot surface reactions with oxygen and water vapour can result in a change of composition of the soot functional groups and reactive centres. As consequence the change of reactivity of the soot particles to other active atmospheric components (OH, O_3, NO_x, NO_y etc.) can be observed during the lifetime of soot in the Earth's atmosphere.

Another plausible reason for the change of soot surface properties is soot surface contact with liquid water condensing from atmospheric water vapour. The presence of water-soluble and hydrophilic compounds and functional groups in the soot composition can aid the surface condensation of liquid water and in changes of the soot surface reactivity after water evaporation.

The objective of this work was a comparative analysis of functional groups containing heteroatoms in the composition of laboratory ethylene, laboratory kerosene and aircraft soot samples. A study of the nature, composition and reactivity of these groups was undertaken.

I. Barnes (ed.), Global Atmospheric Change and its Impact on Regional Air Quality, 129–135.

Investigated samples

1. **Original laboratory ethylene and kerosene soot samples** were prepared by incomplete combustion of ethylene-air and of kerosene vapour-air premixed stoichiometric mixtures and collected on a metal substrate downstream of the combustion zone at near room temperature in the LCSR laboratory, CNRS, France. The collected soot samples were stored in a closed volume at atmospheric conditions. The study of the ethylene soot samples carried out by BET method showed the absence of significant porosity. The microporous volume appears to be 77mm^3/g. The specific surface area was equal to 180m^2/g.
2. **Samples of aircraft soot** were collected at near room temperature on paper filters and on metal substrates mounted in the exhaust stream of an experimental aircraft-engine at SNECMA (France). The engine was placed indoors near the surface.

Aircraft soot samples scraped from the wall of an operable aircraft near to an exhaust pipe of the engine were also studied.

3. Several types of **model samples** were prepared on the basis of original soot samples:

- **model samples of the graphitic and organic fractions of ethylene and aircraft soot** were prepared by organic solvent extraction (dichloromethane, toluene) of the original soot at room temperature with ultrasonic agitation. Filtration was used for the separation of the graphitic fraction of the soot from the solution of the organic fraction.
- **model samples of synthetic oxygen containing groups of different nature** covering the surface of the graphitic fraction of the soot were formed by the reaction of surface of the graphitic fraction of the ethylene soot with different gaseous reactants at known temperature and pressure during a given time.

The following model samples were prepared:

- **model acidic surface groups** containing oxygen atoms only (Barton and Harrison, 1975) were generated by cooling the soot graphitic fraction from a temperature of 1100C to 250C in an atmosphere of dry oxygen and also via it's oxidation in an atmosphere of dry oxygen at a stable temperature in this region.
- **model basic surface groups** containing oxygen atoms only (Boehm and Voll, 1994) were generated via low temperature oxidation of the graphitic fraction of original laboratory ethylene soot in dry oxygen at room temperature (20C) and at 250C.
- **oxygen containing groups** originating from high temperature reactions of the soot surface with stable combustion products (CO, CO_2, H_2O and mixture H_2O+O_2) were produced on the graphitic fraction of original ethylene soot at temperatures between 1150 and 300C.
- **model samples covered by nitrogen oxygen containing surface groups** were generated by reaction of original laboratory ethylene and aircraft soot samples with gaseous NO_2 at room temperature.

The **impact of water vapour on the low temperature oxidation** of the graphitic fraction of the soot surface was studied by exposing model acidic and basic samples at room temperature to an atmosphere of moist air (at humidities of (70-80 %) and 100 %) during a set time.

The **effect of the contact of the soot surface with liquid water** was investigated at room temperature using original laboratory ethylene and aircraft soot.

Experimental techniques

Thermal Desorption Mass Spectrometry, IR spectroscopy and BET techniques were used to investigate the soot samples. The Thermal Programmed Desorption (TPD) device with mass spectral monitoring of gas phase desorption products (Eremenko et al., 1990, Otake and Jenkins, 1993) consists of a double wall reactor containing the soot sample connected to a quadrupole or magnetic mass-spectrometer. The samples were subjected to linear heating in vacuum ($P=10^{-6}$-10^{-7} mm Hg) from room temperature to 1200C with a heating rate of 15 degree/min. The mass of soot sample was usually between 0.2 and 2 milligrams.

The heating of samples in vacuum results in desorption and pyrolysis of physisorbed compounds and in destruction of functional groups containing heteroatoms. Intensities of volatile substances forming in these processes were monitored as the functions of temperature of the sample (thermal desorption curves (TD curves)). The area under the TD curve of a given compound is proportional to the evolved amount of this compound. The presence of several maxima on any given TD curve indicates the existence of several processes (evaporation, destruction of several different types of functional groups) taking place at different temperatures with the evolution of the same compound.

The thermal destruction of oxygen containing functional groups in soot composition appears as the evolution of volatile oxygen containing compounds, mainly as CO, CO_2 and H_2O. The nitrogen containing groups are destroyed with evolution of volatile NO, NO_2, HONO, N_2O, and NH_3 and the sulphur containing groups in soot composition are destroyed with evolution of SO, SO_2. H_2S, and CS_2. In addition volatile organic fragments originating mainly from desorption and destruction of the organic fraction of soot are also detected.

Experimental results and discussion

Oxygen containing groups are more abundant in composition of heteroatoms containing functional groups of soot.

TPD comparative investigations of original laboratory and aircraft soot samples and the same samples subjected to organic solvent extraction show that more than 90% of the oxygen-containing functional groups destroyed during TPD with evolution of CO and CO_2 are functional groups attached to the surfaces of the graphitic fraction of the soot.

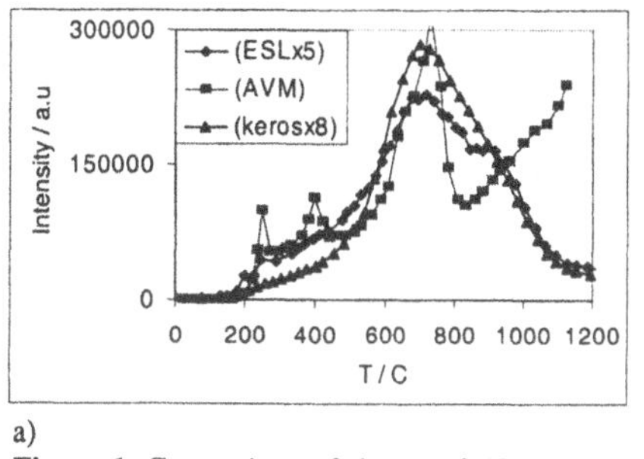

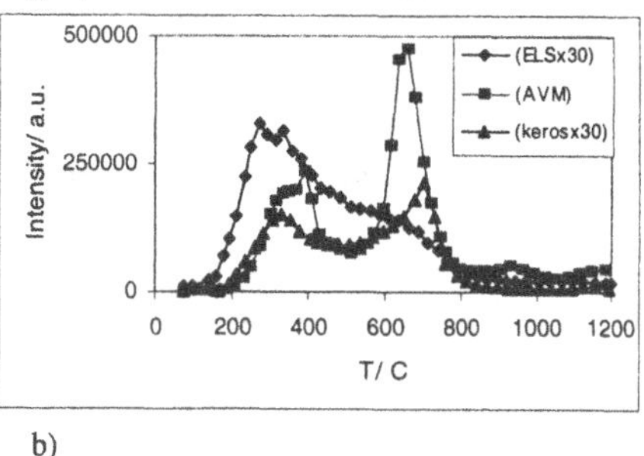

a) b)

Figure 1. Comparison of shapes of TD curves for laboratory ethylene (ESL), laboratory kerosene (keros) and aircraft kerosene (AVM) soot samples.
a) CO evolution; b) CO_2 evolution

Figure 1a shows a comparison of the shapes of TD curves for CO evolution from original soot samples: laboratory ethylene soot (curve (ESL); laboratory kerosene soot (curve (keros)); aircraft soot (curve (AVM)). Figure 1b shows a comparison of the shapes of TD curves for CO_2 evolution for the same samples. These figures demonstrate that CO and CO_2 molecules are evolving in a complex manner. Each of the TD curves contains several maxima. Each maximum characterises a certain type of oxygen containing group being destroyed at a given temperature. At the same time all the curves in figure1a and also figure 1b have a common structure of local maxima with different relative intensities.

The TPD responses of different types of synthetic oxygen containing groups (acidic and basic) formed in different high and low temperature reactions of the oxidation of the surface of the graphitic fraction of soot were studied to facilitate the interpretation of these curves and to understand the origin of the individual local maxima. The comparison of the shapes of TD curves for different types of these synthetic groups with TD curves for original soot samples helps in the identification of the origin of the individual maxima of the TD curves for the evolution of CO and CO_2. Figure 2 shows the positions of these individual maxima with different origin on TD curves for CO_2 and CO evolution from laboratory ethylene soot. The identification of the origin of the individual maxima gives two interesting conclusions:

First, oxygen-containing groups with basic and acidic nature can be present simultaneously on the surface of the graphitic fraction of soot. Figure 3 demonstrates additional experimental results confirming this conclusion. This figure shows TD curves of CO_2 evolution for the model samples of the soot graphitic fraction oxidised in dry oxygen: (A) – by cooling from 1150°C down to room temperature; (A4) - by exposure at 400°C (oxygen containing surface groups with acidic properties are the result of this process); (B6) - by cooling from 250°C to 20°C (oxygen containing surface groups with basic properties are the result of this process). It is clear that curve (A) is a superposition of curves (A4) and (B6).

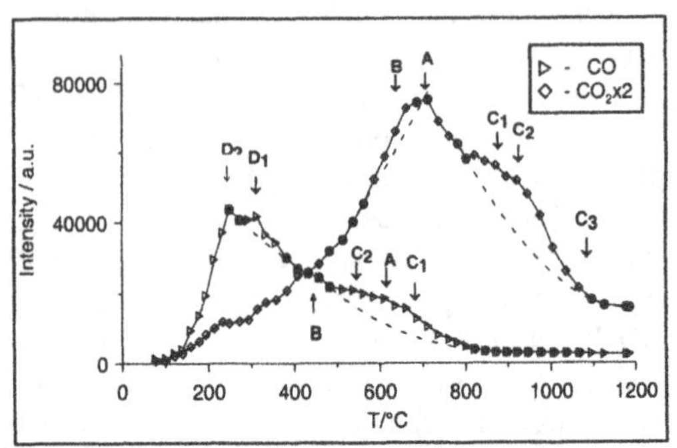

Figure 2. The origin of oxygen-containing surface groups evolving CO_2 and CO during TPD measurement for laboratory ethylene soot.
A - acidic groups formed by high temperature reaction with O_2; B – basic groups formed by low temperature reaction with O_2; C1 – groups formed by high temperature reaction with CO_2; C2 – groups formed by high temperature reaction with H_2O; C3 – groups formed by high temperature reaction with CO; D1 – groups formed by low temperature reaction B with O2 + H2O; D2 – groups formed by low temperature reaction A with O2 + H2O.

It is possible to expect that the ratio between the abundance of acidic and basic surface groups depends not only on the structure of the graphitic fraction of soot but also on the conditions of the soot surface oxidation in a temperature range above 250°C and below 250°C. Apparently the soot particles generated by different combustion sources will have different abundances of acidic and basic groups and, as consequence, different surface properties. This means, that to reproduce soot samples with similar reactivity it is necessary not only to use the same fuel and oxidant and create the same conditions of combustion but it is also necessary to reproduce the same exhaust conditions of combustion products.

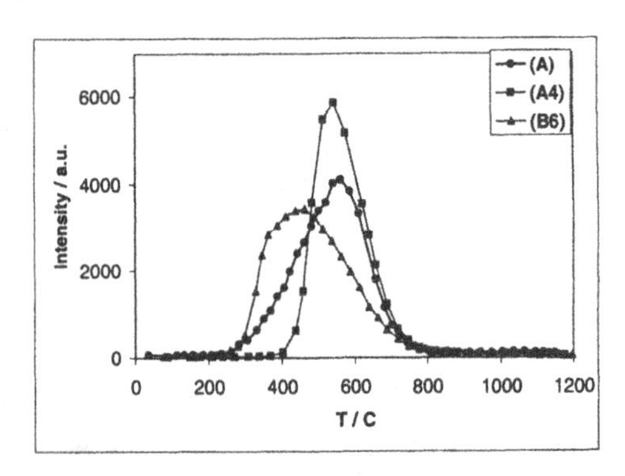

Figure 3. Comparison of the shapes of TD curves of CO_2 evolution for model samples oxidized in dry oxygen at different conditions:
(A) - by cooling in the temperature region 1100°C-20°C;
(A4) - by exposure at 400°C;
(B6) - by cooling in the temperature region 250°C-20°C.

Second, all investigated original soot samples contain oxygen-containing groups which are the result of low temperature oxidation of the soot surface in moist air (compare local maxima of the TD curves for CO_2 evolution at the temperature region near 300C in Figure1b and Figure 2). The abundance of these groups is significant with respect to the abundance of oxygen-containing groups with high temperature oxidation origin. The phenomenon of low temperature oxidation of the soot surface in moist air has the common character.

Model experiments showed that both soot surfaces covered by acidic oxygen-containing groups and covered by basic oxygen-containing groups are able to react with moist air at low temperature and that the increase of air humidity intensifies this process. Low temperature oxidation of the soot surface results in creation of new types of surface groups, which are destroyed during TPD with evolution of CO, CO_2, O_2, and H_2O. The ompositions of new groups are completely different for acidic and basic carbon surfaces. For acidic surfaces the ratio between theamounts of CO and CO_2 molecules evolving during TPD of new groups is equal to10. For basic surfaces this ratio is equal to only 3. Hence, low temperature oxidation of soot surfaces initially covered by basic groups creates the new surface groups more saturated with oxygen than the oxidation of a soot surface initially covered by acidic oxygen-containing groups. H_2O evolution during TPD testifies to the creation of OH surface groups that are reactive, for example, to NO_2 (Laritchev and Petit, 1999).

The kinetic study of the low temperature oxidation of model samples of soot in moist air showed (Laritchev and Petit, 2000) that the reaction time for the creation of oxygen-containing groups being destroyed during TPD with CO_2 evolution is approximately 3 hours for both acidic and basic carbon surfaces at a humidity of 100%. At humidities less than 100% the reaction time may be several orders longer. The generation of new types of oxygen-containing groups will continuously change the soot surface composition and as a consequence the surface chemical activity towards other gas phase reactants present when atmospheric air. At a humidity of 100% the reaction time of about 3 hours is not important in compared with the soot residence time in the atmosphere (about 20 days) but it is comparable with the time of soot sampling for laboratory investigations. Probably the existence of low temperature oxidation is the reason why the reproducibility of the results of laboratory studies of soot reactivity is very difficult. For this reason special precautions should be taken when sampling soot at near room temperature for laboratory studies of soot sample reactivity. It is not improbable that similar low temperature oxidation processes also take place for other types of functional groups or reactive centres on "fresh" soot leaving the exhaust pipe. The reaction time for these reactions can be quite different. Suitable candidates are heteroatom-containing surface groups.

The main fraction of the nitrogen-oxygen atom containing functional groups in the composition of original soot samples was destroyed during TPD with evolution of NO. The comparison of TPD responses in NO for original soot and soot after organic solvent extraction indicates that the major fraction (more than 90%) of these groups are attached to the organic fraction of the soot because they can be removed from soot composition. The origin of these groups is soot surface reaction with products of combustion containing nitrogen-oxygen atoms taking place at temperatures below 200C. Additional oxidation of the soot surface with creation of oxygen-containing groups evolving CO_2 instead CO during TPD is also the result of this process.

The nitrogen-hydrogen atoms containing groups evolving NH_3 during TPD were only observed in the composition of kerosene original soot samples (laboratory and aircraft). They are mainly attached to the soot organic fraction because they can be removed completely during organic solvent extraction. The specific concentration of these groups in the composition of aircraft soot is 10-100 times more than in the composition of laboratory kerosene soot. The amount of these groups correlates with the amount of sulphur atom containing groups in the soot composition and, in principle, its existence may be connected with the existence of strongly acidic sulphur-oxygen containing groups.

Sulphur atom containing groups are destroyed during TPD with the evolution of SO_2, H_2S, and CS_2 (80%, 15% and 5% of the total amount of sulphur atoms, respectively). Organic solvent extraction can not remove sulphur-oxygen containing functional groups from the soot composition. This fact indicates that these groups are mainly attached to the soot graphitic fraction or to inorganic compounds in the soot composition.

The presence of considerable amounts of hydrophilic groups containing heteroatoms in the composition of soot particles favours not only the condensation of water vapour on the soot surface but also makes very probable contact of the soot surface with liquid water. IR studies of liquid water after contact with the surface of soot shows the presence of water-soluble compounds for all original soot samples. The same absorption bands corresponding to absorption of functional groups: hydrophilic –OH (3100-3600 cm^{-1}) and C-O (1650 - 1770 cm^{-1}) groups and hydrophobic –CH aliphatic and aromatic groups (2800 – 3100 cm^{-1}) were observed for original ethylene and aircraft soot samples. The latter is indicative of the presence of soluble organic compounds in soot composition. Further, the spectra of water-soluble compounds for aircraft soot samples contain the adsorption bands for N-H groups, sulphur-containing groups and polyaromatic hydrocarbons (PAH). All water extracts are acidic in nature.

About 90% of nitrogen-oxygen containing functional groups in the composition of all original soot samples can be removed from the soot composition as a result of liquid water contact with the soot surface but at the same time the amount of these groups detected in the water solution after this contact is negligible. It means that contact of liquid water with the soot surface results in hydrolysis of a significant part of the nitrogen-oxygen containing groups in the soot composition and, probably, in generation of products volatile at room temperature, for example, HONO and HNO_3. These volatile products will leave the liquid water-soot system and enter the gas phase.

The major part (more than 90%) of sulphur-oxygen containing groups in the composition of original aircraft soot samples can pass into solution on contact with liquid water. This transition causes the transformation in structure of soluble compounds containing sulphur-oxygen functional groups. The major part (more than 95%) of nitrogen-hydrogen containing groups in the composition of original aircraft soot samples can leave the soot composition during contact with water and pass in solution without change in their structure.

The soot interaction with liquid water can also change the composition of carbon-oxygen containing groups attached to the soot graphitic fraction which evolve CO and CO_2 during TPD. Comparative TPD studies show significant difference in the CO and CO_2 evolution for all original soot samples and the same soot samples after contact with liquid water and indicate that about 50% of carbon-oxygen containing groups are hydrolabile. The shapes of TPD curves for CO and CO_2 evolution from water-treated original aircraft soot samples and for aircraft soot samples scraped from the wall of operable aircraft are very similar. This shows that the considered process of soot interaction with liquid water is realistic and can be observed under real atmospheric conditions. Hence, the contact of soot surfaces with liquid water causes significant changes in the soot functional groups. It should change essentially the reactivity of the soot surface.

The solutions of water-soluble substances present in soot composition have a high ability to precipitate or to produce sediment. On standing the concentration of water-soluble compounds in the solution decreased drastically with time. Visible feculence was observed on the bottom of the vessel. This process is reversible. Subsequent agitation can return the solution to its initial state. The more active participants involved in this process are the compounds containing –OH, -NH, -SO, and -SH functional groups and also PAH.

The presence of water-soluble organic compounds in soot composition supports the presence of molecules of these hydrophilic or highly-polarised compounds in the exhaust stream of combustion products. These molecules may serve as condensation nuclei for water vapour in contrails and for atmospheric aerosol generation.

Conclusion

The Thermal Programmed Desorption technique can be used successfully for identification of the origin of heteroatom functional groups and surface groups in the composition of soot samples. Functional and surface groups containing oxygen atoms are mainly attached to the graphitic fraction in all the original soot samples. These groups are the result of the complex effect of high temperature (above 250C) and low temperature (below 250C) reactions of the soot surface with oxygen of the air and oxygen-containing gaseous combustion products. Oxygen-containing groups with acidic and basic properties can co-exist on the soot surface.

The influence of water vapour was evident both for high and for low temperature oxidation processes. The presence of water vapour significantly intensifies air oxidation of the soot surface at low temperature. Increases in air humidity increase the rate of this process. New types of functional and surface groups are created on the soot surface. The change of soot surface composition can change significantly the soot surface reactivity exhibited by the soot surface to other active components of air during its lifetime in atmosphere. The influence of low temperature oxidation can also significantly change the soot surface properties during the time of soot sampling for laboratory investigations of the soot reactivity if special precautions against this phenomenon are not taken. Nitrogen-oxygen containing functional groups are mainly attached to the soot organic fraction. They were present in the composition of all the investigated original soot samples. These groups are the result of the reaction of the soot surface with nitric oxides at temperatures below 200°C. Nitrogen-hydrogen and sulphur-containing groups were detected only in the composition of laboratory and aircraft kerosene soot samples. The major part of heteroatoms (N, O, S) containing groups in the original soot composition are attached to water-soluble molecules or are able to hydrolyse. Contact of the soot surface with liquid water during the soot atmospheric residence time will drastically change the chemical activity of the soot surface. Hydrolysis of nitrogen-oxygen containing functional groups present in original soot produces volatile nitrogen-oxygen containing compounds.

Acknowledgements

This research was supported by the EC project "CHEMICON" (contract n°: ENV4-CT97-0620). One of the authors (M.N.L.) received partial support from the Russian Foundation for Basic Research (RFBR-00-15-972291). The authors are indebted to Dr. G. Le Bras (CNRS/LCSR, Orleans, France) for fruitful discussions and support of this work.

References

Penner J.E. and T. Novakov, Carbonaceous particles in the atmosphere: A historical perspective to the Fifth International Conference on Carbonaceous Particles in the Atmosphere, *J. Geophys. Res.*, 101, (1996), 19373-19378.

Akhter, M.S., A.R. Chughtai, D.M. Smith, The structure of hexane soot II: Extraction studies, *Applied Spectroscopy*, 39, N1, (1985), 154-167.

Eremenko G.O., M.N. Larichev, I.O. Leypunsky, D.A. Trapeznikov, A.I. Pavlova, Thermal Desorption Mass-Spectrometric (TDMS) Analysis of Carbon Materials Surface, *In MICC-90, (Ed. Fridljander I.N., Kostikov V.I.),* Proceedings of the Moscow Int. Conf. on Composites, Moscow, USSR, (1990). Elsevier, London and New York, 608-612, 1990.

Otake Y., R.G. Jenkins, Characterization of oxygen-containing surface complexes created on a microporous carbon by air and nitric acid treatment, *Carbon*, 31, 1, (1993), 109-121.

Barton S.S., B.H. Harrison, Acidic surface oxide structures on carbon and graphite-I, *Carbon*, 13, (1975), 283-288.

Boehm, H.P., M.Voll, Some aspects of the surface chemistry of carbon blacks and other carbons, *Carbone*, 32, N5, (1994), 759-769.

Laritchev, M.N., J.C. Petit, Study of NOx Reactions with Carbonaceous Surfaces, *Geophysical Research Abstracts*, **1**, (24th General Assembly European Geophysical Society, 19-23 April 1999, The Hague, Netherlands) (1999), 507.

Laritchev, M.N., J.C. Petit, Reactions of ethylene soot with moist air and nitrogen dioxide, *Proc. Joint EC/EUROTRAC-2 Workshop, EC Cluster 4 "Chemical Processes and Mechanisms"*, EPFL, Lausanne, Switzerland, Sept. 11-13, (2000), 213-216.

Poster Contributions

Kinetic and Mechanistic Studies

Investigation of Particle Formation in the Photo-oxidation of Dimethyl Sulphide

Cecilia Arsene[1,2], Ian Barnes[1], Mihaela Albu[1,2] and Karl-Heinz Becker[1]

[1]*Bergische Universität, Physikalische Chemie, Gaußstraße 20, D-42097 Wuppertal, Germany,* [2]*"Al.I. Cuza" Iasi University, Analytical Chemistry, Carol I 11, 6600 Iasi, Romania*

Introduction

Chemical processes involving aerosols can affect the Earth's radiative balance and the oxidative state of the troposphere and, thus, the atmospheric lifetimes of important trace gases such as methane, ozone, non-methane hydrocarbons (NMHC), and dimethyl sulphide (DMS: CH_3SCH_3). Dimethyl sulphide, an important component of the biogeochemical sulphur cycle and the second most important source of sulphur in the atmosphere, is estimated to account for approximately 60 % of the total natural sulphur gas released to the atmosphere in both hemispheres (Andreae, 1990; Bates et al., 1992; Spiro et al., 1992). It is believed that this compound plays an important role in the global sulphur cycle and its emissions represent more than 90 % of the biogenic sulphur emissions from oceans (Andreae and Raemdonck, 1983; Dacey et al., 1987).

It has been postulated that the emission of DMS from the oceans and the subsequent formation of sulphate aerosols can have a significant influence on the Earth's radiation budget and possibly on climate regulation, the so called CLAW hypothesis (Charlson et al., 1987). From the positive correlation between concentrations of DMS and cloud condensation nuclei (CCN) in oceanic air masses observed in some studies, it has been concluded that DMS makes a large contribution to the production of nss-sulphate (nss-SO_4^{2-}) and CCN (Nagao et al., 1999). The general mechanism by which DMS plays a role as a climate modulating species involves the oxidation of the sulphur atom in DMS to the +VI oxidation state followed by heteromolecular nucleation, which leads to the formation of new CCN. Any effect that DMS can have on the climate is considered to be critically dependent on the production of gas phase sulphuric acid (H_2SO_4) and new particles.

Presented here is a detailed study on particle production from the OH-radical initiated oxidation of DMS under different conditions of temperature. The study was aimed mainly at establishing whether or not temperature has a significant influence on aerosol production.

Experimental

The experiments were performed in a temperature regulated 1080 litre reaction chamber (Figure 1) at a total pressure of 1000 mbar synthetic air and temperatures of 284, 295 and 306 $\pm$ 2 K. The UV photolysis of H_2O_2 with 32 low-pressure mercury lamps (Philips TUV 40 W: λmax = 254 nm)

$$H_2O_2 + h\nu(\lambda = 254 \text{ nm}) \rightarrow OH + OH$$

was used for the production of OH-radicals. Mixtures of DMS/H_2O_2/synthetic air were irradiated over 20 - 30 minutes periods. Particle formation was monitored using a Scanning Mobility Particle Sizer (SMPS: TSI Model 3071 A) in series with a Condensation Nuclei Counter (CNC: TSI Model 3022 A). The CNC has an upper range limit for particles detection of 10^7 particles cm^{-3} and the SMPS analyser is capable of detecting particles in the size range of 12 - 640 nm. The SMPS was operated with a 5 minute scan rate with 4 minute of up and 1 minute of down scanning. The system was set to the conditions corresponding to the laboratory air which was used as the sheath air. The pressure in the chamber was maintained at 1000 mbar by continually pressurising with synthetic air.

I. Barnes (ed.), Global Atmospheric Change and its Impact on Regional Air Quality, 141–146.

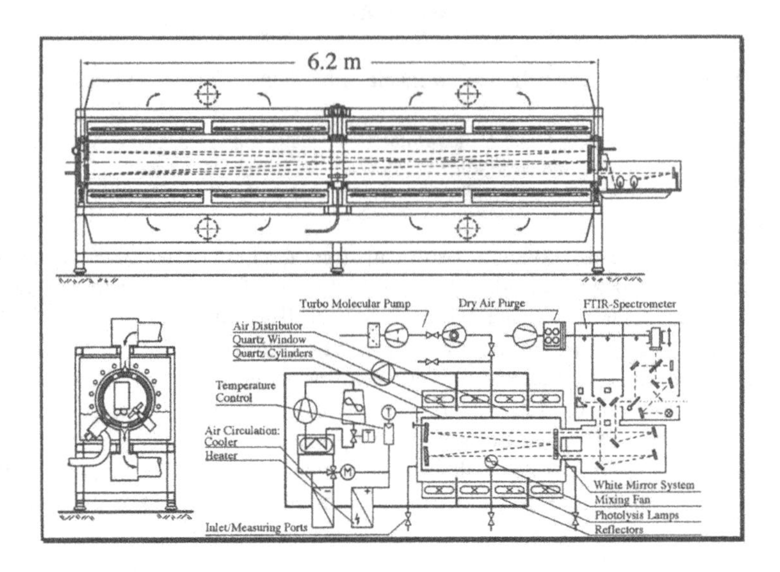

Figure 1: Schematic diagram of the 1080 L quartz glass reactor with FTIR spectrometer.

Results and discussion

Figures 2a and 2b show typical examples of the number and volume distribution, respectively, of the particle formation for an O_2 partial pressure of 200 mbar and temperatures of 284, 295 and 306 K. At each temperature the particle concentration reaches the maximum value after 5 minutes of irradiation. Usually, the rapid rise in number concentration of particles is followed by a rapid decay. It was observed that the maximum number concentration reached was 3.25 x 10^6 particles cm^{-3} at 284 K, 5.1 x 10^6 particles cm^{-3} at 295 K, and up to 5.75 x 10^6 particles cm^{-3} at 306 K. However, the number size distributions in the investigated size range exhibited mostly a monomodal feature with the exception of a few cases where a bimodal feature was observed. The mean diameter for particles resulting from the oxidation of DMS at 284 K was below 250 nm while that for particles formed at 295 and 306 K was under 400 nm. At 284 K the mean diameter for the maximum number of particles was located at about 50 nm. Usually, with increasing irradiation time, the mean diameter corresponding to the maximum number and volume distribution of particles slowly increased, as can be observed in Figure 2a and Figure 2b.

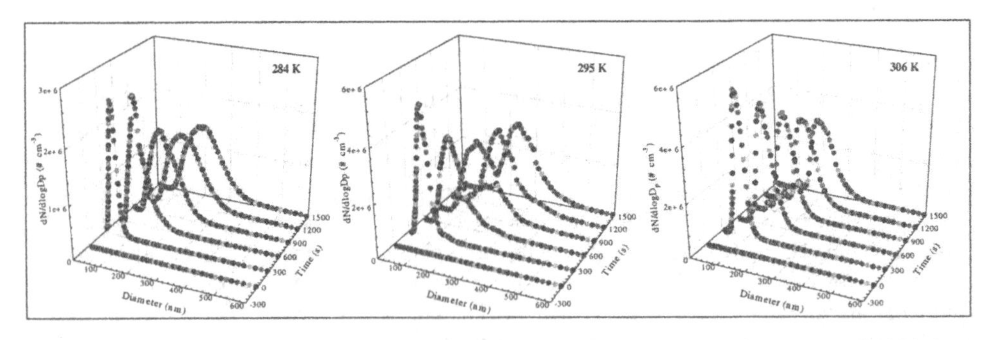

Figure 2a: Number distributions of the particles formed at an O_2 partial pressure of 200 mbar and temperatures of 284, 295 and 306 K.

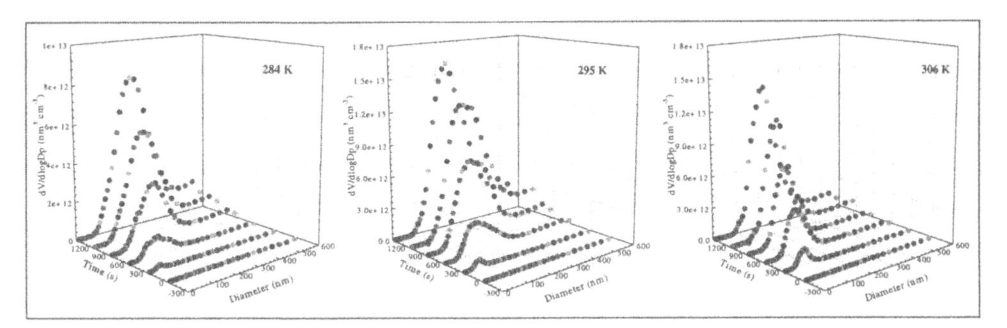

Figure 2b: Volume distributions of the particles formed at an O_2 partial pressure of 200 mbar and temperatures of 284, 295 and 306 K.

Qualitative information concerning aerosol yields in the studied systems were obtained with the help of the gas/aerosol absorption model proposed by Odum et al. (1996). In this model the fractional aerosol yield (Y) is defined as the fraction of a reactive organic gas (ROG) which is converted into aerosol, and is calculated by the expression

$$Y = \frac{\Delta M0}{\Delta[ROG]}, \qquad \text{(E. 1)}$$

where ΔM_0 is the organic aerosol mass concentration ($\mu g/m^3$) produced for a given amount of reacted ROG, ΔROG ($\mu g/m^3$). The expression for the overall aerosol yield of an multi-component system (Y) was defined as being given by the expression:

$$Y = \sum_i Y_i = M_0 \sum_i \left(\frac{\alpha_i K_i}{1 + K_i M_0} \right) \qquad \text{(E. 2)}$$

and the aerosol yield of an individual product (Y_i) by the following expression:

$$Y_i = M_0 \left(\frac{\alpha_i K_i}{1 + K_i M_0} \right) \qquad \text{(E. 3)}$$

where M_0 represent the total amount of organic aerosol, α_i is the yield of product i which is partitioning between gas and particulate phase, and K_i is the partitioning coefficient of the species i.

Figure 3 shows plots of the fractional aerosol yield, as defined by Odum et al. (1996), and the results of fitting the experimental points using the gas/particle partitioning absorption approach of Odum et al. (1996), as a function of the secondary aerosol mass M_0 ($\mu g\ m^{-3}$). Since the molecular composition of the aerosol phase is unknown the aerosol yields were calculated on a mass/mass basis: µg of aerosol formed per µg of DMS reacted. For the calculation the assumption was made that only one secondary product was responsible for particle formation, *i.e.* H_2SO_4, since it is known that in the presence of OH radicals, SO_2, the major oxidation product of DMS, can be oxidised to H_2SO_4, a species which is considered to be a potential source of sulphate particle. Figure 3 shows also that the experimental points could be well fitted using an equation of hyperbolic type:

$$y = {}^{ax}\!/_{(b + x)}. \qquad \text{(E. 4)}$$

The used equation in the fitting procedure of the experimental points allowed us to correlate the parameter α from (E. 3) with the parameter "a" predicted by the calculations using the equation (E. 4), and $1/K_i$ with b, respectively.

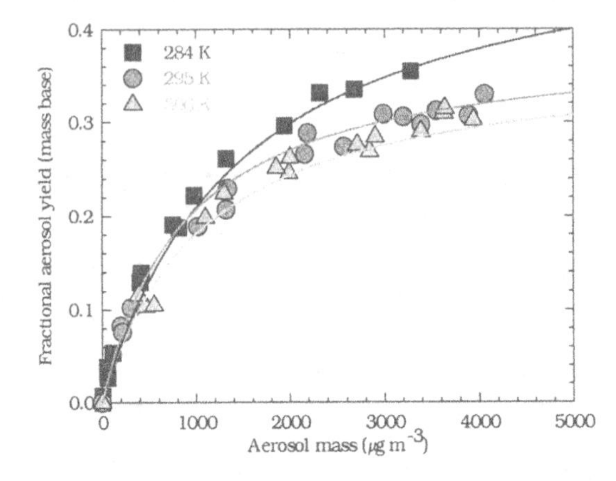

Figure 3: Plots of the fractional aerosol yield as a function of the secondary aerosol mass, M_0 (μg m^{-3}), measured in the present work (closed symbols) and the calculated aerosol yield profiles obtained by using the approach of Odum et al. (1996) (full lines).

Figure 4 shows examples of plots of the aerosol mass concentration versus the consumed amount of DMS. The figure clearly shows that aerosol formation starts after a certain threshold consumption of DMS. The data obtained for the aerosol yields, calculated from the point of the onset of the thresholds, are summarised in Table 1. Also listed are the parameters α (the yield of the product which is partitioning between the gas and particulate phase) and K (partitioning coefficient of the respective species) which were obtained from the calculated fits to the experimental points in Figure 3.

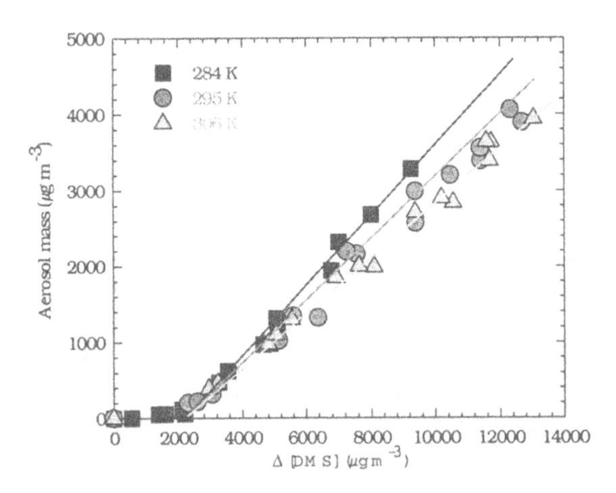

Figure 4: Plots of the aerosol mass concentration as a function of the amount of DMS consumed from experiments performed at an O_2 partial pressure of 200 mbar and different temperatures.

Table 1: Fractional aerosol yields from the OH radical initiated oxidation of DMS and the α, K parameters values predicted by the fit to the points in Figure 3.

Temperature (K)	Fractional aerosol yield (Y)	Parameters values	
	~ 6.5 ppm DMS	α	K
284	0.394	0.418	0.00101
295	0.378	0.389	0.00113
306	0.358	0.363	0.00119

From the results presented in Table 1 it can be observed that the aerosol yields does not show a high variability from one temperature to another. As well, from this table, it can be observed that the values of α, as they were predicted from the calculations, are almost in the same range as the values of the aerosol yield estimated on a mass base. It is possible that in the experimental cases investigated here, and for which we assumed that only one component with low volatility was mostly responsible for particle formation, in the presence of high levels of aerosol mass, the yields of aerosol (Y) will tend towards the yield of the product which is partitioning between the gas and aerosol phase (α).

In the studied systems, evidence was found for the formation of H_2SO_4 as important product in the DMS oxidation. The yields of H_2SO_4 formed in the OH-radical initiated oxidation of DMS were estimated from a knowledge of the concentration-time profile of SO_2 and the OH-radical concentration, assuming that reaction of SO_2 with OH to form H_2SO_4 was the only loss process for SO_2 (the formation yields of SO_2 from the OH-radical initiated oxidation of DMS have shown that SO_2 is the major product of DMS oxidation (Arsene et al., 1999)). However, it is known that production of H_2SO_4 in the oxidation of DMS can be controlled by several factors and the gas phase H_2SO_4 levels will be affected by the concentrations of OH and NO, temperature dependent reaction pathways and aerosol loss processes (Jefferson et al., 1998). By plotting the estimated contribution of the fraction of SO_2 oxidised by OH-radicals to gas phase H_2SO_4 versus the aerosol yields determined on a mass bases a good correlation was observed (Figure 5). In Figure 5 it can be seen that the trends in H_2SO_4 are similar to those observed in plots of the fractional aerosol yields as a function of the aerosol mass concentration. This observation is considered as an indication that H_2SO_4 is the predominant species in the aerosol phase.

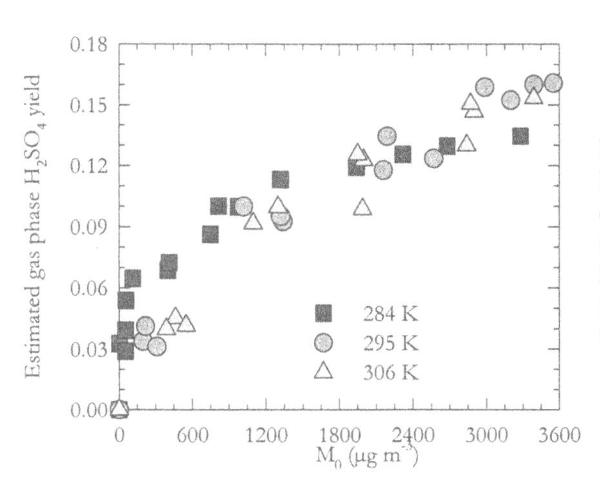

Figure 5: Estimated gas phase H_2SO_4 concentration as a function of the aerosol mass concentration in the OH radical initiated oxidation of DMS at oxygen partial pressure of 200 mbar and different temperatures.

Observations

- In photo-oxidation of DMS, aerosol formation starts after conversion of a certain threshold value of DMS. The diameter for the maximum number of particle corresponds to 50 - 60 nm and usually grows slowly with increasing irradiation period.
- The aerosol formation takes the temperature order 284>295>306 K and the measured aerosol yields show only marginal dependence on temperature (0.36 - 0.40).
- The aerosol yield was found to be a smooth function of the aerosol mass concentration.
- Under the conditions of the experiments it would appear that the aerosol is probably composed mainly of one secondary product (H_2SO_4) formed via reaction of OH radicals with SO_2.

Acknowledgement

Financial support for this work by the European Commission within the EL CID project and the BMBF within the AFS project is gratefully acknowledged.

References

Andreae, M.O.; Ocean-atmosphere interactions in the global biogeochemical sulphur cycle, *Marine Chem.* **30** (1990), 1-29.

Andreae, M.O. and H. Raemdonck; Dimethyl sulphide in the surface ocean and the marine atmosphere. A global view, *Science* **221** (1983) 744-747.

Arsene, C., I. Barnes and K.H. Becker; FT-IR product study of the photo-oxidation of dimethyl sulphide: temperature and O_2 partial pressure dependence, *Phys. Chem. Chem. Phys.* **1** (1999) 5463-5470

Bates, T.S., B.K. Lamb., A. Guenther, J. Dignon and R.E. Stoiber; Sulphur emissions to the atmosphere from natural sources, *J. Atmos. Chem.* **14** (1992) 315-337.

Charlson, R.J., J.E. Lovelock, M.O. Andreae and S.G. Warren; Oceanic phytoplankton, atmospheric sulphur, cloud albedo and climate, *Nature* **326** (1987) 655-661.

Dacey, J.W.H., G.M. King and S.M. Wakeham; Factors controlling emission of dimethyl sulphide from salt marshes, *Nature* **330** (1987) 643-645.

Nagao, I., K. Matsumoto and H. Tanaka; Characteristics of dimethylsulfide, ozone, aerosols, and cloud condensation nuclei in air masses over the north western Pacific ocean, *J. Geophys. Res.* **104**, 11,675-11,693.

Odum, J.R., T. Hoffmann, F. Bowman, D. Collins, R.C. Flagan and J.H Seinfeld; Gas/ particle partitioning and secondary organic aerosol yields, *Environ. Sci. Technol.* 30 (1996) 2580-2585.

Spiro, P.A., D.J. Jacob and J.A. Logan; Global inventory of sulphur emissions with $1° \times 1°$ resolution, *J. Geophys. Res.* **97** (1992) 6023-6036.

Model Study of the Photooxidation of $CH_3SO_2SCH_3$ at Atmospheric Pressure: Thermal Decomposition of the CH_3SO_2 Radical

N. I. Butkovskaya and I. Barnes

Bergische Universität Gesamthochschule Wuppertal, Physikalische Chemie/FB9. 42097 Wuppertal, Germany

Introduction

The knowledge of the rate of CH_3SO_2 decomposition is critical for understanding the distribution of the end products of the tropospheric oxidation of organic sulphur-containing compounds, such as DMS, CH_3SSCH_3 (DMDS) or CH_3SH. A global mechanism of their oxidation was developed a decade ago (Yin et al., 1990). The core of this mechanism is a series of oxidation steps leading to a branching between formation of methane sulfonic acid, CH_3SO_3H (MSA), and sulphate, SO_4^{-2}, originating from SO_2. Since SO_2 and MSA are produced mainly from CH_3SO_2, the competition between CH_3SO_2 decomposition and further oxidation to CH_3SO_3 will determine their yield distribution. Rapid unimolecular dissociation of CH_3SO_2 to CH_3 and SO_2 favours production of sulphuric acid, which leads to formation of new cloud condensation nuclei. Atmospheric MSA is absorbed by pre-existing aerosols and droplets and washed out with precipitation. Therefore, the decomposition-to-oxidation branching of CH_3SO_2 is an important factor in determining the composition of the marine atmospheric aerosol.

To examine the kinetic behaviour of CH_3SO_2 radicals, a study of the photo-oxidation of methyl methanethio-sulfonate (MMTS), $CH_3SO_2SCH_3$ was carried out, using *in sitù* monitoring of the products by FTIR absorption spectroscopy. A numerical integration program incorporating the $CH_3S(O)_x + O_2 \leftrightarrow CH_3S(O)_xOO$ equilibrium was developed to simulate the concentration-time profiles of the products of photolysis under the static conditions of the experiments and to estimate the lifetime of CH_3SO_2 radicals with respect to thermal dissociation.

Experiment

The experiments were performed at room temperature in a 405 l glass cylindrical chamber equipped with a White mirror system coupled to a FTIR Magna 500 spectrometer. The total optical path was 50.2 m. Liquid $CH_3SO_2SCH_3$ (Aldrich, 97%) was injected into the chamber; its concentration was typically 3×10^{13} molecule cm^{-3}. Calibration of MMTS was accomplished by injections of measured volumes into a stream of N_2 passing through a heated reactor port. The MMTS wall loss was measured in dark experiments to be $k_w = 8.7 \times 10^{-4}$ s^{-1}. The composition of the buffer gas, $N_2 + O_2$, was adjusted to have a total pressure in the reactor of 760 Torr. The mixture was exposed to $\lambda = 254$ nm light from three low-pressure mercury lamps for a period of 10-15 min. During the irradiation, IR spectra were continuously recorded with a spectral resolution of 1 cm^{-1} and averaging every 1 min.

Product Yields

The major products observed were SO_2, CH_2O, MSA, CH_3OH, CH_3OOH and DMDS. Other sulphur-containing products, COS and $CH_3SO_2CH_3$, and also HCOOH and CO were produced in substantially less amounts and their contribution to the material balance was negligible. The dependence of the product yields on the oxygen concentration is presented in Table 1. It is worth noting that at high O_2 concentrations nearly every photolyzed MMTS

I. Barnes (ed.), Global Atmospheric Change and its Impact on Regional Air Quality, 147–152.

molecule is transformed into MSA. At low $[O_2]$ the consumption of MMTS and the yields of most products are practically invariable with the change of oxygen concentration.

Table 1. Product yields from the photolysis of $CH_3SO_2SCH_3$ at atmospheric pressure and different O_2 concentrations after 10 min irradiation.

$[O_2]$ 10^{16}	[CH3SOO]/ /[CH3S]$_{eq}$	$[MMTS]_0$ 10^{14}	Δ[MMTS]/ /$[MMTS]_0$	CH_2O	CH_3OH	CH_3OOH	SO_2	DMDS	MSA	Total S	Total C
N_2	6×10^{-4}	0.32	0.21	0.49	0.27	0.16	1.03	0.07	no	0.59	0.53
1.85	0.003	0.42	0.19	0.71	0.19	0.24	1.13	0.06	0.05	0.65	0.66
3.87	0.006	0.32	0.19	0.82	0.15	0.26	1.20	0.06	0.08	0.70	0.72
7.26	0.010	0.41	0.20	0.81	0.10	0.24	1.14	0.05	0.16	0.69	0.70
33.7	0.044	0.19	0.24	0.90	0.05	0.26	1.15	0.05	0.33	0.79	0.82
134	0.20	0.25	0.53	1.06	0.03	0.10	1.31	0.01	0.60	0.95	0.90
510	0.69	0.36	0.59	0.62	0.01	0.06	0.61	0.01	0.80	0.74	0.78

General Mechanism

The general reaction scheme for the system, based largely on the mechanism proposed by Yin with co-workers, is shown below. The detected products are enclosed by frames. The scheme does not include some radical recombination reactions, such as $CH_3S + CH_3SO \rightarrow CH_3S(O)SCH_3$, or possible hydrogen transfer reactions, such as $CH_3S(O)_xOO + HO_2 \rightarrow CH_3S(O)_xOOH + O_2$, which are responsible for the major part of "missing" products. Unidentified bands in the IR spectra are consistent with such products. Possible secondary reactions of MMTS with peroxy radicals are also not shown. RO_2 designates all the peroxy radicals present in the system (CH_3O_2 and HO_2 at low $[O_2]$ with the appearance of considerable concentrations of $CH_3S(O)_xOO$ (x = 0-2) at higher $[O_2]$).

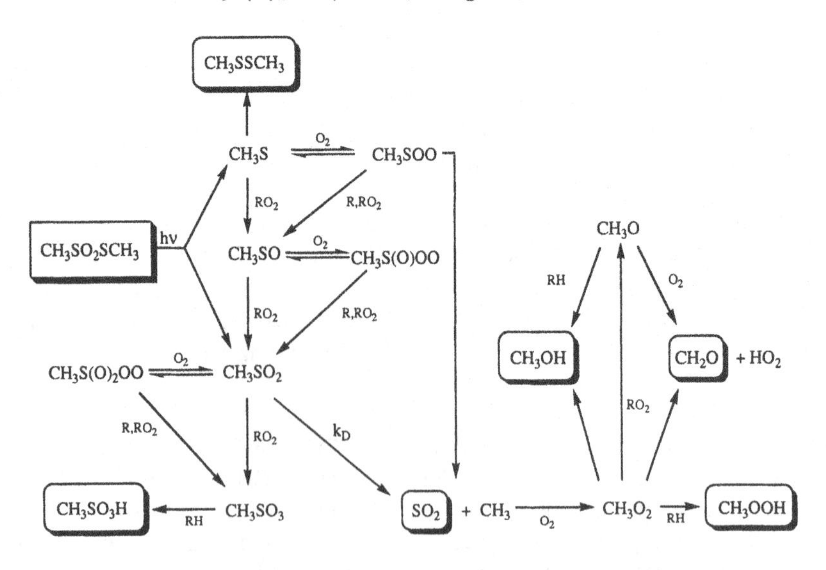

Figure 1. Reaction pathways for the photo-oxidation of $CH_3SO_2SCH_3$.

To describe quantitatively the oxidation kinetics, we assume that the CH_3S radical reversibly adds to O_2, forming the CH_3SOO adduct (Turnipseed et al., 1992) and that CH_3SOO thermally converts to CH_3 + SO_2 as supported by our recent results from a study on the photo-oxidation of DMDS. The second column in the Table 1 shows the equilibrium $[CH_3SOO]/[CH_3S]$ ratio derived from an extrapolation of the low temperature data of Turnipseed et al. to 298 K, giving $K_c = 1.4 \times 10^{-19}$ cm^3 $molecule^{-1}$. Simulation of the MMTS photo-oxidation system was made assuming that similar equilibria with oxygen also exist for CH_3SO and CH_3SO_2 radicals.

Model Calculations

The concentration-time profiles measured during the photolysis of $CH_3SO_2SCH_3$ at 298 K and $[O_2]$ = 0.6, 10 and 150 Torr are shown as scattered-point plots in Figure 2. Results of the numerical calculations are represented by the solids lines. In the present model, the equilibration between the CH_3S and CH_3SOO concentrations, as derived from the extrapolation to 298 K equilibrium constant value, was inserted into every step of the numerical integration of the kinetic system. This was possible due to the fast establishment of the equilibrium, which according to Turnipseed et al. takes place very rapidly at 298; within the scale of 3 μs. Therefore, a calculation time step $\Delta t \gg 3$ μs had to be used during the integration. For the period from t = 0 to the time of attainment of the steady-state concentrations of intermediate species a value of $\Delta t = 2 \times 10^{-4}$ s was chosen, thereafter Δt was increased to 1×10^{-3} s in order to speed up the calculations.

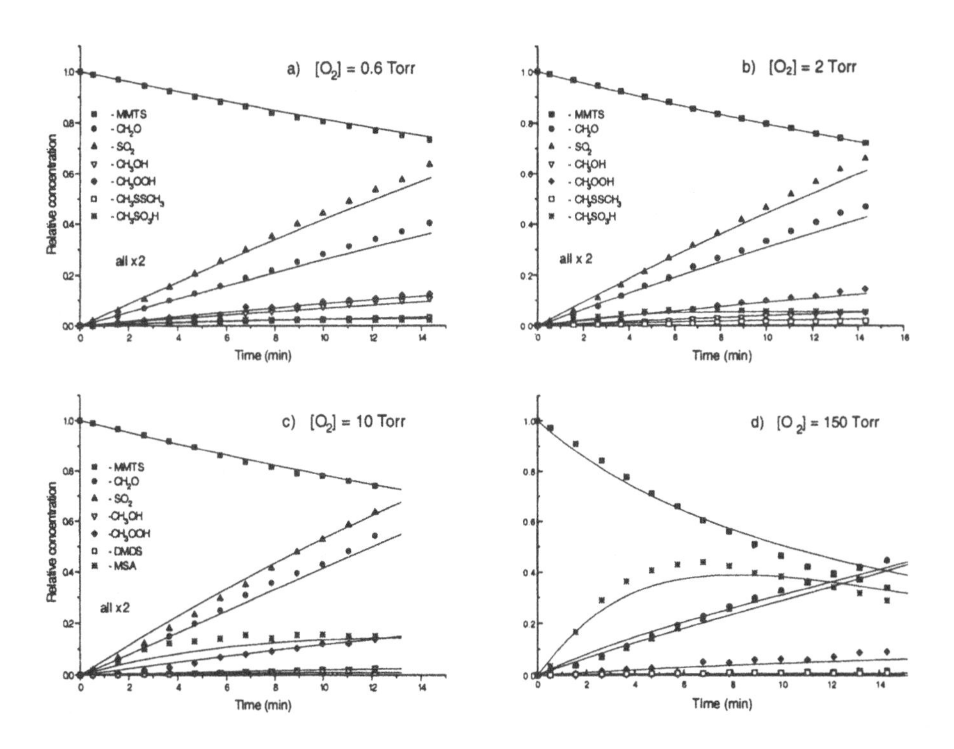

Figure 2. Concentration-time profiles of the reactant and products at different $[O_2]$.

Due to the simultaneous registration of DMDS and CH_3OOH, the steady-state concentrations of CH_3S and CH_3O_2 radicals could be controlled during the calculation making the system stiff. Some independent results from model studies of CH_3SH and DMDS photo-oxidation systems performed in our laboratory were used, as, for example, the rate coefficients for $CH_3S(O)_x$ + RO_2 reactions estimated from the CH_3SH study, and the frequency of the CH_3SOO conversion to CH_3 + SO_2 from the DMDS study. The full scheme used for simulation included 130 reactions and 25 species.

At low $[O_2]$, when $CH_3S(O)_xOO$ chemistry is not important, the reaction kinetics can be described by the above scheme. In this case MSA is produced from CH_3SO_3 formed from the oxidation of CH_3S and CH_3SO_2 radicals. The best fit of the observed SO_2 and MSA profiles corresponds to k_D of about 0.4 s^{-1} assuming that the decomposition of CH_3SO_3 is negligible (Figure 2a,b).

Table 2. Sensitivity coefficients for the photo-oxidation of $CH_3SO_2SCH_3$ at t = 10 min.

Reaction[a]	k[b]	DMDS	SO_2	CH_3OOH	MSA	DMDS	SO_2	CH_3OOH	MSA
		$[O_2]$ = 0.6 Torr				$[O_2]$ = 150 Torr			
1	1.0×10^{-12}	-0.030	-0.024	-0.023	-0.035	-0.007	-0.003	-0.002	-0.002
2	2.4×10^{-11}	-0.479	0.049	-0.084	0.058	-0.075	-0.001	-0.020	0.004
3	6.0×10^{-12}	-0.011	-0.005	-0.050	0.016	0.004	-0.004	-0.010	-0.004
4	1.0×10^{-12}	0.077	-0.034	-0.128	0.249	0.010	-0.010	-0.034	-0.003
5	2.4×10^{-11}	-0.750	0.113	-0.010	0.041	-0.292	0.017	-0.051	0.018
6	6.0×10^{-12}	0.047	0.021	0.044	0.029	0.028	-0.006	-0.018	-0.014
7	1.0×10^{-12}	0.149	-0.053	-0.199	0.418	0.066	-0.018	-0.099	-0.009
8	**0.4**	**-0.201**	**0.115**	**0.360**	**-0.666**	**0.120**	**0.009**	**0.676**	**-0.374**
9	8	-0.022	0.002	0.012	0.004	-0.793	-0.075	0.107	-0.125
10	1.4×10^{-19}	-0.031	0.004	0.016	0.018	-1.560	-0.004	0.087	-0.022
11	7.0×10^{-20}	-0.008	0.002	0.008	0.027	-0.035	0.017	0.033	0.021
12	4.0×10^{-20}	0.000	-0.001	-0.002	0.034	-0.199	0.451	-0.514	0.394
13	1.2×10^{-13}	-0.010	0.003	0.008	0.063	-0.978	0.214	-0.313	0.543

[a] Reactions: $CH_3S + CH_3SO_2 \rightarrow CH_3SO_2SCH_3$ **(1)**; $CH_3S + HO_2 \rightarrow CH_3SO + OH$ **(2)**; $CH_3SO + HO_2 \rightarrow CH_3SO_2 + OH$ **(3)**; $CH_3SO_2 + HO_2 \rightarrow CH_3SO_3 + OH$ **(4)**; $CH_3S + CH_3O_2 \rightarrow CH_3SO + CH_3O$ **(5)**; $CH_3SO + CH_3O_2 \rightarrow CH_3SO_2 + CH_3O$ **(6)**; $CH_3SO_2 + CH_3O_2 \rightarrow CH_3SO_3 + CH_3O$ **(7)**; $CH_3SO_2 \rightarrow CH_3 + SO_2$ **(8)**; $CH_3SOO \rightarrow CH_3 + SO_2$ **(9)**; $CH_3S + O_2 \leftrightarrow CH_3SOO$ **(10)**; $CH_3SO + O_2 \leftrightarrow CH_3S(O)OO$ **(11)**; $CH_3SO_2 + O_2 \leftrightarrow CH_3S(O)_2OO$ **(12)**; $CH_3SO_2SCH_3 + CH_3S(O)_xOO \rightarrow$ adduct **(13)**.

[b] Bimolecular rate constants (1-7,13) are in cm^3 moilecule^{-1} s^{-1}; unimolecular rate (7,8) constants are in s^{-1}; equilibrium constants (10-12) are in cm^3 moilecule^{-1}.

Coefficients $s > 0.1$ are underlined for reader's convenience.

The rapid decay of MMTS at elevated O_2 can be explained by the secondary reactions with $CH_3S(O)_xOO$ peroxy radicals. The most probable mechanism is electrophylic addition of the terminal oxygen of the radical to the non-oxidised sulphur atom of MMTS, followed by reaction with O_2 giving CH_3SO_3, CH_3SO_2 and $CH_3S(O)_xO$. The channel giving CH_3SO_2, CH_3O, SO_2 and $CH_3S(O)_xO$ is also possible. It's worth noting that for alkyl peroxy radicals such reactions are less exothermic by $\approx$36 kcal mol^{-1}. These secondary reactions provide fast generation of MSA as shown in Figure 2c for an intermediate oxygen concentration. Under these conditions the simulation has additional degrees of freedom such as manipulation of the mechanisms of the MMTS reactions and the equilibrium concentrations for $CH_3S(O)OO$ and

$CH_3S(O)_2OO$. Nevertheless, since the decay of MMTS is tied to the $CH_3S(O)_xOO$ total concentration, it was possible to show that attributing $K_c(CH_3SO_2)$ a value of one order of magnitude higher or lower than $K_c(CH_3S)$ required a change of k_D from 0.6 to 0.15, respectively, to fit the SO_2/CH_2O and MSA kinetics. The calculation for the oxidation in air presented in Figure 2d corresponds to the suggested mechanism with $k_D = 0.3\ s^{-1}$. In this calculation the $[CH_3S(O)OO]/[CH_3SO]$ and $[CH_3S(O)_2OO]/[CH_3SO_2]$ equilibrium ratios were assumed to be equal to that for CH_3S.

Table 2 presents the results of the sensitivity analysis for the present chemical system. Sensitivity coefficients for DMDS, SO_2, CH_3OOH and MSA were directly calculated as $s = [C(t,k+\delta k)-C(t,k-\delta k)]/2\delta C(t,k)$, where C is the product concentration, changing the rate constants by $\delta = 5\%$. The selected line corresponds to the frequency of CH_3SO_2 decomposition. We see that at low $[O_2]$ a change in k_D acts upon all major products; at the same time, production of these species depends on the rate of oxidation of CH_3SO_x by HO_2 and CH_3O_2. In air, the $CH_3SO_2 + O_2 \leftrightarrow CH_3S(O)_2OO$ equilibrium is very important, since it also determines the consumption of MMTS.

The rate constant obtained for the thermal decomposition of CH_3SO_2 is substantially smaller than the value of $510\ s^{-1}$ from the study of Ray et al. (1996). This can be explained by the fact that the CH_3SO_2 radicals generated in the study of Ray et al. by the $CH_3SO + NO_2$ reaction underwent a partial prompt decomposition as was later confirmed by Kukui et al (2000). Our value is also lower than $k_D = 11\ s^{-1}$ recommended by Yin et al. (1990); calculations with $k_D = 11\ s^{-1}$ give practically negligible MSA production independent of the mechanism of secondary MMTS reactions. The low decomposition rate is in agreement with the statement of Barone, Turnipseed and Ravishankara (1995) that CH_3SO_2 is quite stable, with the suggested $k_D \approx 3\ s^{-1}$, which also agrees with a recent report giving an upper limit of $k_D < 10\ s^{-1}$ as determined by direct observation of CH_3SO_2 radical by chemical ionization mass-spectrometry (Le Bras et al. in ELCID (2001)).

It was also found that the decay of MSA in the reactor is probably attributable to reversible formation of dimers rather than loss to the wall. Calculation of the CH_3SO_3H diffusion coefficient using the method of Fuller et al. (1969) for 1 atm and T = 298 K, gives $D_{MSA} = 0.10\ cm^2\ s^{-1}$.

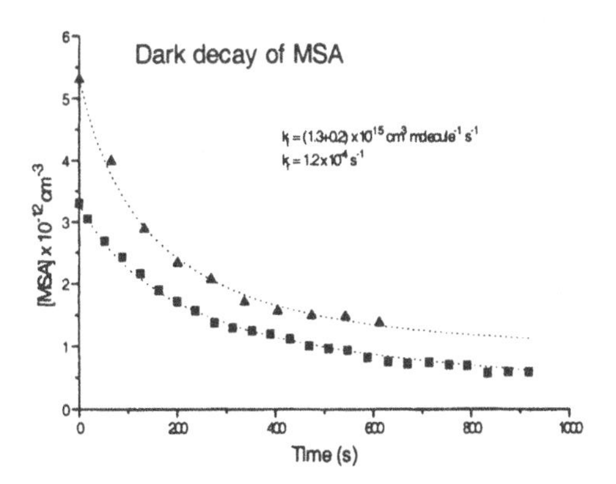

Figure 3. Concentration-time profiles of the dark decay of MSA. Fitted curves correspond to reversible formation of dimers.

The corresponding upper limit for the rate of wall loss is about $k_w = 1.8 \times 10^{-3}\ s^{-1}$, as follows from the diffusion equation $x^2 = 4Dt$ with x = 15 cm as the characteristic reactor size. Figure 3 shows two runs of dark experiments for different MSA concentrations produced in the course of the photo oxidation of MMTS. The initial decay is faster than the diffusion to walls, the concentration reaches a certain quazi stationary level, and then slowly asymtotically declines to zero. Fitting of the curves was made using the analytical solution for the MSA + MSA $\leftrightarrow$ $(MSA)_2$ chemical process. The obtained rate constants for the addition and decomposition reactions are $(1.3\pm0.2) \times 10^{-15}$ cm^3 $molecule^{-1}$ s^{-1} and 1.2×10^{-4} s^{-1}, respectively. During the modelling of the oxidation of MMTS, combination of dimer formation and wall losses was used to describe the MSA kinetic profiles.

Conclusions

The generally accepted oxidation scheme of CH_3S and CH_3SO_2 radicals assuming $CH_3S(O)_x + O_2 \leftrightarrow CH_3S(O)_xOO$ equilibria can reproduce the kinetics of formation of all the major products formed in the course of the photo-oxidation of $CH_3SO_2SCH_3$.

The high yields of methane sulphonic acid observed from the photo-oxidation of $CH_3SO_2SCH_3$ under atmospheric conditions corresponds to a rate coefficient for the thermal decomposition of CH_3SO_2 radicals of $k_D = 0.4\pm0.2$ s^{-1}.

References

Barone S. B., A. A. Turnipseed and A. R. Ravishankara; Role of adducts in the atmospheric oxidation of dimethyl sulfide, *Faraday Discuss.* **100** (1995), 37-54.

ELCID (Evaluation of the Climate Impact of Dimethyl Sulfide): Report (2001) obtainable from barnes@uni-wuppertal.de

Fuller E. N., K. Ensley and J. C. Giddings; Diffusion of Halogenated Hydrocarbons, *J. Phys. Chem.* **73** (1969), 3679.

Kukui A., V. Bossoutrot, G. Laverdet and G. LeBras; Mechanism of the reaction of CH_3SO with NO_2 in relation to atmospheric oxidation od dimethyl sulfide: experimental and theoretical study, *J. Phys. Chem. A* **104** (2000), 935-946.

Ray A., I. Vassalli, G. Laverdet and G. LeBras; Kinetics of the thermal decomposition of the CH_3SO_2 radical and its reaction with NO_2 at 1 Torr and 298 K, *J. Phys. Chem.* **100** (1996), 8895-8900.

Turnipseed A. A., S. B. Barone and A. R. Ravishankara; Observation of CH_3S addition to O_2 in the gas phase, *J. Phys. Chem.* **96** (1992), 7502-7505.

Yin F., D. Grosjean and J. H. Seinfeld; Photooxidation of dimethyl sulfide and dimethyl disulfide. I: Mechanism development, *J. Atmos. Chem.* **11** (1990), 309-364.

Atmospheric Oxidation of Selected Primary Alcohols

Fabrizia Cavalli, Ian Barnes and Karl Heinz Becker

Bergische Universität, Physikalische Chemie, Gaußstraße 20, D-42097 Wuppertal
barnes@uni-wuppertal.de

Introduction

Oxygenated organic compounds are currently under consideration as potential replacements for the traditional aromatic hydrocarbon and chlorinated organic solvents. The drive behind the search for replacements is the need for more environmentally acceptable solvents with respect to ambient air quality, i.e. compounds which will help to reduce the tropospheric O_3 and photo-oxidant burden. However, even the use of oxygenated replacements will lead to the release of these compounds to the atmosphere where they will undergo photo-oxidative degradation processes, the most important and often dominant being reaction with the OH radical (Atkinson, 1994). Accurate kinetic data and mechanistic information on the atmospheric fate of replacement oxygenated solvents are, therefore, essential components in any attempts to reliably assess the ozone forming potential of these substances and their possible contribution to photochemical air pollution in urban and regional areas.

Alcohols are a class of highly volatile oxygenates, which are finding increasing use both as solvents and fuel additives. The compounds investigated here, 1-butanol and 1-pentanol find already a variety of applications in chemical industry. The alcohol 1-butanol (Ullman's Encyclopedia, 1993) is used principally in the field of surface coating either direct as a solvent or converted into derivatives. It is an excellent thinner and useful for regulating the viscosity and improving the flow properties of varnishes; it has also numerous application in the plastic and textile sectors. 1-Pentanol, a high volume chemical with production exceeding 1 million pounds annually, is particularly useful as a solvent in several syntheses (Ullman's Encyclopedia, 1993); it is also employed as an extracting agent and as a starting material for the production of lubricant additives. Furthermore, 1-pentanol finds an important application in the pharmaceutical industry and in the production of several esters that are applied in different fields.

While several OH + alcohols bimolecular rate coefficients are well established, the atmospheric oxidation mechanism of alcohols has received relatively little attention and its representation in atmospheric models has substantial uncertainties. To extend both the atmospheric kinetic and mechanistic data bases for alcohols, kinetic and product investigation of the room temperature gas-phase reactions of OH radicals with 1-butanol and 1-pentanol are here presented.

Experimental

The two experimental systems used for the present studies have been described in detail elsewhere (Cavalli *et al.*, 2001; Becker, 1996) and are only very briefly outlined here.

Kinetic studies. Kinetic experiments on the reactions of OH radicals with 1-butanol and 1-pentanol were carried out at 298 ± 2 K and 740 Torr total pressure of synthetic air in a 480 L cylindrical Duran glass chamber (Cavalli *et al.*, 2001). A White mirror system mounted inside the reactor and coupled, by an external mirror system, to a Fourier Transform-Spectrometer (Nicolet Magna 550) enables the *in situ* monitoring of both reactants and products by long path infrared absorption using a total path length of 51.6 m and a resolution of 1 cm^{-1}. The rate coefficients for the gas-phase reactions of OH radicals with 1-butanol and 1-pentanol were determined using a relative rate method (Barnes *et al.*, 1982); cyclohexane was employed as reference compound.

I. Barnes (ed.), Global Atmospheric Change and its Impact on Regional Air Quality, 153–158.

Product studies. All of the product experiments were performed in the European Photoreactor *EUPHORE* (Becker, 1996), a large-volume outdoor smog chamber, which is part of the Centro de Estudios Ambientales del Mediterraneo (CEAM), in Valencia/Spain. The facility consists of two identical half spherical fluorine-ethene-propene (FEP) foil chambers mounted on aluminum floor panels covered with FEP foil; each chamber having a volume of ca. 195 m^3. An FT-IR spectrometer (NICOLET Magna 550, 1 cm^{-1} resolution) coupled to a White mirror system (base length 8.17 m, total optical path length 554.5 m) was used to monitor reactants and products. The reactant mixtures were also analysed, during the experiments, by gas chromatography with photo-ionisation detection (GC-PID). In addition, carbonyl compounds were sampled from the chamber using solid phase 2,4-dinitrophenyl hydrazone (DNPH)-Silica cartridges. Hydrazones, formed by derivatisation, were separated and quantitatively measured by HPLC.

Results and Discussion

Kinetic Studies. Using a relative kinetic technique and employing cyclohexane as reference compound, the following rate coefficients have been measured: k(OH + 1-butanol) = $(8.24 \pm 0.84) \cdot 10^{-12}$ cm^3 $molecule^{-1}$ s^{-1} and k(OH + 1-pentanol) = $(1.11 \pm 0.11) \times 10^{-11}$ cm^3 $molecule^{-1}$ s^{-1}. The results of the present kinetic study are in excellent agreement with previous determinations reported in the literature (Wallington and Kurylo, 1987; Nelson *et al.*, 1990).

Product Studies. Analysis of the products formed in the OH-initiated oxidation of 1-butanol showed the formation of butanal, propanal, ethanal and formaldehyde among the products. In addition, a product eluted at or very close to the retention time of 4-hydroxy-2-butanone. While the HPLC retention time of this product closely matched that of an authentic standard of 4-hydroxy-2-butanone, the residual absorptions in the FT-IR product spectra could not easily be compared with those of the authentic standard, because of the weak residual signals. Based on the HLPC analysis and the possible reaction channels, it is believed that this product observed by HPLC is 4-hydroxy-2-butanone.

Similarly, product analysis on the OH-initiated oxidation of 1-pentanol showed the formation of pentanal, butanal, propanal, ethanal and formaldehyde among the products. In addition, a product has been tentatively identified as 5-hydroxy-2-pentanone on the basis of a comparison of its HPLC retention time and its infrared spectrum with that of the authentic compound.

In both systems, the primary aldehydic products could undergo further reaction with OH radicals, photolysis and dilution losses; therefore, to derive the formation yields of the aldehydic products, their measured concentrations had to be corrected for these secondary reactions. Corrections have been performed using the mathematical procedure described by Tuazon (Tuazon *et al.*, 1986).

The corrected values for the formation yields are given in Table 1.

Table 1: Product Formation Yields

1-BUTANOL + OH

Products	Molar yields
Butanal	0.518 ± 0.071
Propanal	0.234 ± 0.035
Ethanal	0.127 ± 0.022
Formaldehyde	0.434 ± 0.024
4-Hydroxy-2-butanone	0.05 ± 0.01
C-yield	0.92 ± 0.16

1-PENTANOL + OH

Products	Molar yields
Pentanal	0.405 ± 0.082
Butanal	0.161 ± 0.037
Propanal	0.081 ± 0.019
Ethanal	0.181 ± 0.042
Formaldehyde	0.251 ± 0.013
C-yield	0.70 ± 0.14

In summary, the carbonyl and the hydroxycarbonyl products observed and quantified in this work account for 92 ± 16 %C of the overall OH radical reaction with 1-butanol and for 70 ± 14 %C of the overall OH radical reaction with 1-pentanol.

General Mechanistic Considerations based on the product yield data

The OH radical reaction with alcohols proceeds by H-atom abstraction from the various $-CH_3$, $-CH_2-$, and -OH groups present in the molecule. In particular, the present product studies strongly suggest that the main degradation pathway of primary alcohols is hydrogen abstraction from the $-CH_2-$ group at the α position to the -OH functional entity due to the activation effect of this group. Furthermore, the reaction pathways involving the H-atom abstraction from the $-CH_2-$ groups at the β, γ, etc. positions gradually lose in importance since the activating effect lessens with increasing distance from the functional group. Finally, the H-atom abstraction from the $-CH_3$ and -OH group are excepted to be of negligible importance (SAR estimation).

H-atom abstraction from the $-CH_2-$ group at the α position to the -OH functional group produces an α hydroxy alkyl radical which reacts rapidly with oxygen to form the corresponding aldehyde in unit yield.

H-atom abstraction from the $-CH_2-$ group at the β, γ, etc. position to the -OH functional entity produces the corresponding (β, γ, etc.) hydroxy alkyl radicals which react rapidly under atmospheric conditions solely with O_2 to form (β, γ, etc.) hydroxy alkyl peroxy radicals, RO_2. The further reaction of the RO_2 radicals with NO leads to the formation of the corresponding alkoxy radicals, RO. Hydroxy alkoxy radicals can then react with O_2, unimolecularly decompose into shorter-chain species, or, if possible, isomerise.

Based on the results of the present product studies, the dominant reaction pathway of the β, γ, etc. hydroxy alkoxy radicals, $RO^{\bullet}$ appears to be decomposition to form aldehydes and hydroxy carbonyl compounds.

In conclusion, the first generation oxidation products of the two investigated primary alcohols are mainly aldehydes and, to a minor extent, hydroxy aldehydes and hydroxy ketones; their further reactions with OH radicals and photolysis have an important influence on ozone formation and on the chemistry and transport of NO_y in the troposphere.

From the observed oxidation products and their distribution, detailed atmospheric degradation mechanisms have been constructed for the OH radical initiated oxidation of 1-butanol and 1-pentanol (Cavalli *et. al.*, 2001, Cavalli *et al.* 2001 submitted) (Scheme 1 and 2, respectively). The Schemes1 and 2 show the reaction pathways for the alkoxy radicals formed after H-atom abstraction from the $-CH_2-$ groups at the α, β, γ, etc. positions to the -OH functional group; the relative importance of the reaction pathways indicated in the schemes refer to the overall reaction of alcohols + OH.

From the experimental data and mechanistic considerations discussed above it is proposed, for the purposes of modelling oxidant formation in urban air masses, that the atmospheric chemistry of 1-butanol can be represented by:

$$CH_3CH_2CH_2CH_2OH + OH + 0.58\ NO \longrightarrow$$
$$0.55\ CH_3CH_2CH_2C(O)H + 0.23\ CH_3CH_2C(O)H + 0.05\ CH_3CH(O)CH_2CH_2OH + 0.13\ CH_3C(O)H + 0.02\ OHCH_2CH(O)H + 0.45\ HCHO + 0.95\ HO_2 + 0.54\ NO_2 + 0.04\ RONO_2$$

and the atmospheric chemistry of 1-pentanol can be represented by the following equation

$$CH_3CH_2CH_2CH_2CH_2OH + OH + 0.78\ NO \longrightarrow$$
$$0.44\ CH_3CH_2CH_2CH_2C(O)H + 0.16\ CH_3CH_2CH_2C(O)H + 0.02\ OHCH_2CH_2CH_2C(O)CHOH + 0.09\ CH_3CH_2C(O)H + 0.02\ OHCH_2CH(O)H + 0.18\ OHCH_2CH_2C(O)H + 0.05\ CH_3C(O)CH_2CH_2CH_2OH + 0.18\ CH_3C(O)H + 0.3\ HCHO + 0.87\ HO_2 + 0.72\ NO_2 + 0.06\ RONO_2$$

Scheme 1: Atmospheric degradation mechanism for the OH radical initiated oxidation of 1-butanol

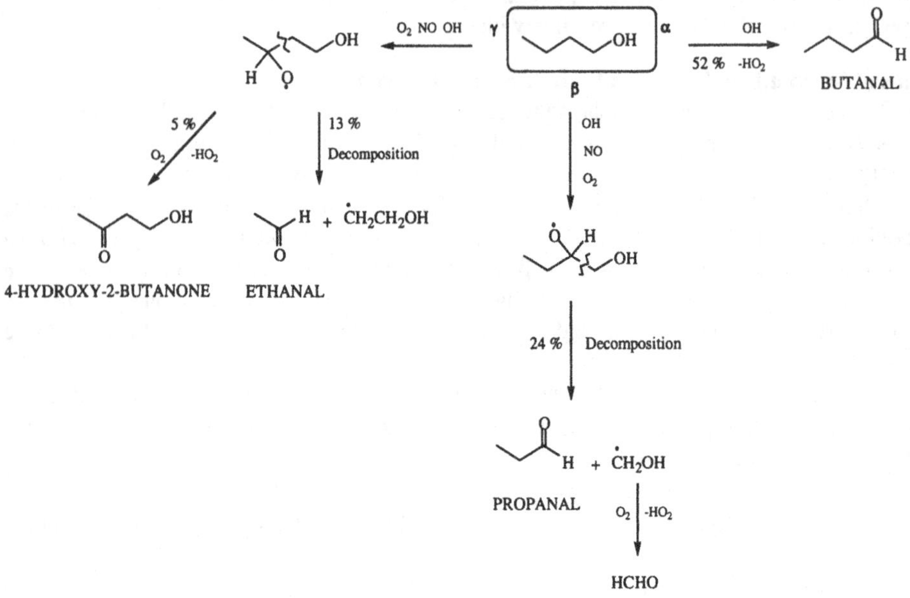

OH
52 % $-HO_2$
BUTANAL
O_2 NO OH
γ α β
5 %
O_2 $-HO_2$
13 %
Decomposition
OH
NO
O_2
4-HYDROXY-2-BUTANONE
ETHANAL
H + $\dot{C}H_2CH_2OH$
24 %
Decomposition
H + $\dot{C}H_2OH$
PROPANAL
O_2 $-HO_2$
HCHO

Scheme 2: Atmospheric degradation mechanism for the OH radical initiated oxidation of 1-pentanol.

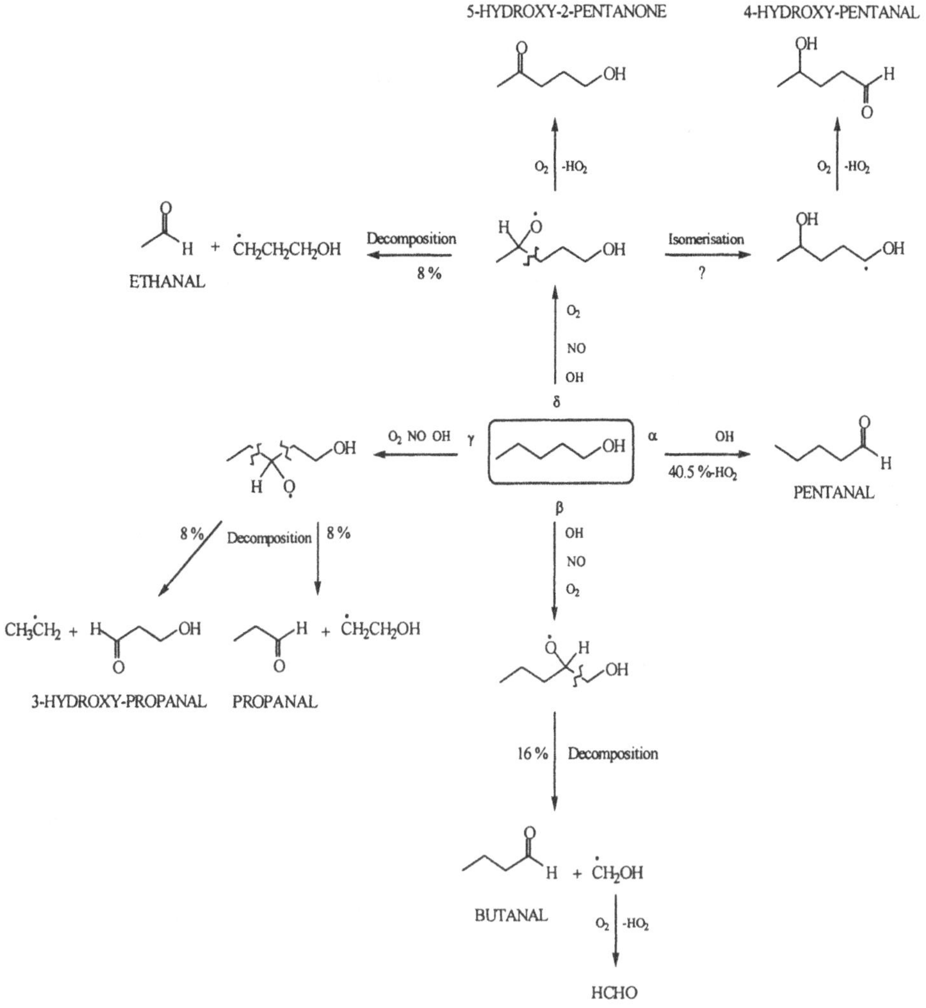

5-HYDROXY-2-PENTANONE

4-HYDROXY-PENTANAL

O_2 | $-HO_2$

O_2 | $-HO_2$

H + $\dot{C}H_2CH_2CH_2OH$ Decomposition 8 %

ETHANAL

Isomerisation ?

O_2 NO OH

δ

O_2 NO OH γ

α OH 40.5 %-HO_2

PENTANAL

β OH NO O_2

8 % Decomposition 8 %

$CH_3\dot{C}H_2$ + H ... OH

H + $\dot{C}H_2CH_2OH$

3-HYDROXY-PROPANAL PROPANAL

16 % Decomposition

H + $\dot{C}H_2OH$

BUTANAL

O_2 | $-HO_2$

HCHO

Acknowledgements

The authors thank the technical staff of CEAM for their assistance with the measurements in *EUPHORE* and the financial support by the "Generalitat Valenciana". F. Cavalli gratefully acknowledges the European Commission for funding her research at the Bergische Universität Wuppertal within the framework of the Environmental and Climate Research and the Specific RTD Programme. Support of the work within the 4th Framework Programme (EUROSOLV) is also gratefully acknowledged.

References

Atkinson R.; Gas-Phase tropospheric chemistry of organic compounds, *J. Phys. Chem. Ref. Data*, Monograph No. 2, (1994).

Barnes I., V. Bastian, K. H. Becker, E. H. Fink and F. Zabel; Reactivity studies of organic substances towards hydroxyl radicals under atmospheric conditions, *Atmos. Environ.* **16** (1982) 545-550.

Becker K. H.; The European Photoreactor EUPHORE. *Final Report of the EC-Project* (1996).

Cavalli F., I. Barnes and K. H. Becker; FT-IR kinetic and product study of the OH radical-initiated oxidation of 1-pentanol, *Environ. Sci. Technol.* **34** (2001) 4111-4116.

Cavalli F., H. Geiger, I. Barnes and K. H. Becker; FT-IR kinetic, Product and modelling study of the OH-initiated oxidation of 1-butanol, *submitted to Environ. Sci. Technol.* (2001).

Nelson L., O. Rattigan, R. Neavyn and H. Sidebottom; Absolute and relative rate constants for the reaction of hydroxyl radicals and chlorine atoms with a series of aliphatic alcohols and ethers at 298 K, *Int. J. Chem. Kin.* **22** (1990) 1111-1126.

Tuazon E. C., H. Mac Leod, R. Atkinson and W. L. P. Carter; α-Dicarbonyl yields form the NO_x-air photooxidations of a series of aromatic hydrocarbons in air, *Environ. Sc. and Technol.* **20** (1986) 383-387.

Ullman's Encyclopedia of Industrial Chemistry; Fifth, Completely Revised Edition, VCH, Weinheim **A19** (1993) 49-60.

Wallington T. J. and M. J. Kurylo; The gas phase reactions of hydroxyl radicals with a series of aliphatic alcohols over the temperature range 240-440 K, *Int. J. Chem. Kinet.* **19** (1987) 1015-1023.

Interaction of Ozone with Sodium Chloride – a Possible Additional Source of Chlorine in the Atmosphere

A.V. Levanov, E.E. Antipenko, A.V. Zosimov, V.V. Lunin

Chemistry Department of Lomonosov Moscow State University

Introduction

The purpose of the work is to investigate the possibility of chlorine gas emission under interaction of ozone with sodium chloride solutions. This relates to the search of real but not yet known sources of chlorine in the atmosphere.

Experimental Part

The reactor used in the work is a glass cylinder with a porous glass plate as its bottom, through which gases were fed in. During the experiments the reactor was filled with 170 ml of 1M NaCl solution. The experiments were performed at room temperature. Initial gas mixtures contained typically (unless otherwise noted) O_3 0.2 – 1.9 vol.%, CO_2 18.2 or 50 vol.%, O_2 the rest.

Determination of chlorine in the exit gases was performed as follows. Ozone was removed from the gas mixture by passing it through the oven. Chlorine was quantitatively trapped with KI solution with the formation of the I_3^- ion, which was determined photometrically. To prevent oxidation of KI with oxygen, the absorbing solution was buffered with 0.01M $NaHCO_3$.

Chlorine gas emission was not detected on passing ozonised oxygen (1.5 – 4 vol.% O_3) through sodium chloride solutions in water (pH $\geq$7). In our experiments the detection limit of Cl_2 in the gas phase was about 0.03 mcmole/l. Chlorine gas emission did take place, if the initial gas mixture contained carbon dioxide in addition to ozone. The explanation to these facts is given below.

The quantity of chlorine accumulated in the trap and the pH of the solution in the reactor were measured as functions of time. Typical experimental results are given in Figures 1 and 2 in comparison with mathematical simulation results.

As the dependencies of the Cl_2 quantity versus time were straight lines, it was possible to determine stationary concentrations of chlorine in the emitted gas. These concentrations (also in comparison with mathematical simulation results) are given in Figure 3 as a function of O_3 concentration in the initial gas mixture.

Mathematical Simulation of Chlorine Emission Kinetics on Interaction of O_3 + CO_2 with NaCl Solution

As a first stage, a calculation of the equilibrium composition of active chlorine solutions as a function of pH was performed. The calculation revealed that in the solution ClO^-, HOCl, Cl_2, Cl_3^- and in the gas phase Cl_2 (also HOCl at higher pH) should be taken into account. Cl_2O can be neglected both in the solution and in the gas phase.

The processes listed in Tables 1 and 2 were included in the model. [X] and C_X designate concentration of particle X in the solution and in the gas phase over it correspondingly. The simulation was performed for 1M NaCl solution and room temperature. [Cl^-] = 1M is constant, because only a small part of Cl^- (less than 0.1% in our experiments) is oxidized to active chlorine. C_{CO2}, [CO_2], C_{O3}, [O_3] are constant, because they are fed into the reactor at constant rate.

I. Barnes (ed.), Global Atmospheric Change and its Impact on Regional Air Quality, 159–162.

Table 1. Chemical reactions in the solution.

Reaction	Rate Constant *or* Equilibrium Constant (20 °C)
$Cl^- + O_3 \rightarrow ClO^- + O_2$	$k_1 < 0.18$ L/mole*min (Hoigné *et al.*, 1985); $k_1 =$ 0.013 L/mole*min (0 °C), 0.038 L/mole*min (9.5 °C) (I=0.3) (Yeatts and Taube, 1949)
$ClO^- + O_3 \rightarrow Cl^- + 2O_2$ $ClO^- + 2O_3 \rightarrow ClO_3^- + 2O_2$	$W=k_2[ClO^-][O_3]$, $k_2 + k_3 = 7200$ L/mole*min $W=k_3[ClO^-][O_3]$ (Hoigné *et al.*, 1985)
$HOCl = H^+ + ClO^-$	$K_1 = [H^+][ClO^-]/[HOCl] = 4*10^{-8}$ mole/L (Morris, 1965)
$Cl_2 + H_2O = H^+ + ClO^- + HOCl$	$K_2 = [H^+][Cl^-][HOCl]/[Cl_2] = 3.6*10^{-4}$ mole^2/L^2 (Connick and Chia, 1959)
$Cl_2 + Cl^- = Cl_3^-$	$K_3 = [Cl_3^-]/[Cl_2][Cl^-] = 0.19$ L/mole (Wang *et al.*, 1994)
$CO_2 + H_2O = H^+ + HCO_3^-$	$K_4 = [H^+][HCO_3^-]/[CO_2] = 4.2*10^{-7}$ mole/L (Nikolskiy, 1965a)

Table 2. Henry's Law Constants for Equilibrium X (gas phase) = X (solution), $H_X = [X]/C_X$ (20 °C, 1M NaCl) (estimated on the basis of literature data).

X	O_3	CO_2	Cl_2
H_X	0.17 (Horváth *et al.*, 1985)	0.71 (Nikolskiy, 1965b)	1.5 (Yakimenko *et al.*, 1976)

The active chlorine concentration can be represented as the product of $[ClO^-]$ and the term that depends only on $[H^+]$.

$$S_{Cl} = [ClO^-] + [HOCl] + [Cl_2] + [Cl_3^-] =$$
$$= [ClO^-] + [H^+][ClO^-]/K_1 + [H^+]^2[Cl^-][ClO^-]/K_1K_2 + K_3[H^+]^2[Cl^-]^2[ClO^-]/K_1K_2 =$$
$$= [ClO^-]*\{1 + [H^+]/K_1 + \frac{[H^+]^2[Cl^-]}{K_1K_2}(1+K_3[Cl^-])\} = [ClO^-]*A.$$

$$A \equiv 1 + [H^+]/K_1 + \frac{[H^+]^2[Cl^-]}{K_1K_2}(1+K_3[Cl^-]).$$

Therefore the variables are $[ClO^-]$ and $[H^+]$.

The problem is formulated as follows.

$$\begin{cases} \dfrac{d\{[ClO^-]\cdot A\}}{dt} = k_1[Cl^-][O_3] - (k_2+k_3)[ClO^-][O_3] - \dfrac{\upsilon}{V}C_{Cl2} \\ K_4[CO_2] = [H^+]^2 + [H^+]^2[ClO^-]/K_1 + 2\dfrac{[H^+]^3[Cl^-]}{K_1K_2}[ClO^-](1+K_3[Cl^-]+\dfrac{\upsilon t}{VH_{Cl2}}) \\ C_{Cl2} = \dfrac{[H^+]^2[Cl^-][ClO^-]}{H_{Cl2}K_1K_2} \qquad t = 0, [ClO^-] = 0 \end{cases}$$

$\upsilon = 0.367$ L/min (gas flow rate), $V = 0.17$ L (solution volume in the reactor).

We neglected the volume of bubbles compared to the volume of liquid in the reactor, and HOCl compared to Cl_2 in the gas phase.

The problem was solved numerically.

The rate constant k_1 was varied to provide the minimal difference between experimental and calculated data. $k_1 = 0.15$ L/mole*min was found. The results of mathematical simulation are given in Figures 1 to 3 in comparison with experimental data.

The emission of chlorine gas from sodium chloride solutions is due to the following reactions taking place in the solution: $Cl^- + O_3 \rightarrow ClO^- + O_2$ and $ClO^- + H^+ = HOCl$, $HOCl + H^+ + Cl^- = Cl_2 + H_2O$. The role of carbon dioxide is that it forms H^+ ions in the solution.

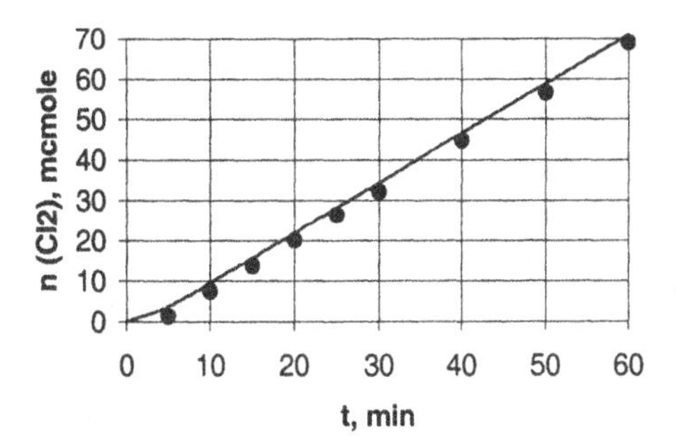

Figure 1. Quantity of emitted chlorine gas as a function of time (points – experiment, line – calculation). Composition of initial gas mixture: O_3 1.86 vol%, CO_2 50 vol.%, O_2 the rest.

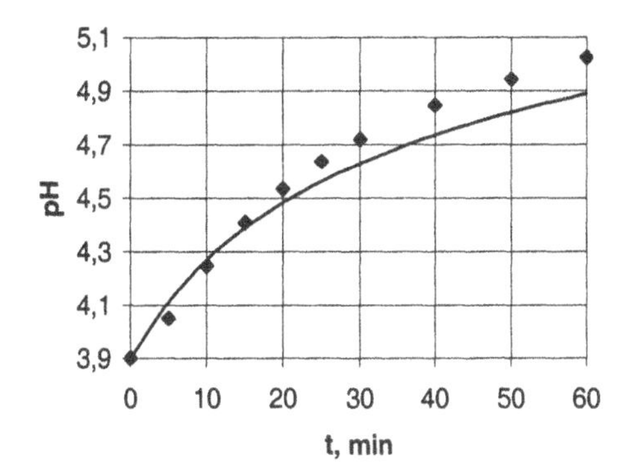

Figure 2. pH of the solution in the reactor as a function of time (points – experiment, line – calculation). Composition of initial gas mixture: O_3 1.86 vol%, CO_2 50 vol.%, O_2 the rest.

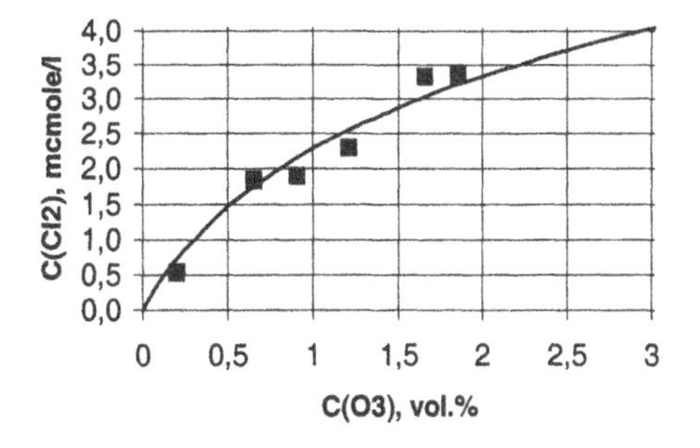

Figure 3. Cl_2 concentration in the emitted gas as a function of O_3 concentration in the initial gas (points – experiment, line – calculation). Composition of initial gas mixture: O_3, CO_2 50 vol.%, O_2 the rest.

Conclusion

1. Chlorine gas is emitted into the gas phase on interaction of ozone with sodium chloride solutions.
2. Chlorine emission was detected experimentally at $pH \leq 5.1$. This pH value was created by addition of CO_2 to the initial gas mixture.
3. A mathematical model that explains chlorine emission on interaction of O_3 with NaCl solutions has been developed.
4. The rate constant k_1 = 0.15 L/mole*min (at 20 °C and ion strength 1 mole/L) for the reaction in water solution $Cl^- + O_3 \rightarrow ClO^- + O_2$ has been determined with the help of the model. The value is in good agreement with literature data.
5. There is a chlorine source in the atmosphere due to the reaction $Cl^- + O_3 \rightarrow ClO^- + O_2$. The interaction O_3 + NaCl can take place as an interaction of atmospheric ozone with ocean surface or aerosol of seawater drops.

References

Connick, R.E. and Y.-T. Chia; The hydrolysis of chlorine and its variation with temperature, *J. Am. Chem. Soc.* **81** (1959) 1280-1284.

Hoigné, J., H. Bader, W.R. Haag and J. Staehelin; Rate constants of reactions of ozone with organic and inorganic compounds in water – III – inorganic compounds and radicals, *Water Res.* **19** (1985) 993-1004.

Horváth, M., L. Bilitzky and J. Hüttner, Ozone. Amsterdam – London – New York – Tokyo (1985).

Morris, J. Carrell; The acid ionization constant of HOCl from 5 to 35 °C, *J. Phys. Chem.* **70** (1966) 3798-3805.

Nikolskiy, B.P. (ed.), Spravochnik khimika (Chemist's handbook). **3**, Moscow (1965a) 106-107.

Nikolskiy, B.P. (ed.), Spravochnik khimika (Chemist's handbook). **3**, Moscow (1965b) 316-321.

Wang, T.X., M.D. Kelley, J.N. Cooper, R.C. Beckwith and D.W. Margerum; Equilibrium, kinetic and UV-spectral characteristics of aqueous bromine chloride, bromine, and chlorine species, *Inorg. Chem.* **33** (1994) 5872-5878.

Yakimenko, L.M. and I.M. Pasmanik, Spravochnik po proizvodstvu khlora, kausticheskoy sody i osnovnykh khlorproduktov (Handbook on production of chlorine, caustic soda and basic chlorine products). Moscow (1976) 162-164.

Yeatts, L.R.B. and H. Taube; The kinetics of the reaction of ozone and chloride ion in acid aqueous solution, *J. Am. Chem. Soc.* **71** (1949) 4100-4105.

Transition Metal Ions in The Processes Of Acid Rain Formation

Alexander N. Yermakov, Anatoliy P. Purmal
ul. Kosygina 2, N.Semenov Institute of Chemical Physics of RAS, 117829, Moscow, Russia

Abstract

Redox dissolution of a nuclei of which a part is insoluble Fe_2O_3 favors sulfite oxidation in the atmosphere from a thermodynamic and kinetic point of view. Coupled with UV light absorption, the uptake of HO_2 radicals from the gas through the reaction $FeOH^{2+} + O_2^{\cdot-}$ and the presence in the droplets of manganese ions, the redox dissolution results in a non-linear rise of the dynamics of the redox cycling of iron and faster removal of SO_2 from the gas.

Introduction

Despite the low liquid water content in the atmosphere, 10^{-7} - 10^{-6} (Behra and Sigg, 1990; Sedlak *et al.*, 1997) it serves as an ideal reaction media for SO_2 removal from the gas phase (Schwartz and Freiburg, 1981). Among the factors differentiating the oxidation in the atmosphere and that occurring under laboratory conditions the following can be outlined:

- *$OH^{\cdot}$, $HO_2^{\cdot}$ and $NO_3^{\cdot}$ radicals' fluxes bombard the droplets and uptake of H_2O_2, and O_3 from the gas.*
- *Absorption of UV light by components dissolved in the aqueous media.*
- *The presence of transition metal ions forms synergistic pairs (Fe + Mn).*
- *Self-influencing of sulfite oxidation rate induced by changes in pH, [S(IV)] and $[Fe]_{sol}$.*
- *A conjugation of atmospheric aqueous phase oxidation processes, for instance simultaneous oxidation of sulfite and $CH_2(OH)_2$).*

The paper aims to show a link between the unavoidable dissolution of Fe_2O_3 as a part of the nuclei and sulfite oxidation rate (Desboeufs and Losno, 1999). An important looks cooperation between reduction and oxidation processes involving iron species (Warnek, 1990). The reduction provided by a sum of UV light absorption (Graedel *et al.*, 1986) and uptake of HO_2 radicals (Matthijsen *et al.*, 1995) together with the oxidation of ferrous ions which can be accelerated by traces of Mn(II) present in the droplets enhances the dynamics of iron redox cycling (Yermakov and Purmal, 2001) leading to a faster removal of SO_2 from the gas phase.

The loss of SO_2 in the gas is not the sole reason that slows down its oxidation in the droplet phase. Acidity growth of the media following removal of SO_2 from the gas looks important bearing in mind that $[HSO_3^-] \approx H_{SO_2}P_{SO_2}K_I/[H^+]$. Here H_{SO_2} and P_{SO_2} are the Henry law coefficient and partial pressure of SO_2, expressed respectively′, in units of M atm^{-1} and atm. The $K_I = 1.4\times10^{-2}$ M (Jacob, 1986) is the equilibrium constant of the $SO_{2(aq)} \Leftrightarrow HSO_3^- + H^+$ (I) ionization of $SO_{2(aq)}$. The pH drops from 6 to 3 resulting in decrease in $[HSO_3^-]$ by a factor of 10^3. The fall in sulfite content in the course of say a reaction occurring in a tube would significantly cut it off regardless of the presence in the system of any catalyst. Under field conditions, however, the oxidation may continue because of an increase in the ferric ion content provided by dissolution both $Fe^{II}O$ and $Fe_2^{III}O_3$ as parts of nuclei (Hoffmann, 1991):

$$FeO + 2H^+ \Leftrightarrow Fe^{2+} + H_2O, \qquad K_{II} \approx 10^{13.7} \qquad (II)$$

Fe^{II}, however, itself is not a true catalyst for the oxidation (Ziajka *et al.*, 1994). It gives the catalytically active form, namely ferric ions (Huss *et al.*, 1982), through oxidation of the ferrous ions by atmospheric oxidants such as HO_2, H_2O_2, $SO_5^{\cdot-}$, HSO_5^-. It is expected also that

I. Barnes (ed.), Global Atmospheric Change and its Impact on Regional Air Quality, 163–165.

the fraction of FeO in nuclei is negligible compare with that given by Fe_2O_3 (Dedik and Hoffmann, 1992). Its solubility is extremely small even in acidic solutions because the process is thermodynamically unfavorable:

$$Fe_2O_3 + 6H^+ \Leftrightarrow 2Fe^{3+} + 3H_2O, \qquad \Delta G_{III} \approx +13 \text{ kJ} \qquad \text{(III)}$$

The presence of sulfite enhances the solubility of ferric ions due to redox dissolution:

$$Fe_2O_3 + HSO_3^- + 3H^+ \Leftrightarrow 2Fe^{3+} + SO_4^{2-} + 2H_2O, \qquad \Delta G_{IV} \approx -117 \text{ kJ} \qquad \text{(IV)}$$

To check weather the redox process proceeds a number of model runs were performed to detect Fe^{2+} as a product of the redox reaction. Its production was monitored by addition in the system of α,α'-dipyridil forming with a Fe(II) a colored complex $Fe^{II}(Dyp)_3$ with $\varepsilon_{max} = 8.75\times10^3$ M^{-1} cm^{-1} at λ_{max} = 525 nm. Optical density ($OD_{\lambda=525}$) was measured after introduction of a portion of sulfite into a tube containing an amount of a powder of Fe_2O_3 (130 mg). To provide a constant pH an acetic buffer was used, $[AcO^-] = 0.1$ M. The time over which the sulfite contacted with the powder was kept constant at 1.5 hour. A first series of the runs were made in the absence of sulfite selecting pHs over the range 3-5. No Fe^{2+} production in reaction (III) was detected. In all cases a linear dependence of $OD_{\lambda=525}$ on time was observed. Contrary to that in these runs with sulfite ($[HSO_3^-] = 2\times10^{-2}$ M, pH = 2.8 - 6) a growth of $OD_{\lambda=525}$ was detected. The measured OD is found to be proportional to a weight of the Fe_2O_3 and to $[HSO_3^-]^n$ with $n < 1$. A linear correlation between the extent of the reduction of ferric ions of interest and the weight of the Fe_2O_3 reflects likely the proportionality of ferric ions dissolution rate and a surface of the solid. The fact that $n < 1$ shows probably the saturation in adsorption of HSO_3^- with an increase of $[HSO_3^-]$. The Table below shows the correlation between pH and the extent of the reduction (III) expressed in a form of $OD_{\lambda=525}$:

pH	5. 1	4.7	4.2	3.7	3.5	3.0
$OD_{\lambda=525}$	0.035	0.04	0.05	0.063	0.08	0.13

The results illustrate a chance of increasing of concentration of ferric ions served as a true catalyst based on thermodynamic ($K_{III} \approx 10^{20}$ M) and kinetic points of view. Both ferric and ferrous ions co-exist in the cloud droplets. The distribution between reduced and oxidized forms is a function of daytime, temperature, and liquid water content in the gas.

Acidification of the cloud droplets shifts the equilibrium $FeOH^{2+} \Leftrightarrow Fe(OH)_2^+$ to the left so accelerating both thermal and external channels of chain carrier production (Warnek *et al.*, 1996):

$$FeOH^{2+} + HSO_3^- \Leftrightarrow FeOHSOH^+ \Rightarrow Fe^{2+} + H_2O + SO_3^{\cdot-}$$
$$FeOH^{2+} \xrightarrow{h\nu} Fe^{2+} + OH^{\cdot}$$
$$OH^{\cdot} + HSO_3^- \Rightarrow SO_3^{\cdot-} + H_2O$$

These in turn lead to in-cloud formation of HSO_5^- that enters into reaction with Fe^{2+} producing two new chain carriers for the one that is consumed in the process (Gilbert and Stell, 1990):

$$HSO_5^- + Fe(II) \Rightarrow Fe^{3+} + SO_4^{\cdot-} + OH^-$$

UV light absorption together with uptake of HO_2 results in a faster reduction of ferric ions compared with that occurring in their absence. Following the photo-dissociation of hydroxo-complexes of ferric ions, production in the droplets of OH radicals together with their flux input, significantly enhances the concentration of HSO_5^-. The enhancement is especially large in the presence of manganese ions. The reaction $SO_5^{\cdot-} + Mn(II) \xrightarrow{H^+} HSO_5^- + Mn(III)$, $k \approx 10^8\ M^{-1}\ s^{-1}$ (Berglund *et al.*, 1995) becomes dominating in the production of HSO_5^- starting with the lowest [Mn(II)]. As a sequence the HSO_5^- + Fe(II) rate is increased so leading to increasing concentrations of chain-carriers and *visa-versa*. Thus both reduction and oxidation of iron species accelerates significantly the dynamics of the iron cycle in the atmosphere making the removal of SO_2 faster. In the nighttime, except for $NO_3^{\cdot}$, there are no external fluxes of atmospheric radicals. Under the conditions a reduction of ferric ions by $HO_2^{\cdot}$ radical arisen in a parallel oxidation of methylene glycol (McElroy et al., 1991) together with intrinsic one $FeOH^{2+} + HSO_3^-$ looks important.

Acknowledgement

Support of this work by the Russian Fund of Basic Researchers (00-05-64029, 01-02-16172) and by the Ministries of Education and Atomic Industry of Russian Federation (6-26) is gratefully acknowledged

References

Behra P., and L.Sigg; Evidence for redox cycling of iron in atmospheric water droplets, *Nature* 344 (1990) 419-421

Dedik A.N., and P.Hoffmann; Chemical characterization of iron in atmospheric aerosols, *Atmos. Envir.* 26A (1992) 2545-2548.

Desbouufs K., and R.Losno; Trace metal supply and acid-base reactions in clouds and rain water, *in CMD Aachen Germany EC/EUROTRAC II joint workshop*, Eds. R.Vogt and G.Axelsdotter (1999) 150-153.

Gilbert B.C., and J.K.Stell; Mechanisms of peroxides decomposition. An ESR study of the reactions of the peroxomonosulfate anion ($HOOSO_3^-$) with Ti^{III}, Fe^{II}, and α–oxygen-sabstituted radicals, *J. Chem. Soc. Perkin Trans.* 2 (1990) 1281-1288.

Graedel T.E., M.L.Mandich, and C.J.Weschler; Kinetic model studies of atmospheric droplet chemistry2. Homogeneous transition metal chemistry in raidrop, *J. Geophys. Res.* 91 (1986) 5205-5221.

Jacob D.J.; Chemistry of OH in remote clouds and its role in the production of formic acid and peroxomonosulfate, *J. Gephys. Res.* 91 (1986) 9807-9826.

Hoffmann H., P.Hoffmann , and K.H.Lieser; Transition metals in atmospheric aqueous samples, analytical determination and speciation, *Fresenius J.Anal. Chem.* 340 (1991) 591-597

Huss A., P.K.Lim, and C.A.Eckert; Oxidation of aqueous sulfur dioxide. 1. Homogeneous manganese(II) and iron(II) catalysis at low pH, *J. Phys. Chem.* 86 (1982) 4224-4228.

Matthijsen J., P.J.H.Builtjes, and D.L.Sedlak; Cloud model experiments of the effect of iron and cooper on tropospheric ozone under marine and continental conditions, *Meteor. Atmos. Phys.* 57 (1995) 43-60.

McElroy W.J., and S.J.Waygood; Oxidation of formaldehyde by the hydroxil radical in aqueous solution, *J. Chem. Soc. Farad. Trans.* 87 (1991) 1513-1521.

Mirabel V., G.A.Salmon, R.van Eldik, C.Winckier, K.J.Wannowius, and P.Warnek; *Review of the activities and achievements of the EUROTRAC Subproject HALIPP*. 2. Warneck P. (Eds) Berlin. Springer. (1996). 7-74.

Sedlak D., J. Hoigné, M.D.David, R.N.Colvile, E.Seyeffer, K.Acker, W.Weipercht, J.A.Lind, and S.Fuzzi; The cloud water chemistry of iron and cooper at Great Dan Fell, *Atmos. Envir.* 31 (1997) 2515-2526.

Schwartz S.E., J.E. Freiberg; Mass-transport limitation to the rate of reaction of gases in liquid droplets: Application to oxidation of SO_2 in aqueous solutions; *Atmos. Envir.* 15 (1981) 1129-1142.

Yermakov A.N., and A.P.Purmal; A nature of synergism between manganese and iron ions in sulfite oxidation, *Chem. Phys. Reports* (2002) in press.

Warnek P.; Chemistry and photochemistry in atmospheric water drops, *Ber. Bunsenges Phys. Chem.* 96 (1992) 455-460.

Ziajka J., F.Beer, and P.Warnek; Iron-catalyzed oxidation of bisulfite aqueous solution: evidence for a free radical chain mechanism, *Atmos. Envir.* 28 (1994) 2549-2552.

The Role of HCl and HBr in Heterogeneous Processes of Stratospheric Ozone Decomposition

Savilov S.V. and Yagodovskaya T.V.
Moscow State University, Department of Chemistry, Vorobjevi gori, 119899 Moscow
savilov@inorg.chem.msu.ru

Introduction

Today, probably, there is no necessity to speak about the great importance of the stratospheric ozone for the human life. It is enough only to recollect, what public resonance the report about the occurrence of the ozone holes above our planet caused [1]. Almost at once after that event a series of the various normative acts regulating the manufacture of substances, potentially suitable for ozone destruction, was accepted. They were devoted, first of all, to CFCs, which are widely used in various fields of industry. However, the question on the final mechanism of ozone layer destruction above the Earth still even now remains open.

It is known, that the natural process of ozone destruction always accompanies to the process of its synthesis. At the same time, in the connection with the presence of various anthropogenic impurities in the atmosphere, nitrogen, chlorine and hydrogen cycles of ozone destruction are stressed [2]. Fifteen years from that moment have passed, but this question has not lost topicality. The universal theory explaining formation of these local centers of seasonal ozone destruction does not exist for today. It is possible to connect this phenomenon either with human activity (chlorine-containing substances, nitrogen oxides etc.) [3] or with natural emission of biosphere, of processes occurring in terrestrial atmosphere and in terrestrial nucleus [4, 5, 6]. The decision of this question is extremely important today whereas the transition of the industrial enterprises from the application of CFCs in cooling systems etc. to the other materials requires the large financial investments, that is especially important in the conditions of modern Russia.

The common point in the first and second hypotheses about ozone layer destruction is the fact that both of them don't deny the contribution of heterogeneous reactions with participation of stratospheric ozone with other compounds on the particles of polar stratospheric clouds, and role of atmospheric circulation processes in the transport of ozone-destroying substances.

Chlorine cycle of stratospheric ozone destruction

Heterogeneous reactions on the border line between gas and solid state play an important role in physical and chemical processes occurring in the Earth's atmosphere. The classical theory of stratospheric ozone destruction explains this process in the following way: chlorine exists in the atmosphere mainly as its natural reservoirs - HCl, HOCl and $ClONO_2$. On the surface of the particles of the polar stratospheric clouds these compounds participate in the reactions, listed below [7]:

$$ClONO_2 + HCl \rightarrow Cl_2 + HNO_3,$$
$$HCl + HOCl \rightarrow Cl_2 + H_2O,$$
$$ClONO_2 + H_2O \rightarrow HOCl + HNO_3.$$

The molecular chlorine formed undergoes photolysis in the atmosphere. Radicals of "active" chlorine form in this process. They react with ozone according to the following reactions [8], which results in ozone destruction:

$$2\,(Cl + O_3 \rightarrow ClO + O_2),$$

I. Barnes (ed.), Global Atmospheric Change and its Impact on Regional Air Quality, 167–171.

$$2\,ClO + M \rightarrow Cl_2O_2 + M,$$
$$Cl_2O_2 + h\nu \rightarrow ClO_2 + Cl,$$
$$ClO_2 + M \rightarrow Cl + O_2 + M,$$

where M is a third body substance. The "active" chlorine radicals can be formed as a result of the photolysis of CFCs and CH_3Cl.

Bromine cycle of stratospheric ozone destruction.

The Br/BrO pair have the most importance in this cycle. The basic reactions according to the published literature data [9] are:

$$Br+O_3 \rightarrow BrO + O_2,$$
$$BrO+O \rightarrow Br+O_2,$$
$$BrO+NO_2+M \rightarrow BrONO_2,$$
$$BrONO_2+h\nu \rightarrow Br+NO_3,$$
$$NO_3+h\nu \rightarrow NO+O_2,$$
$$NO+O_3 \rightarrow NO_2+O_2.$$

The combined cycle of stratospheric ozone destruction.

This cycle is not stressed as an individual cycle in many works: it belongs either to the chlorine, or to bromine mechanisms. However, it is a terminological question. The combined cycle can be realized by two ways [10].
The first of them occur according to the reactions:

$$BrO+ClO \rightarrow BrCl+O_2,$$
$$BrCl+h\nu \rightarrow Br+Cl,$$
$$Br+O_3 \rightarrow BrO+O_2,$$
$$Cl+O_3 \rightarrow ClO+O_2.$$

However, the much more effective way is the following one:

$$BrO+ClO \rightarrow Br+ClOO,$$
$$ClOO+M \rightarrow Cl+O_2,$$
$$Br+O_3 \rightarrow BrO+O_2,$$

The first way occurs, as a rule, at height of 15-35 km, while the second one occurs between 20 and 30 km.

Everything, discussed above, has a photochemical background. However, recent experiments on the simulation of the chlorine cycle of stratospheric ozone decomposition [11], shows the possibility of direct interactions between HCl and O_3 at low temperatures and pressures. It is possible to assume, that the reaction of ozone with HBr will occur similarly. The present work is devoted to detailed research of these processes.

For the investigation of the products of these reactions the low-temperature IR-spectroscopy technique can be applied. Of course, the realization of these experiments has some restrictions with the reaction conditions, but it is suitable for a qualitative estimation. Quantum-chemical calculations can also be very useful in this case.

Experimental

The experiments were carried out with the use of vacuum flowing gas-discharged device. As starting compounds HCl, HBr, O_2 and H_2O (chemically pure grade) were used.

Ozone was synthesized in the oxygen discharge tube and was cooled down to 77 K. Hence, it did not have oxygen impurities.

Distilled water was placed on the walls of a low-temperature trap by means of equable wetting of its surface at room temperature. Then the trap was connected with the vacuum device, was cooled by liquid nitrogen and was pumped out down to 10^{-2} mm Hg. The occurrence of a thin ice film on the trap surface was observed.

At the temperature 293K HCl and HBr vapours with various concentration moved to the trap. They contacted with the ice surface at 77K. The supply of HCl or HBr vapours was done by several doses.

The system was interacted with ozone at a pressure of 10^{-2} mm Hg. The initial ozone pressure was 2 mm Hg. Unreacted ozone was eliminated by pumping. Investigations of interactions between components of the system were carried out at various temperatures.

Several experiments were also made in the following sequence. Ozone contacted with the ice, then water was placed on its surface. After that the system was frozen by liquid nitrogen and contacted with hydrogen chloride or hydrogen bromide.

Results and discussions

Passing to discussion of the results obtained, it is necessary to note, that the reaction of HCl and HBr with ozone at low values of temperature and pressure have not been studied before. There are literature data [12] about thermo-chemical effects and kinetic parameters for the processes occurring on irradiation of mixtures of hydrogen chloride with ozone at temperatures and pressures about 480-1300K and 4-8 atmospheres, respectively. In the case described HCl does not react with ozone directly, it serves only as a catalyst of ozone decomposition. Nevertheless chlorine oxide ClO is formed, but indirectly - interaction of atomic chlorine (product of reaction of oxidation of HCl by atomic oxygen) with ozone.

On injecting gaseous ozone to the trap with HCl adsorbed on the ice surface at 77K an immediate interaction with ozone was observed. It resulted in the formation of a red condensate. The pressure in the device fell and reached 1 mm Hg, probably, because of molecular oxygen formation. After evacuating the system at 77K, the pressure remained constant. Increasing the temperature up to 223K resulted in the appearance of a ring of deep red colour above the brightened solid condensate in the low-temperature trap. The pressure increased again, probably there was a formation of chlorine oxide with formula Cl_2O_6 from the products of primary interaction of the system ice – HCl (adsorbed) with ozone. On the basis of the data, published in [13], it was possible to make a conclusion that the original condensate basically consisted of Cl_2O and (or) ClO_2. When the temperature increased, Cl_2O_6 and (or) others chlorine oxides were formed. It is necessary to note, that the reaction of ozone with HCl (adsorbed on ice) on the walls of the trap without the ice film proceeds with formation of the same red condensate at 77K [11], however, much more slowly. It proves that HCl state in the later experiment mentioned differs from the state of HCl molecule, adsorbed on the ice surface, under the similar conditions.

It was found that a highly developed ice surface and low concentrations of the hydrochloric acid are beneficial to the reaction.

The IR-spectra of system investigated show that reaction occurs. This evidenced by the existence of a number of intensive absorption peaks, which can be assigned to the Cl-O vibrations in chlorine oxides with the formulas Cl_2O, and ClO_2. However, because of the peaks in the region 1477 cm^{-1} and 1442 cm^{-1}, which are related to the O-O vibrations in the structure of ClO_2 [14, 15], it is possible to imply primary formation of this oxide under the reaction conditions. ClO_2 has two isomers, the structures and energy are given in a figure 1, based on the literature data [16] (energies are given with the zero point correction).

(a) ClOCl (C_{2v}), E=-610,541500 a.u. *(b)* ClOO, E=-610,545413 a.u.

Figure 1. Structures and energies of Cl_2O and ClO_2

As shown, the isomer (b) is energetically more favourable, but according to [17], it is kinetically unstable when irradiated. Since the reaction occurs at low temperatures and obviously by means of a non-radical mechanism (without any irradiation), it is possible to consider isomer (b) (which certainly will undergo the further transformations under stratospheric conditions) as the product.

The interaction of a thin layer of hydrogen bromide, adsorbed by ice, with ozone at low temperatures also proceeds immediately. Br_2O is formed initially, which undergoes further oxidation to BrO_2. This process is promoted by increasing the temperature. In the reactions of HCl and HBr with ozone there are some distinctions especially with decreasing ice surface and acid concentration roles in the latter case.

In the case, when HCl contacted with ozone, previously adsorbed on the ice, the described interaction was not observed. Apparently, it is connected with the formation of ozone hydrates, which protect it from reaction with hydrogen chloride.

According to the literature [18], when HCl comes on the ice surface of PSC's particles, it undergoes ionization with formation of Cl^- and H_3O^+. The exit of the chloride ion to the surface from the bulk ice crystal is the result of this multistage process. This ion is surrounded by water molecules and H_3O^+ ions. This process is the process of active chlorine ion formation.

Thus, our current ideas about HCl and HBr as stratospheric chlorine and bromine reservoirs, according to the experiments described, appeared to be without validity. These compounds are capable of reacting with ozone on the surface of ice particles of polar stratospheric clouds. Under doubt is the other fact, which was obvious till now – i.e. the statement that CFCs are the basic source of active chlorine in stratosphere. Recent works specify that sodium chloride from the marine water can occupy their place [19, 20]. However, it is necessary to note, that the reactions of the chlorine ion, resulting from NaCl solutions and from HCl, adsorbed on the ice, with ozone have absolutely various nature of chlorine ion existence conditions. The data presented allow a dramatic change to be made to the current chlorine and bromine cycle theories. The application of the data obtained to the theory of the chlorine cycle of the stratospheric ozone decomposition is shown on the figure 2.

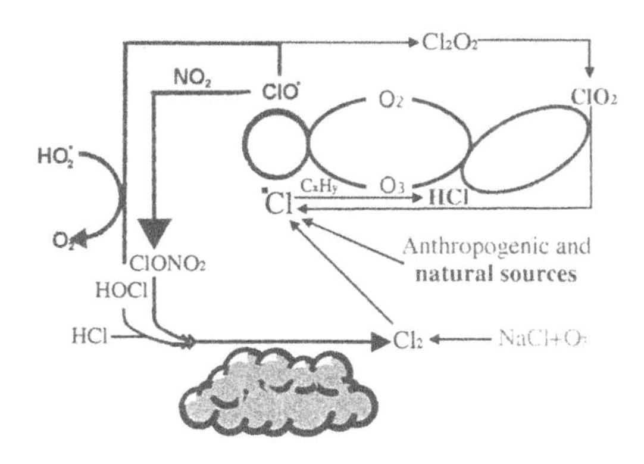

Figure 2: The new scheme of the chlorine cycle of the stratospheric ozone decomposition

Acknowledgements

This work has been supported by FCP "Integratsia" N 467. The authors thank Dr., Prof. Valeriy. V. Lunin for helpful discussions.

References

1. Farman J.C., Gardiner B.G., Shanklin J.D. *Nature.* **315** (1985) 207.
2. van Loon G.W., Duffy S.J. Environmental Science. Oxford University press, 2000.
3. Wayne R.P. Chemistry of the atmospheres. Oxford University Press. 2000.
4. Sivorotkin V.L. Riftogenez I ozonoviy sloy. Moscow.: Geoinformmark, 1996.
5. Marakushev O.O. *Vestnic RAN.* **68** (1998) 813.
6. Butler J.H. *Nature.* **403** (2000) 260.
7. Zondlo M.A., Barone S.B., Tolbert M.A. *J. Phys. Chem. A.* **102** (1998) 5735.
8. Molina L.T., Molina M.J. *J. Phys. Chem.* **91** (1987) 433.
9. Lary D.J. *J. Geoph. Res. D.* **102D** (1997) 21515.
10. Lary D.J. *J. Geoph. Res. D.* **101D** (1996) 1505.
11. Yagodovskaya T.V., Gromov A.R, Zosimov A.V., Lunin V.V. *Zh. fiz. him.* **73** (1999) 662.
12. Park C. *J. Phys. Chem.* **81** (1977) 499.
13. Nikitin I.V. Himiya kislorodnih soedineniy galogenov. Moscow: Nauka. 1986.
14. Muller H.S., Willner H. *J. Phys. Chem.* **97** (1993) 10589.
15. Johnsson K., Engdhl A., Nelander B. *J. Phys. Chem.* **97** (1993) 9603.
16. Beltran A., Andres J. *J. Phys. Chem. A.* **103** (1999) 3078.
17. Pursell C.J., Conyers J., Denison C. *J. Phys. Chem.* **100** (1996) 15450.
18. Gertner B.J., Hynes J.T. *Science.* **271** (1996) 1563.
19. Behnke W., Elend M., Kruger U., Zetzsch C. *Proc. of the EC/EUROTRAC-2 joint workshop.* September 11-13, 2000, Switzerland.
20. Abbatt J.P., Waschewsky G.C.G. *J. Phys. Chem. A.* **102** (1998) 3719.

Comparison of the NO_3 Heterogeneous Uptake on Dry and Humidified NaX (X=Cl, Br) Salts

Vladislav V. Zelenov and Elena V.Aparina

Institute of Energy Problems of Chemical Physics (Branch), Russian Academy of Sciences, Chernogolovka 142432, Moscow District, Russian Federation; zelenov@binep.ac.ru

Introduction

Sea salt is one of the dominant components of atmospheric aerosols (Brimblecombe, 1986). An interest in heterogeneous reactions on sea salt aerosols with participation of nitrogenated matter arose after the Cl^- deficit in aerosol particles was found. In especially polluted areas, for small-sized particles the deficit reaches up to 100%. At the same time, extremely high Cl_2 concentrations have been monitored in coastal air and MBL regions (Spicer *et al.*, 1998). At sunrise, this may correspond to Cl concentrations up to 10^5 cm^{-3} (Spicer *et al.*, 1998) and be competitive with OH as a main atmospheric oxidant. Once a new atmospheric oxidizing agent had been revealed, a number of studiess concerning a possible mechanism of Cl_2 formation followed. The reaction mechanisms between NO_2, NO_3, N_2O_5, HNO_3, $ClONO_2$, etc. and sea-salt aerosols have been investigated. The nitrate radical ranks a special place among these oxidants. The heterogeneous reaction of NO_3 with sea-salt particles

$NO_3(g)+NaCl(s)\rightarrow NaNO_3(s)+Cl$, $\Delta_{r1}H^o_{298}=-9.5$ kJ/mol (R1),

$NO_3(g)+NaBr(s)\rightarrow NaNO_3(s)+Br$, $\Delta_{r2}H^o_{298}=-68.7$ kJ/mol (R2)

can be a direct nonphotolytic source of halogen atoms(Seisel *et al.*, 1997).The nighttime concentration of NO_3 in the troposphere may reach up to 2×10^9 cm^{-3} (Platt *et al.*, 1980), therefore, the reactions (R1), (R2) can make an important contribution to the oxidizing potential of the troposphere at night when other oxidants like $O(^1D)$ and OH are absent. Despite sea salt being poor in NaBr (0.124%), it can be a potential source of the strong ozone destroying agent (Sander and Crutzen, 1996).

The reaction mechanism of an interaction between NO_3 and aqueous solutions of ionic salts as well as its application to atmospheric chemistry has been already considered by Rudich (Rudich *et al.*, 1998). A reactive uptake of NO_3 on solid NaCl (Seisel *et al.*, 1999; Zelenov *et al.*, 2001; Seisel *et al.*, 1997; Gratpanche *et al.*, 1999; Gershenzon *et al.*, 1999) and NaBr (Zelenov *et al.*, 2001; Gratpanche *et al.*, 1999; Gershenzon *et al.*, 1999) has been studied for different salt presentations. Based on the data obtained, the uptake coefficients $\gamma_{NO3}^{NaCl}=(2\text{-}5)\times10^{-2}$, $\gamma_{NO3}^{NaBr}=0.1\text{-}0.2$ are independent of the NO_3 volume concentration in the range of $[NO_3]=10^{10}-10^{12}$ cm^{-3}. Bromine atoms have been established as the primary products of the reaction (R2). Chlorine atoms as products of the reaction (R1) have not been detected. Based on the temperature dependence of γ as well as the γ dependence on $[NO_3]$ in the range of $[NO_3]=10^{10}-10^{14}$ cm^{-3} (Zelenov *et al.*, 2001), the mechanism of the NO_3 reactive uptake on NaCl and NaBr has been determined, and includes elementary rate coefficients and their temperature dependencies.

In the course of its lifetime, the sea-salt aerosol undergoes a modification from saturated aqueous solution to dry crystallite. Nevertheless, there are no systematic data on the dependence of γ on relative humidity to date, except the data of Rossi et al (Seisel *et al.*, 1999), where the dependence is found to be absent at low $[H_2O]\leq 10^{13}$ cm^{-3}.

To summarize briefly, there are several problems that still remain unsolved: (i) why Cl atoms are absent among the reaction products, (ii) whether uptake coefficient depends on relative humidity at high $[H_2O]$, (iii) what are the primary products of the reaction between NO_3 and humidified sea-salt aerosol?

I. Barnes (ed.), Global Atmospheric Change and its Impact on Regional Air Quality, 173–179.

Experimental section

Flow reactor coupled to mass spectrometer. The setup is designed for investigation of gas-phase and heterogeneous reactions. It consists of a high-resolution low-energy electron-impact ionization mass spectrometer combined with several replaceable flow reactors. The distinguishing features of the mass spectrometer are: working resolution of 20,000; the mass range of 1 – 4000 Da; molecular beam sampling with phase-sensitive detection; the unique home-made ion source with an electron impact from an indirectly heated LaB_6 cathode specially designed for a molecular beam.

Coated insert flow tube reactor. The reactor is described elsewhere (Gershenzon *et al.*, 1999). In brief, the helium carrier gas flow with added NO_3 radicals runs along an outer wide glass tube (1.3 cm inner diameter and 50 cm long) covered with a Teflon film. A thin quartz rod (1.3 mm outer diameter) covered with a salt of interest can be slide along the axis of the outer tube to vary the exposure time of the salt to the NO_3 flux. The NO_3 radicals are transported into the reactor by a He flow from a thermocell where they are formed under N_2O_5 thermal decomposition. A source of N_2O_5 is a Teflon ampoule filled with N_2O_5 frozen and placed in a cryostat. A variation in the NO_3 concentration inside the reactor is achieved by varying the temperature of the cryostat from -93° C to -80° C. At the same time, the cryostat serves as a low-temperature trap for HNO_3 that can be produced in the synthesis of N_2O_5. An additional He/H_2O flow runs along the full length of the outer tube. A molecular beam sampling into the mass spectrometer is realized through an orifice at the top of an alumina inlet cone fixed at the butt of the outer reactor tube. A salt was deposited on the rod by dipping it in aqueous non-saturated solution followed by drying the rod in an oven at 200° C and keeping it in vacuum for a day.

Metering procedure. The NO_3 concentration was established in every experiment by titration of NO_3 with NO via the fast $NO + NO_3 \rightarrow 2NO_2$ reaction. Before each experiment, the mass spectrometer was calibrated using absolute concentrations of stable substances (NO, NO_2, N_2O_5, HCl, $ClONO_2$, $BrONO_2$, Br_2, H_2O). For substances with overlapping mass spectra, the intensities of their parent ions were measured, using low-energy electron-impact ionization near the ionization threshold. Some of substances, e.g. $ClONO_2$ and $BrONO_2$, do not have parent ions. In this case their concentrations were determined by comparison between some residual mass-spectral line intensities and their reference mass spectra obtained during the calibration procedure.

Data treatment. An uptake coefficient γ was calculated by measuring a relative change in the NO_3 signal intensity without a rod, I^0_{NO3}, and with the rod inserted as a unit, I^r_{NO3} based on the relation

$$\ln(I^0_{NO3} / I^r_{NO3}) = \frac{\gamma \cdot c_{NO3}}{d_R} \cdot \frac{d_r}{d_R} \cdot t \qquad (1)$$

where c_{NO3} is a mean molecular velocity of NO_3 radicals; d_r=0.13 cm is the outer diameter of the rod; d_R=1.3 cm is the inner diameter of the reactor tube; $t=L_r/v_l$ is an exposure time of the rod to the NO_3 flux; L_r is the length of the rod; v_l is the linear velocity of the He flux. The relation (1) is based on the well-known equation for the first order of uptake

$$d\,[NO_3] / dt = k_w \cdot [NO_3] \qquad (2)$$

under conditions when $d_r \ll d_R$ and a reactive uptake is mainly determined by kinetic transfer rather than by diffusion.

Data on the reaction product formation are presented in terms of some partial NO_3 uptake coefficient, γ_P, leading to a yield of the given product P with concentration [P] at the sampler. Such a presentation is based on the formal equation for the product formation in the heterogeneous reaction

$$d[P]/d\tau = k_w(\gamma_p/\gamma)\cdot[NO_3(\tau)] \qquad (3)$$

where k_w is the rate coefficient for NO_3 heterogeneous uptake defined in eqn.(2). An expression for γ_p follows from integration of eqn.(3) over an interval $\tau[0, t]$

$$\gamma_p = ([P]/[NO_3]_0)\cdot\gamma/(1-\exp(-k_w t)) \qquad (4).$$

Such a presentation is very convenient, as it allows us to follow a material balance between the reactant consumption and the product formation at any moment.

Results and discussion

The mechanisms of NO_3+NaX (X=Cl,Br) reactions were studied at $[NO_3]=(1 - 10)\times10^{12}$ cm^{-3} and $[H_2O]=(2 - 1000)\times10^{12}$ cm^{-3}. During every initial treatment of fresh and humidified NaX by a NO_3 flux, the $\gamma(t)$ is found to decrease exponentially in time down to some steady-state γ which is determined by the specific salt, the NO_3 volume concentration, and that of H_2O. A reason for the initial γ decrease is likely to be the same as that for HNO_3 uptake on NaCl solid (Beichert and Finlayson-Pitts, 1996). What all our steady-state γ have in common is that they increase directly proportional to $[H_2O]$, beginning with some initial γ_{dry} that corresponds to "dry" conditions for a given $[NO_3]$. In the present work, a mechanism of only steady-state uptake is considered.

Figures 1 and 2 show dependencies of both total uptake coefficient for NO_3 and the partial ones on H_2O volume concentration. Linear regressions of the data in Figure 1 give: $\gamma=(0.5\pm0.1(2\sigma))\times10^{-3}+(1\pm0.15)\times10^{-18}[H_2O]$, $2\gamma_{NO2}=(0.45\pm0.06)\times10^{-3}$, $2\gamma_{ClONO2}=(0.78\pm0.1)\times10^{-18}[H_2O]$, $2\gamma_{HCl}=(0.34\pm0.04)\times10^{-18}[H_2O]$. The same procedure for data in Figure 2 gives: $\gamma=(3.1\pm0.2)\times10^{-3}+(2\pm0.3)\times10^{-18}[H_2O]$, $2\gamma_{BrONO2}=(1.8\pm0.2)\times10^{-3}+(1.7\pm0.3)\times10^{-18}[H_2O]$, $\gamma_{totalBr}=(1.5\pm0.1)\times10^{-3}-(1.3\pm0.8)\times10^{-19}[H_2O]$. The last $\gamma_{totalBr}$ combines 3 contributions, $\gamma_{totalBr}=\gamma_{Br}+2\gamma_{Br2}+2\gamma_{NO2}$, with relation of 0.1:0.25:0.65 between them. The last two of them are an outcome of secondary processes that transform initial Br atoms into Br_2 and NO_2 because of heterogeneous recombination and fast secondary gas-phase $Br+NO_3\rightarrow BrO+NO_2$ reaction. Doubling in some partial uptake coefficients corresponds to formation of only one molecule of the product per two NO_3 radicals spent.

The data in Figures 1,2 indicate validity of the mass balance, i.e. $\gamma=2\gamma_{NO2}+2\gamma_{ClONO2}+2\gamma_{HCl}$ in Figure 1 and $\gamma=\gamma_{totalBr}+2\gamma_{BrONO2}$ in Figure 2. Additionally, they show that some of the products do not depend on humidity whereas the other ones are exactly determined by it.

Reactive uptake of NO_3 on dry solid NaCl and NaBr has been already described elsewhere (Gershenzon *et al.*, 1999). To summarize briefly, the uptake is supposed to proceed via several consecutive steps including(e.g. for NO_3 uptake on NaCl):

1. Reversible adsorption of NO_3 on the defect sites that have the surface concentration equaled to $f\cdot z_0^{NaCl}$, where $z_0^{NaCl}=6.4\times10^{14}$ cm^{-2} is a surface density of active sites; f=0.2 is a fraction of the total z_0 having defects (Dai and Ewing, 1993)
$$NO_3(g)+NaCl(s)\Leftrightarrow z_s', \qquad k_{ad}, k_d' \qquad (R1a).$$
2. Unimolecular decomposition of surface complex
$$z_s'\rightarrow Cl(ad)+NaNO_3(s), \qquad k_{R1b}, \qquad (R1b).$$
3. Secondary heterogeneous reaction transforming Cl(ad) into NO_2(ad)
$$NO_3(g)+Cl(ad)\rightarrow NO_2(ad)+ClO(ad) \qquad (R3).$$
4. Reversible desorption of the products.

An analytical representation of γ in terms of elementary constants follows from the mechanism given above

$$\gamma_{dry}^{NaCl} = \frac{2k_{R1b}\cdot f}{k_d'}\cdot\left(1+\frac{[NO_3]}{n_{th}'}\right)^{-1} \qquad \text{with} \quad n_{th}' = \frac{k_d'\cdot z_0^{NaCl}}{k_{ad}} \qquad (5)$$

where $k_{R1b}=\nu_{R1}\cdot\exp(-E_{R1} / RT)$ is the rate constant for unimolecular decomposition (R1b); $k_d'=\nu_d'\cdot\exp(-Q_{ad}^{NaCl}(NO_3) / RT)$ is desorption rate constant, $k_{ad}=c_{NO3} / 4$ is adsorption rate constant; n_{th}' is threshold volume concentration of NO_3; E_{R1}=56.7 kJ/mol; $Q_{ad}^{NaCl}(NO_3)$=58 kJ/mol, $\nu_{R1}=0.78\times10^9$ s^{-1}, $\nu_d'=1.4\times10^{11}$ s^{-1}.

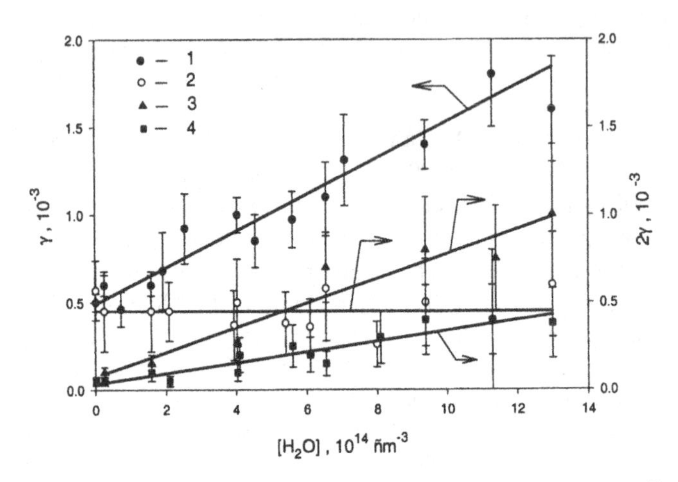

Figure 1. Reactive uptake of NO_3 on NaCl and the product formation at p=2 Torr, $[NO_3]=3.9\times10^{12}$ cm^{-3}, t=0.15 s. Symbols are experimental data; 1, uptake coefficient for NO_3; 2,3,4, partial uptake coefficients leading to NO_2, $ClONO_2$, and HCl formation, respectively. Solid lines are linear regressions.

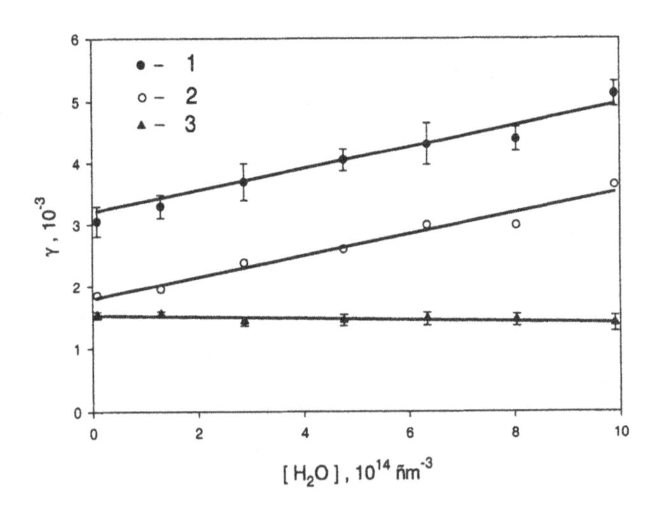

Figure 2. Reactive uptake of NO_3 on NaBr and the product formation at p=1 Torr, $[NO_3]=2.3\times10^{12}$ cm^{-3}, t=0.075 s. Symbols are experimental data; 1, uptake coefficient for NO_3; 2,3, partial uptake coefficients leading to $BrONO_2$ and Br atoms formation, respectively. Solid lines are linear regressions.

A similar mechanism of NO_3 uptake on dry solid NaBr is presented by

$$NO_3(g)+NaBr\cdot2H_2O\Leftrightarrow z_s'', \quad k_{ad}, k_d'' \qquad (R2a)$$

$$z_s''\rightarrow Br(g)+NaNO_3(s), \quad k_{R2b} \qquad (R2b)$$

$$NO_3(g)+Br(g)\rightarrow NO_2(g)+BrO(g), \quad k_{R4}(298\ K)=1.6\times10^{-11}\ cm^3\ s^{-1} \qquad (R4)$$

$$Br(g)\rightarrow 0.5\ Br_2(ad)\Leftrightarrow 0.5\ Br_2(g), \quad k_w^{Br}$$

$$\gamma_{dry}^{NaBr} = \frac{(1+\beta)\cdot k_{R2b}\cdot f}{k_d''}\cdot\left(1+\frac{[NO_3]}{n_{th}''}\right)^{-1} \qquad \text{with} \quad n_{th}'' = \frac{k_d''\cdot z_0^{NaBr}}{k_{ad}} \qquad (6)$$

where $k_{R2b}=\nu_{R2}\cdot\exp(-E_{R2}\ /\ RT)$; $k_d''=\nu_d''\cdot\exp(-Q_{ad}^{NaBr}(NO_3)\ /\ RT)$; E_{R2}=58.2 kJ/mol; $Q_{ad}^{NaBr}(NO_3)$=61.8 kJ/mol; $\nu_{R2}=8.4\times10^9$ s^{-1}, $\nu_d''=0.73\times10^{11}$ s^{-1}; the numerical value of β is determined by a contribution into NO_3 consumption from secondary reactions. In particular, β=0.35 under conditions given in Figure 2.

A distinction between probable aggregative states of halogen atoms formed in reactions (R1), (R2) follows from a diagram of the energy transfer for reactive system along the reaction coordinate. Enthalpies of (R1) and (R2), $\Delta_{r1}H^o_{298}$ and $\Delta_{r2}H^o_{298}$, correspond to formal transition between initial and final states of the reactive system. In fact, the reaction (R1b) or (R2b) starts only after the NO_3 radical is adsorbed. Thus, true enthalpy of reaction $NO_3(ad)/NaCl(s) \rightarrow Cl(g)/\ NaNO_3(s)$, $\Delta_rH^o_{298}=\Delta_{r1}H^o_{298}+Q_{ad}^{NaCl}(NO_3)$=+48.5 kJ/mol indicates that Cl atoms can not leave the surface during unimolecular decomposition. They are formed in an adsorbed state on the defect sites. Adsorption energy of Cl atoms can be evaluated by comparison between their characteristic lifetime on the surface, $k_d^{-1}=\nu_d^{-1}\exp(Q_{ad}^{NaCl}(Cl)\ /\ RT)$, and the time interval between reactive collisions of NO_3 radicals with Cl adatom, $\tau_{col}=4z_0^{NaCl}\ /\ \gamma\ [NO_3]\cdot c_{NO3}$. Since nobody saw Cl atoms or Cl_2 as the products of their recombination, a relation $k_d^{-1}\geq\tau_{col}$ is always valid under experimental conditions. Based on the $\gamma=5\times10^{-2}$ at $[NO_3]=10^{10}$ cm^{-3} (Seisel *et al.*, 1997) as the most reliable one under tropospheric conditions and at normal preexponential factor $\nu_d=10^{13}$ s^{-1}, the last inequality gives $Q_{ad}^{NaCl}\geq87.2$ kJ/mol with characteristic lifetime $k_d^{-1}\geq162$ s.

In contrast to Cl(g) formation, the reaction $NO_3(ad)/NaBr(s)\rightarrow Br(g)+NaNO_3(s)$, $\Delta_rH^o_{298}=\Delta_{r2}H^o_{298}+\ Q_{ad}^{NaBr}(NO_3)$=–6.9 kJ/mol is a slightly exothermic one. Just in the course of unimolecular decomposition (R2b), Br atom may leave the surface. Adsorption energy of Br atom is much lower as compared with that of Cl atom, *viz.* $Q_{ad}^{NaBr}(Br)\geq79.8$ kJ/mol with $k_d^{-1}\geq0.1$ s. A lower $Q_{ad}^{NaBr}(Br)$ is an outcome of Br uptake on some fine energy traps because the deepest of them are already occupied by the particles that have a higher gas-phase concentration.

The main assumption that allows us to explain all the experimental dependencies and to propose a kinetic model for NO_3 uptake on humidified surface is a correlation between reactive uptake and the surface number density of water. At present, the adsorption isotherms of water on NaCl and NaBr are well known. At low coating factor, the surface number density of H_2O is known to be directly proportional to H_2O volume concentration: $\theta_{H2O}=0.5\cdot[H_2O]\ /\ n_{th}(H_2O)$, where $n_{th}(H_2O)=1.2\times10^{17}$ cm^{-3} for NaCl and 6×10^{16} cm^{-3} for NaBr are the proportionality constants between volume concentration, $[H_2O]$, and an average fraction of the surface covered with adsorbed H_2O, θ_{H2O}. Physical adsorption is a reversible process that takes place on the crystal surface free of defects.

The component part of the γ in Figure 1,2 dependent on $[H_2O]$ is supposed to be determined by uptake of NO_3 on vacant sites generated by physisorbed water. A mechanism of NO_3 uptake on NaCl can be proposed as the following steps

1. Reversible competitive adsorption of NO_3 and H_2O on NaCl surface sites free of defects, i.e. $(1-f)\cdot z_0^{NaCl}$

$$NO_3(g)+NaCl(s)\Leftrightarrow NO_3\text{–}NaCl(s),\ H_2O(g)+NaCl(s)\Leftrightarrow H_2O\text{–}NaCl(s).$$

2. Reversible competitive adsorption of NO_3 and H_2O on supplementary adjacent sites followed by the surface complex formation

$$NO_3(g)+H_2O\text{–}NaCl(s)\Leftrightarrow NO_3\text{–}NaCl\text{–}H_2O(s),\ H_2O(g)+NO_3\text{–}NaCl(s)\Leftrightarrow NO_3\text{–}NaCl\text{–}H_2O(s).$$

3. Unimolecular decomposition of the surface complex leading to Cl_{ad} formation

$$NO_3\text{–}NaCl\text{–}H_2O(s)\rightarrow NaNO_3+Cl_{ad}\text{–}H_2O,\quad k_{R5} \qquad (R5).$$

4. Secondary reaction leading to $ClONO_2$(ad) formation followed by reversible desorption
$NO_3(g)+Cl_{ad}-H_2O \rightarrow ClONO_2(ad)-H_2O$, $ClONO_2(ad)-H_2O \Leftrightarrow ClONO_2(g)+H_2O(ad)$.
5. Partial hydrolysis of $ClONO_2$ (Timonen *et al.*, 1994) as a pathway of HCl formation
$ClONO_2(ad)-H_2O \rightarrow HOCl(ad)+HNO_3(ad)$, $HNO_3(ad)-NaCl \rightarrow HCl(g)+NaNO_3(s)$ (R6).

Based on the mechanism, a particular expression for γ under conditions given in Figure 1 is

$$\gamma_{hum}^{NaCl} = \frac{k_{R5} \cdot z_0^{NaCl}}{k_{ad}} \cdot \frac{[H_2O]}{n_{th}(H_2O)} \cdot \frac{(1-f)}{[NO_3]} \quad (7).$$

The γ proves to be directly proportional to $[H_2O]$ and inversely proportional to $[NO_3]$, that corresponds to experimental dependencies. Based on eqn. (7), the elementary rate constant $k_{R5}=7.2\pm1.1\ s^{-1}$ can be calculated from the slope of dependence of γ on $[H_2O]$ given in Figure 1.

A mechanism of NO_3 uptake on NaBr humidified is similar to that of NO_3/NaCl. The lack of HBr among the products is likely to be related with the tenfold lower reactivity of HNO_3 with NaBr in the reaction that is similar to (R6).

The expression for γ is identical to the one given by eqn. (7), replacing k_{R5} by k_{R7}

$NO_3-NaBr-H_2O(s) \rightarrow NaNO_3+Br(ad)-H_2O$, k_{R7} (R7),

and by using $n_{th}(H_2O)=6\times10^{16}\ cm^{-2}$ for NaBr. Based on experimental dependence of γ on $[H_2O]$ in Fig.2 and analytical expression for γ_{hum}^{NaBr} given by eqn. like (7), elementary rate constant k_{R7} is calculated to be of $5.5\pm0.8\ s^{-1}$. The values of k_{R5} and k_{R7} proved to be virtually identical. This means that the rate of unimolecular decomposition of the $NO_3-NaX-H_2O$ surface complex is weakly dependent on a type of ionic lattice but is mainly determined by a structure and energetics of the very complex.

As follows from the data in Figure 2, the NO_3 uptake on dry and humidified NaBr gives different aggregate states of initial Br atoms. For dry NaBr, the NO_3 uptake is sure to occur with an allowed transition of the Br atom into the gas phase. But the allowance is quite miserable, i.e. ~ 6.9 kJ/mol. A slight variation in some parameters of uptake on humidified surface, e.g. heat of solvation, can lead to a forbidden transition of Br atom into the gas phase. This can lead in turn to a long-term uptake of Br atoms on the deep energy traps similar to those for Cl_{ad} atoms.

Acknowledgment. The authors are grateful to Dr. M.J.Rossi for valuable comments and useful discussions. The authors wish to thank the Russian Foundation for Basic Research for funding this work under grant № 00-03-32999.

References

Beichert P. and B.-J.Finlayson-Pitts; Knudsen cell studies of the uptake of gaseous HNO_3 and other oxides of nitrogen on solid NaCl: the role of surface-adsorbed water, *J.Phys.Chem.* **100** (1996) 15218-15228.

Brimblecomb P.; Air composition and chemistry, *Cambridge Univ. Press* (1986).

Dai D.J. and G.E.Ewing; Induced infrared absorption of H_2, HD, and D_2 physisorbed on NaCl films , *J. Chem Phys.* **98** (1993) 5050-5058.

Gershenzon M.Yu., S.D.Il'in, N.G.Fedotov, Yu.M.Gershenzon, E.V.Aparina, and V.V.Zelenov; The mechanism of reactive NO_3 uptake on dry NaX (X=Cl, Br), *J. Atmos. Chem.* **34** (1999) 119-135.

Gratpanche F. and J.-P. Sawerysyn; Uptake coefficients of NO_3 radicals on solid surfaces of sea salts, *J. Chim. Phys.* **96** (1999) 213-231.

Platt U.F., A.W.Winer, H.W.Bierman, R.Atkinson, and J.N.Pitts Jr.; Measurement of nitrate radical concentrations in continental air, *Environ. Sci. Technol.* **18** (1984) 365-369.

Rudich Y., R.K.Talukdar, and A.R.Ravishankara; Multiphase chemistry of NO_3 in the remote troposphere, *J. Geophys. Res.* **D103** (1998) 16,133-16,143.

Sander R. and P.J.Crutzen; Model study indicating halogen activation and ozone destruction in polluted air masses transported to the sea, *J. Geophys. Res.* **D101** (1996) 9121-9138.

Seisel S., F.Caloz, F.F.Fenter, and H. Van den Bergh; The heterogeneous reaction of NO_3 with NaCl and KBr: a nonphotolytic source of halogen atoms, *Geophys. Res. Lett.* **24** (1997) 2757-2760.

Spicer C.W., E.G.Chapman, B.-J.Finlayson-Pitts, R.A.Plastridge, J.M.Hubbe, J.D.Fast and C.M.Berkowitz; Unexpectedly high concentrations of molecular chlorine in coastal air, *Nature* **394** (1998) 353-356.

Timonen R.S., L.T.Chu, M.-T.Leu, and L.F.Keyser; Heterogeneous reaction of $ClONO_2(g)+NaCl(s) \rightarrow Cl_2(g)+$ $NaNO_3(s)$, *J.Phys.Chem.* **98** (1994) 9509-9517.

Zelenov V.V., E.V.Aparina, Yu.M.Gershenzon, M.Yu.Gershenzon, and S.D.Il'in; Kinetic mechanisms of uptake of atmospheric trace gases on the sea-salt surface. II. NO_3 uptake on NaX (X=Cl, Br) dry salts, *Chemical Physics Reports* **20** (2001) in press.

Global and Regional Air Pollution Studies

Air Pollution in Turkey

Tuncay DOGEROGLU

Anadolu University Environmental Engineering Department, Iki Eylul Campus 26470 Eskisehir-TURKEY
E-mail:tdoverog@anadolu.edu.tr

Introduction

As is the case with all environmental problems, two primary causes of air pollution in Turkey are urbanisation and industrialisation. Among the developments contributing to air pollution in the cities are incorrect urbanisations for the topographical and meteorological conditions, low quality fuel and improper combustion techniques, a shortage of green areas, and an increase in the number of motor vehicles. Industrial sources of air pollution in Turkey are mainly the fertiliser, iron and steel, paper and cellulose, sugar, cement, textile, petrochemical, pesticide and leather industries and power plants.

In order to protect the atmosphere, the Government of Turkey promotes policies and programmes in the areas of energy efficiency, environmentally sound and efficient transportation, industrial pollution control, and management of toxic and other hazardous waste. The Ministry of Environment (MOE) is in fact primarily responsible for decision-making related to protection of the atmosphere and the Ministry of Health (MOH) is responsible for transboundary atmospheric pollution control. Studies on air pollution have been carried out by the universities in Turkey and the Scientific & Technical Research Council (TUBITAK). Most of the studies have concentrated on mostly the chemical composition of atmospheric aerosols in rural sites, monitoring traffic-origined heavy metal pollution, industrial air pollution monitoring and emission inventories.

The Air Quality Control Regulation, which came into force on 2 November 1986, has not been revised in the light of the Agenda 21. According to this national legislation, parallel with German TA Luft 1985, much detailed technological sets of emission limitations are available for industrial activities. Presently, however the air quality limits (Table 1) are subject to a radical change by the MOE. For this reason of great interest in the international activities and trade, WHO criteria and EC directives were also being closely followed in many case studies.

Table 1: Air quality limits in Turkey in comparison to WHO, EC and U.S. regulations (AQPR, 1986; Elbir et al., 2000)

Air quality parameters	Turkish Limits ($\mu g/m^3$)*		WHO guidelines ($\mu g/m^3$)	EC regulations ($\mu g/m^3$)	U.S. regulations ($\mu g/m^3$)
	Short term	Long term			
NO_2	300 (200)	100 (80)	200 (1 h) 40 (annual)	50 (long term)	100 (annual)
SO_2	400 (250)	150 (100)	125 (24 hrs) 50 (annual)	125 (24 hrs) 50 (long term)	365 (24 hrs) 80 (annual)
CO	30000 (10000)	10000 (5000)	10000 (8 hrs) 30000 (1 h)	-	10000 (8 hrs) 40000 (1 h)
O_3	240 (1 h)	-	120 (8 hrs)	-	235 (1 h) 157 (8 hrs)
PM_{10}	300 (200)	150 (100)	-	80 (annual)	50 (annual) 150 (24 hrs)
(*)The values in the parenthesis indicate probable revision values					

I. Barnes (ed.), Global Atmospheric Change and its Impact on Regional Air Quality, 183–188.

Turkey signed the Geneva Convention on Long- Range Transboundary Air Pollution in 1979, the Montreal Protocol in 1991, and the London and Copenhagen Amendments in 1995. Although Turkey is not a party to the United Nations Framework Convention on Climate Change (UNFCCC), studies on the estimation of greenhouse gases which is an obligation to the parties of the convention have been started.
The project called "Establishment of National Environmental Database Project System to Prevent Pollution in Turkey" has started at 13 July 1999 by a protocol signed by the MOE and TAI Tusas Aerospace Industries, Inc. The website of this project (http:/www.ucvt.gov.tr/english/ TJeindex.html) covers all the information related with the proposed project including the searchable form of the products mentioned above and project forms to be filled on line (Balta, 2000).

General information about Turkey

Geographical properties. Turkey is located between Asia, Europe and Africa which are the old continents and acts as a natural bridge connecting these continents and also Black sea and Mediterranean Sea (Fig 1.). It has a total surface area of 77 797 127 ha and is bounded by the Black Sea in the north, by the Aegean Sea in the west, and by the Mediterranean Sea in the south. Because of its varying geographical conditions Turkey shows the characteristics of the big continent within its ecological structure, climate, plant cover and topology. About 35 percent of surface area is suitable for various types of agriculture.

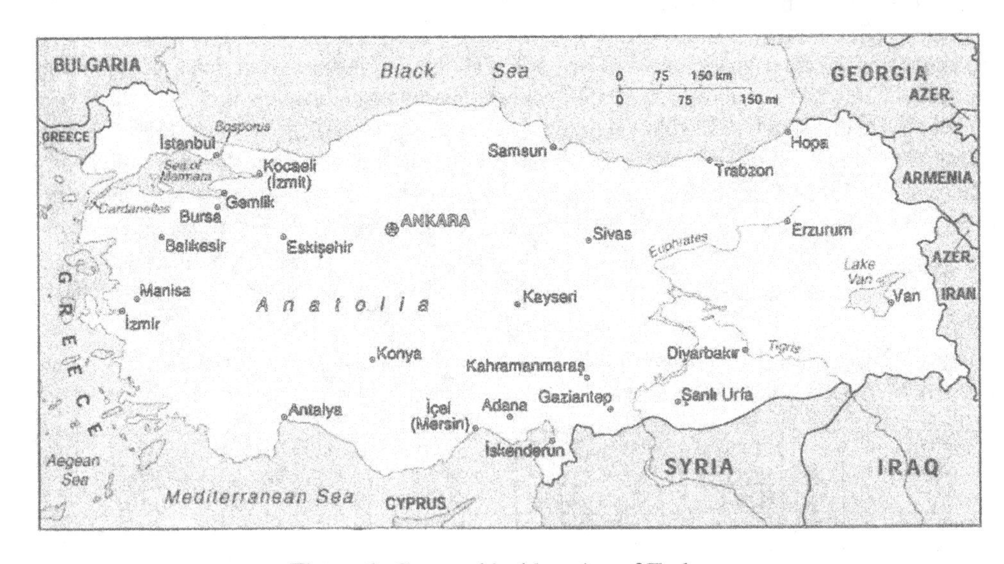

Figure 1: Geographical location of Turkey

Topological properties. The topological structure of Turkey is generally sloping, highlanded and mountainous. The slope of 62.5 percent of the land is greater than 15 percent.
Climatological properties. Turkey is generally located within the sub-tropical climatic line, winters being relatively mild and summers being long, warm and dry. The average annual temperature varies between 18-20 °C on the south coast, falls to 14-15 °C on the west coast, and finally in the interior areas, fluctuates between 4-18 °C. The south coast of Turkey is usually warm during winter; the mean temperature varies between 8-12 °C. The winters are

not severe on the north and west coast of Turkey, the mean January temperature varies from 5-7 °C. East Anatolian and interior parts of Turkey are subject to cold winters, because they are shielded from the moderating effect of the sea breezes by the coast line mountains. Average temperatures over these are between 0 and –10 °C in winter.
Turkey is subject both to continental type climate distinguished by dry summers. Generally, in Turkey heavy rainfalls are observed on the slopes of mountains facing seas. But, moving towards interior areas the rainfalls gradually becomes less. Autumn marks the start of rainy season which continues until late spring on the Marmara, Mediterranean and Eagean coast receives rain throughout the year. In the interior areas and Southeast Anatolia rainfall mostly occurs in spring. On the other, in the east of Turkey the winters are usually dry and rainfall occurs in spring and summer.
Demographical properties. The first reliable information on Turkey's population dates from 1927 when the country had a population of about 13.5 million inhabitants. From that time on, Turkey has experienced a very rapid increase of population, the lowest annual increase rate being observed in the period of 1940-1945 at about 10%. In 2000, population was 65 666 677 and the annual population growth rate was 1.27%.
Economy. Turkey has a dynamic economy that is a complex mix of modern industry and commerce along with traditional village agriculture and crafts. It has a strong and rapidly growing private sector, yet the state still plays a major role in basic industry, banking, transport and communication. Its most important industry-and largest exporter-is textiles and clothing, which is almost entirely in private hands.
Industry. Industrialisation is the main condition of achieving economical progress and development, both in developing countries. In Turkey, industrialisation policies gained importance following the introduction of five-year development plans when changes in social and cultural structures developed parallel to industrial development.
Energy consumption. Turkish energy consumption has risen dramatically over past 20 years. From just 1.0 quadrillion Btu in 1980, Turkey's domestic energy consumption has nearly tripled, reaching a level of 2.9 quadrillion Btu in 1998. Although this is still low relative to similar-sized countries such as Germany (13.8 quadrillion Btu), France (10.0 quadrillion Btu), and Poland (3.5 quadrillion Btu), Turkey's upward trend may mean it will surpass these countries in the future. Turkey's per capita energy consumption was 45.6 million Btu in 1998, compared to 350.7 million Btu in U.S., 177.3 million Btu in Russia, 168.6 million Btu in Germany, and 90.8 million Btu in Poland. Of Turkey's total energy consumption, fully 50% is used by the industrial sector, with residential at 27%, transportation at 16.4%, and commercial 6.7%. Oil accounts for 43.9% of this consumption, with coal at 26.7% and natural gas at 13.2% but rising. Although analysts have said that Turkey's continually increasing energy consumption is needed to power the country's developing economy, environmental critics believe that Turkey's economic policies have encouraged energy waste. Because the Turkish energy sector is mainly state-owned, critics charge that the government's pricing policy has encouraged the inefficient use of energy. Experts calm that about 22% of energy generated in Turkey is lost because of inefficient distribution and relay systems. In turn, they argue, this energy waste has necessitated the accelerated growth in energy demand and imports.

Emission Inventories for Turkey

Turkey's carbon emissions have risen in line with the country's energy consumption. Since 1980, Turkey's energy-related carbon emissions have jumped from 18 million metric tons annually to 47.1 million metric tons in 1998. Greenhouse gases emission inventories of Turkey have been estimated by State Institute of Statistics using by the IPCC (Intergovernmental Panel on Climate Change) guidelines.

According to the calculations of Turkey greenhouse gas emissions, CO_2 equivalents of direct greenhouse gases estimated as 200.7 million tonnes for 1990 and 271.2 million tonnes for 1997. Direct greenhouse gas emissions per capita were calculated as 3550 kg in 1990 and as 4340 kg in 1997; these values are below world and OECD average emission values per capita (Akcasoy and Onder, 1999).

Despite upward trends in recent years, Turkey still has the lowest energy-related CO_2 emissions per capita and energy consumption per capita among IEA countries. In 1998, Turkey's carbon emissions per capita were 0.7 metric tons (compared to the U.S. value of 5.5 metric tons).

Table 2 indicates that Turkish emission inventory for some pollutants with respect to main sector categories in 1995. When the total emissions were distributed between cities, Istanbul was shown to be responsible for one-fifth to one-fourth of all air pollutant emissions in Turkey although one-seventh of the Turkish population live in Istanbul (Elbir et al., 2000). Emissions of some European countries, Turkey and US. In 1990 and 1998 were given as shown in Table 3.

Table 2: Turkish emission inventory with respect to main categories (in 1995) (1000 Tonnes/y) (Elbir et al., 2000)

Sector	PM	SO_x	NO_x	NMVOC	CO
Domestic heating	849 (40.41%)	337 (18.96%)	63 (10.68%)	228 (34.00%)	259 (11.33%)
Industry-combustion	598 (28.45%)	451 (25.36%)	59 (10.01%)	6.6 (0.98%)	25 (1.09%)
Industry process	324 (15.44%)	143 (8.07%)	55 (9.47%)	239 (35.73%)	633 (27.71%)
Power generation	310 (14.73%)	846 (47.59%)	168 (28.71%)	17 (2.57%)	17 (0.77%)
Traffic	21 (0.97%)	-	241 (41.12%)	179 (26.72%)	1350 (59.10%)
Total	2101	1778	585	670	2285

Table 3: Emissions of some countries in 1990 and 1998 (1000 tonnes/y) EMEP, 2000; Akcasoy and Onder, 1999)

	France		Germany		Turkey		UK		US	
Pollutant	1990	1998	1990	1998	1990	1998	1990	1998	1990	1998
SO_x	1268	837	5321	1292	833	1288	3736	1615	21463	17622
NO_x	1877	1652	2709	1780	670	851	2788	1753	21815	22083
NMVOC	2535	827	3225	625	524	632	2445	350	18992	4477
CH_4	2870	827	5571	625	173	-	3677	350	31055	4477
NH_3	813	827	765	625	-	-	366	350	3929	4477

Air quality monitoring studies in Turkey

Since 1985 annual and monthly averages of SO_2 and particulate matter (smoke) (PM) from daily integrated monitoring network run by the MOH have been published officially by the Presidency of State Institute of Statistics in Turkey (DIE, 1992; DIE, 1995). Measurements were made with acidimetric titration for SO_2 and refractometric evaluations on 24-hrs integrated dust filter samples in accordance with WHO-recommended measurement methods. Currently, at 175 stations in 76 settlements the levels of SO_2 and PM in ambient air are being monitored under the organisation of MOH and precautions are being taken when necessary. Fig. 2 and 3 indicate that for SO_2 and PM trends for some metropolitan cities in Turkey between 1985 and 2000 respectively.

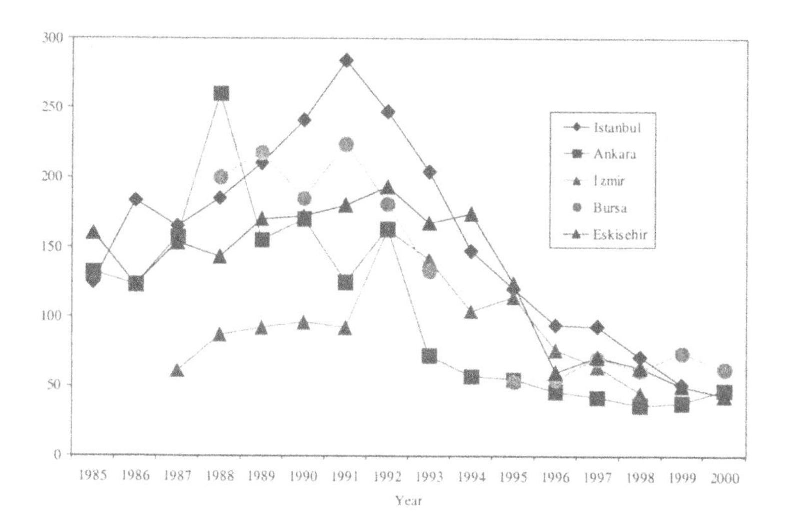

Figure 2: Annual SO_2 averages for some metropolitan cities in Turkey

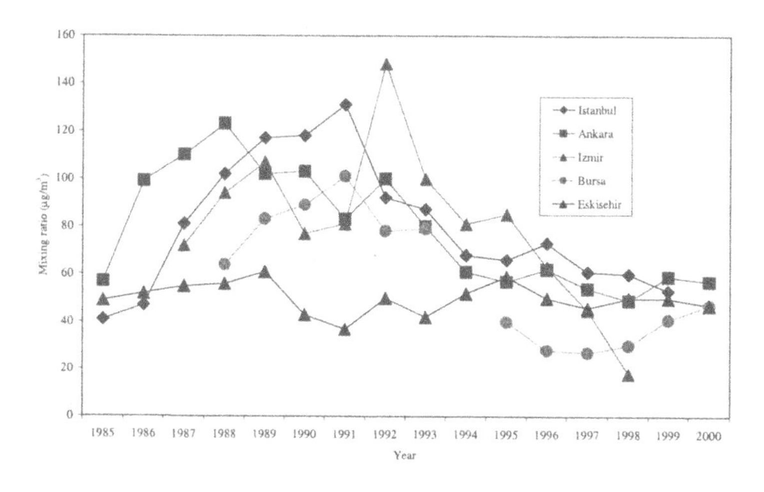

Figure 3: Annual PM averages for some metropolitan cities in Turkey

Besides monitoring of the local air pollution the regional monitoring activities within the framework of the Cooperative Programme for monitoring and evaluation of the Long Range

Transboundary Air Pollution, EMEP carried out at one station with measurements of 20 parameters since 1991.

In addition, concentrations of O_3, NO_x, SO_x and particulate matter were measured in the Uludag National Park, the highest mountain of west Turkey. The aims of these measurements were to track the transportation of the main air pollutants from Europe to Turkey and to investigate the influence of acid rain on the forest of the Uludag National Park (Ozer and Aydin, 1997).

Results and Discussion

For improving the air quality and better protection of community health and ecological balances some policies and programs are definitely needed:

- Air quality monitoring studies have to cover some other parameters such as nitrogen oxides, ozone and VOC for definition of air pollution originated from traffic.
- It is necessary to take some precautions for energy preserving (using central heating system, of renewing the inefficient energy distribution and relay systems etc.)
- It is necessary to improve bad quality coals which is used for energy production for domestic heating.
- Industrial air pollution controls have to be much more efficient. Recognising these issues, the Turkish federal government and municipalities have taken several measures.
- The Air Quality Control Regulation revising studies urgently has been finished.

References

Air quality protection regulation (AQPR), *Official Gazette* 19269, Ministry of Environment (MOE) Ankara (1986).

Akcasoy K. and Onder, F. 1970-2010 yillari arasinda Turkiye sera gazi emisyonlarinin istatistiksel degerlendirilmesi, in: Elbir T. Bayram A. (eds), *Hava Kirlenmesi ve Kontrolu Ulusal Sempozyumu*, Izmir 27-29 Eylul (1999) 15-22.

Balta, D.A.; Ulusal çevre veri tabani sisteminin olusturulmasi projesi, *IV.Cevre Surasi Tebligleri*, T.C. Cevre Bakanligi, Izmir 6-8 Kasim (2000) 181-207.

Bayar, A.B.; Activities on air pollution monitoring in Turkey, *Environmental Research Forum* **7-8** (1997) 635-640.

DIE (State Institute of Statistics Prime Ministry Republic of Turkey), *Environmental Statistics-Air Pollution*, Ankara (1992).

DIE (State Institute of Statistics Prime Ministry Republic of Turkey), *Environmental Statistics-Air Pollution*, Ankara (1995).

Elbir T., A. Müezzinoglu, and Bayram A; Evaluation of some air pollution indicators in Turkey, *Environment International* **26** (2000) 5-10.

Ozer U., R.Aydin and H. Akcay; Air pollution profile of Turkey, *Chemistry International* **19** (1997) 190-191.

http://www.bartleby.com/151/m245.html; Turkey. The World Factbook.2000

http://www.worldbank.org/depweb/beyond/global/chapter10.html; Beyond Economic Growth Meeting the Challenges of Global Development.

http://www.eia.doe.gov/emeu/cabs/turkenv.html, United states Energy Information Administration, Turkey: Environmental Issues, March 2000.

http://www.emep.int/emis_tables Tables of anthropogenic emissions in the ECE region. Pollutants: sulphur, nitrogen oxides, ammonia, NMVOC, carbon monoxide, methane, carbon dioxide, persistent organic pollutants, and heavy metals.

Background State and Trends of Sulfur Dioxide and Sulfate Aerosols in the Atmosphere over Russia.

V.I. Egorov, S.G. Paramonov, B.V. Pastuhov

Institute of Global Climate and Ecology
Glebovskaya 20-B, Moscow, 107258, Russia.

Introduction

Sulfur dioxide and sulfate aerosols play an important role in the atmospheric chemistry and radiation balance of the Earth's atmosphere (Mitchell *et al.*, 1995). The negative impact of sulfur dioxide and sulfate aerosols on the natural environment state in Europe are well known facts.

Sulfur dioxide is emitted to the atmosphere from anthropogenic (fossil fuel combustion, industrial emissions) and natural (volcanic activity, oxidation of biogenic sulfur compounds) sources. For the north hemisphere the contribution of anthropogenic sources of sulfur dioxide in their total emission is approximately 90%. The major sources of sulfate aerosols in the atmosphere are the processes of sulfur dioxide oxidation, erosion of the soil by the wind and emission of seawater aerosols (Cullis and Hirschler, 1980).

The increase in fossil fuel consumption for power production between 1950-1970 caused the growth of sulfur dioxide emissions to the atmosphere and pollution of large territories by the long-range transport of polluted air masses from urbanized areas to the background regions (Acidification..., 1982, Hill *et al.*, 1970).

The observed large scale negative effects of atmospheric pollution on vegetation, acidification of a natural media by acid deposition, forest damage in Europe made it necessary to organize in the 1970ies systematic studies of atmospheric pollution at background and remote regions.

Regular observations of background atmospheric pollution in the territory of the former USSR were begun in the 1980ies at the network of Integrated Background Monitoring System (IBMS). The development of the IBMS was within the framework of the Global Environment Monitoring System (GEMS) and the UNESCO Program "Man and Biosphere". The IBMS observational stations located at the biospheric reserves (BR) territories at the distance 80-100 km from the large industrial sources of atmospheric pollutants. Absence of local anthropogenic sources of air pollution as well as the absence of economical activity within the regions of the observational stations made it possible to assess the background regional level of air pollution (Izrael and. Rovinsky, 1992).

The working program of the IBMS stations includes measurements of sulfur dioxide, sulfate aerosols, nitrogen dioxide, heavy metals, polyaromatic hydrocarbons, organochlorine pesticides and precipitation chemistry.

Methods of observations

At the IBMS stations daily air sampling is conducted. The data on the daily-mean concentrations are later used for assessing the background state of atmospheric pollution in the form of monthly-mean, seasonal mean (April - September, October – March) and annual mean concentrations.

Significantly low background levels of sulfur dioxide and sulfate aerosols puts demands on the application of filter and impregnated solid sorbents for sampling. To define

I. Barnes (ed.), Global Atmospheric Change and its Impact on Regional Air Quality, 189–194.

the sulfur dioxide content in a sample the West-Gaeke method was used. Sulfate aerosols in the air are defined using air sampled on filters with subsequent measurement of sulfate concentrations by a turbidimetric analysis technique (Rovinsky and Wiersma, 1987).

Results of observations

In order to obtain more complete information on the air pollution by sulfur dioxide and sulfate aerosols outside of the urbanized territories (background pollution), analysis of their spatial distribution will utilize not only data from systematic observations of the IBMS stations, but also results of field experiments in remote districts located at extremes from anthropogenic pollution sources (e.g. polar areas, open regions of oceans). The results of the analysis of the obtained data allow estimates at regional and global levels of the background levels of sulfur dioxide and sulfate aerosols in air. The results of systematic and field observations of sulfur dioxide and sulfate aerosols are represented in Table 1.

The data, presented in the Table 1, give evidence on the irregularity of the spatial distribution of sulfur dioxide and sulfate aerosols in the lower atmosphere over the territory of Russia. The highest concentrations of measured substances are observed in a central part of the European territory of Russia (ETR), Kola peninsula. The high values of concentrations are caused by the influence of long-range atmospheric transport from large sources of sulfur dioxide located in these districts. The elevated levels of sulfate aerosols in a central part of the ETR can also be influenced by natural sources (soil erosion by wind in steppe territories).

The districts of Baikal Lake (east coast), polar territories of Asia and the Caucasus region are characterized essentially by lower levels of atmospheric pollution. It is worth noting, that a similar spatial distribution in the territory of Russia is observed for other anthropogenic contaminants (3,4-benzpyrene, lead aerosols, nitrogen dioxide). The results of the analyses of the monitoring data shows a high correlation relationship between sulfur dioxide and 3,4-benzpyrene at a cold season of a year for most of the monitoring stations.

The observed data of background contents of sulfur dioxide and sulfate aerosols collected by the authors in the open regions of Pacific and Indian oceans, testify to a non-uniform latitudinal distribution of defined substances. A moderate increase of sulfur dioxide and sulfate aerosols concentrations in the atmospheric boundary layer of the Pacific ocean (transect 50°N – 40°S, 160°E) is observed at the equatorial region (10°N-10°S) (Huebert *et al.*, 1998).

The most important source of sulfur dioxide input into the atmosphere above the oceans is the oxidation of dimethylsulfide (DMS), emitted from of the ocean surfaces. The subsequent oxidizing of sulfur dioxide up to sulfate, the processes of ocean water aerosols emission are the most important sources of sulfate aerosols in the atmosphere over open areas of oceans and seas.

Except for the districts of Barguzin (Baikal region) and Ust-Lena (Laptev sea) reserves, the regional background levels of sulfur dioxide and sulfate aerosols in the atmosphere over the territory of Russia exceeds acceptable levels.

Results of the analysis of long-term observations of sulfur dioxide and sulfate aerosols show that the inter-annual variations of the atmospheric contamination level have a scale from 4 to 6 months (warm and cold periods of the year) (Figures 1 and 2). The highest concentrations of sulfur dioxide are registered in the cold period of the year (October - March). The most probable reasons for the observed inter-annual variations of sulfur dioxide in the atmosphere are the increased consumption of fossil fuel used for heating during the cold period and a decrease of the rate of dry and wet deposition of this substance onto a underlying surface (snow) in comparison with the warm period of the year (ground, vegetation). The collateral influence of the indicated factors causes an increase of sulfur dioxide in the atmosphere during the cold period.

The data from the observations show, that the highest difference in sulfur dioxide concentration levels in warm and cold periods is registered in districts with higher degrees of urbanization (central and southeast districts ETR). In the districts of Barguzin, Ust-lena and Caucuses BR the inter-annual variability is not so significant.

In contrast to sulfur dioxide, the highest contents of sulfate aerosols in the atmosphere in the Russia territory is observed, as a rule, during the warm period of the year and is caused by the higher intensity of their sources (soil erosion by the wind, processes oxidizing sulfur dioxide). Inter-annual variability of sulfate aerosols is less in comparison with the respective alterations of sulfur dioxide.

One of the important tasks of background monitoring is the estimation of long-term changes of pollution levels in the atmosphere. The measurement data for the sulfur dioxide and sulfate aerosols content testify to the large scale decrease of their concentrations in air over the last 10-12 years (Figures 1, 2). The decreasing tendency of the level sulfur dioxide is also reflected in measurements for a number of Europe countries. On the basis of the presented data it is also worth noting the decrease of the inter-annual amplitude of the level changes for the period 1995-1999 (WMO, 2000).

The most probable reason for the decrease of the sulfur dioxide and sulfate aerosols background levels in air is the decrease in industrial production in Russia for the indicated period. According to available estimations, the total emission of pollutants into the atmosphere has decreased approximately by a factor of 2 over the last 10 years in Russia.

References

Acidification today and tomorrow. Stockholm: Swedish Ministry of Agriculture, Environment 82 Committee, (1982) 231 p.

Cullis C.F., M.M. Hirschler. Atmospheric sulfur: natural and man-made sources. *Atmos. Environ.*, **14** (1980) 1263-1278

Hill A., A.E. Heggestad, S.N. Lenson. Recognition of air pollution injury to vegetation. *Air Poll. Contr. Assoc.* (1970) 49-71.

Huebert B. J., S. Howel, P. Laj. Observation atmospheric Sulfur Cycle on SAGA-3. *Journ. Geoph. Res.* v.98, No. **9**, 16985-16997

Izrael Yu. A., F. Ya. Rovinsky. Integrated background monitoring of environmental pollution in mid-latitude Eurasia. WMO Publ., TD No.**434** (1992) 103 p.

Mitchell J.F.B., T.C. Johns, J.M. Gregory. Climate response to increasing levels of greenhouse gases and sulphate aerosols. *Nature*, **376** (1995) 501-504

Rovinsky F. Ya., G. B. Wiersma. Procedures and methods for integrated global background monitoring of environmental pollution. WMO publ., TD No. **178** (1987) 131 p.

WMO WDCCG Data Report 2000, Japan Meteorological Agency, WMO. No.**23**, 426-443.

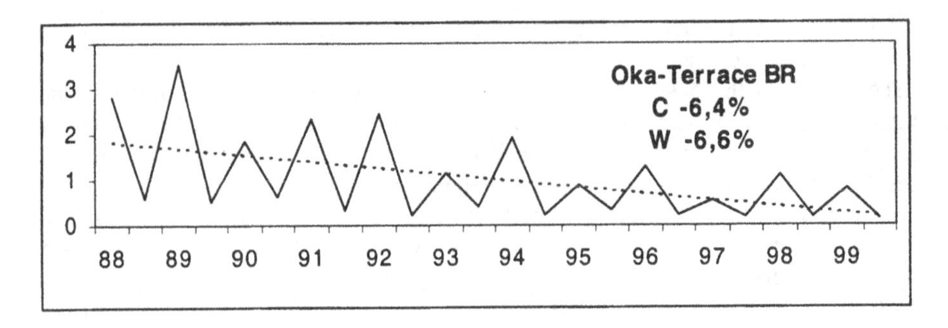

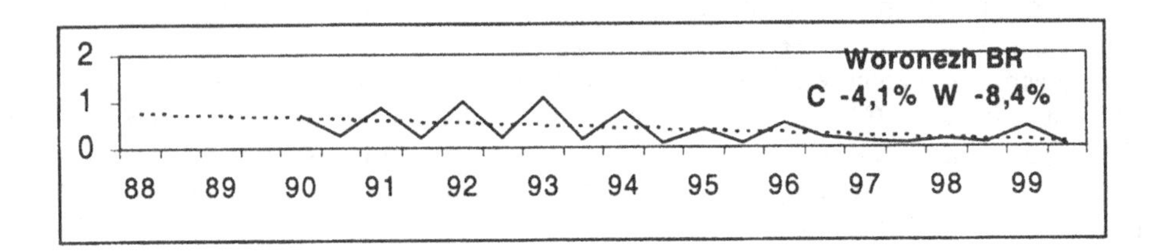

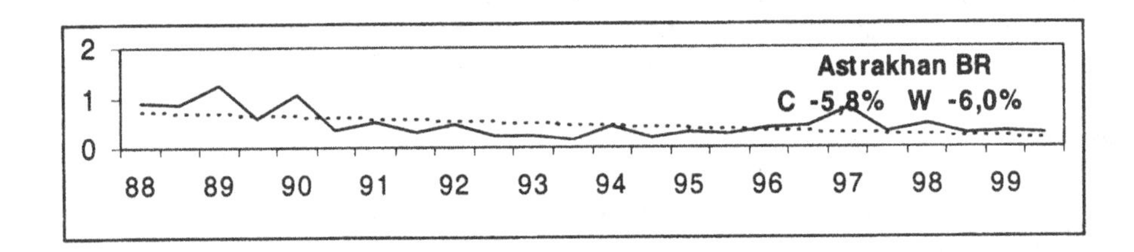

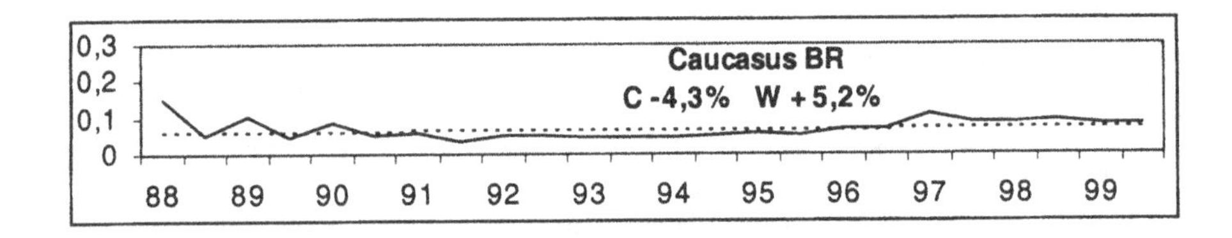

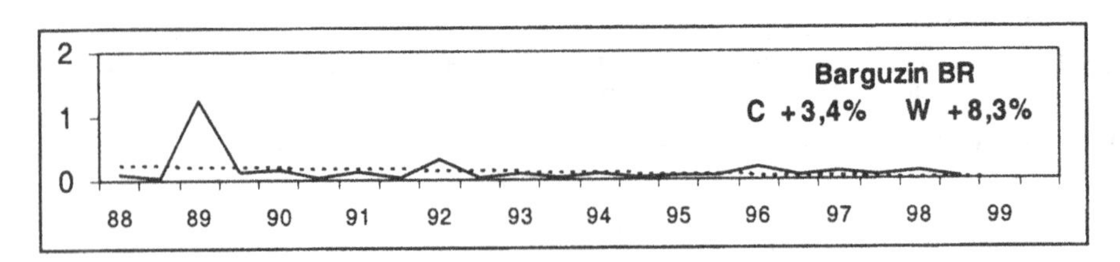

Figure 1: Seasonal variations and trends of sulfur dioxide (ug/m^3) in the atmosphere of background regions at cold (C) and warm (W) seasons.

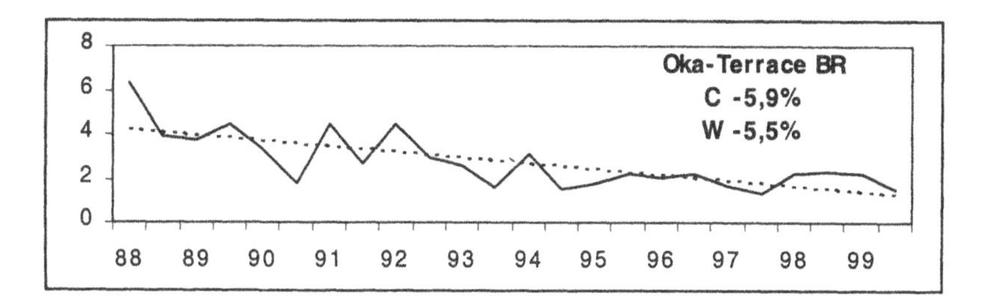

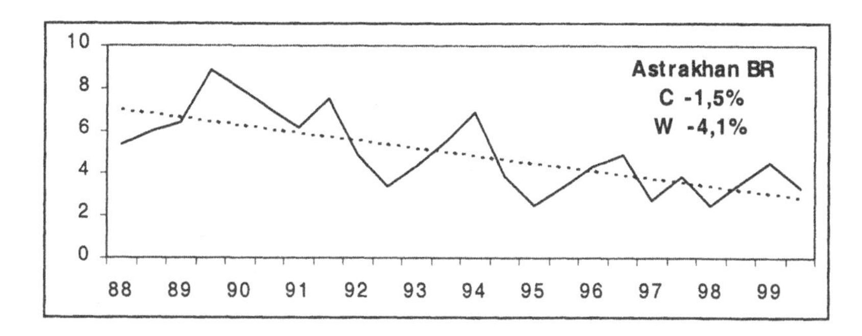

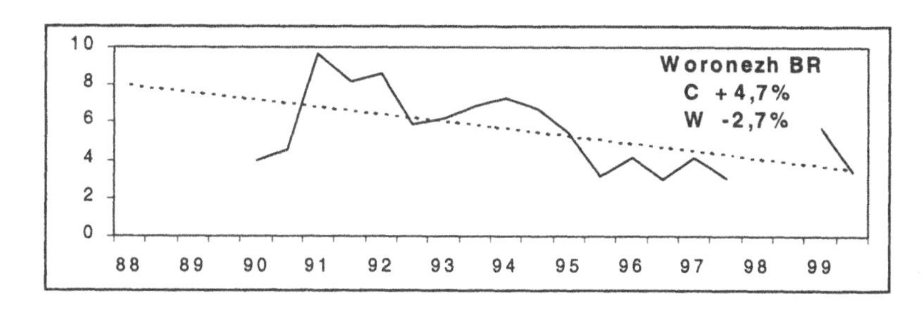

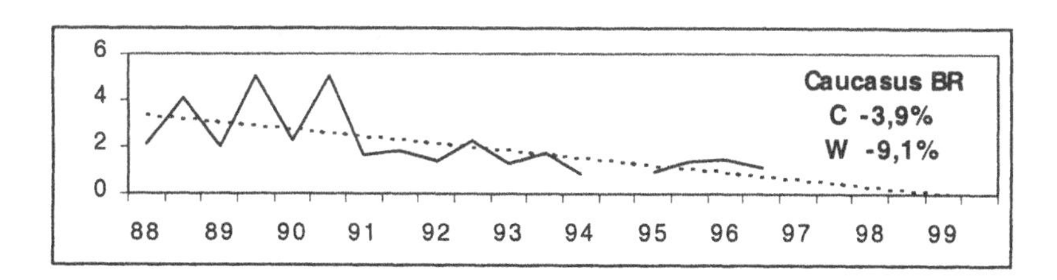

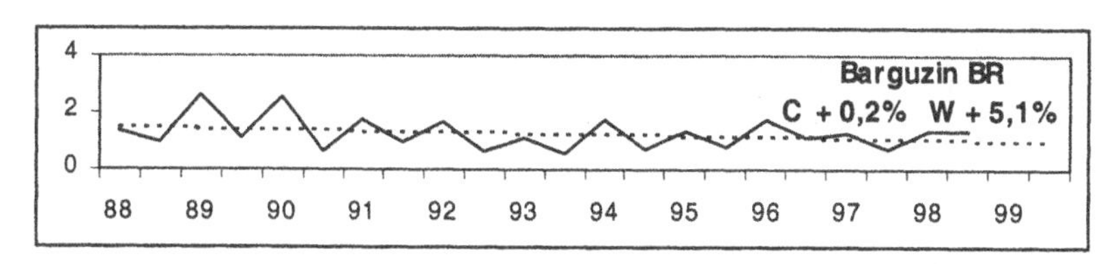

Figure 2: Seasonal variations and trends of sulphate aerosols (ug/m^3) in the atmosphere of background regions at cold (C) and warm (W) seasons.

Table 1: Background levels (μg/m^3) of sulfur dioxide and sulfate aerosols annual mean values in the atmosphere

SUBSTANCE	BIOSPHERIC RESERVE, POINTS AND REGIONS OF OBSERVATIONS									
	EUROPEAN PART OF RUSSIA						ASIAN PART OF RUSSIA		OCEAN	
	Astrakhan	**Woronezh**	**Caucasus**	**Oka-Terrace**	**Central Forest**	**Janiskosky**	**Barguzin (Baikal lake)**	**Ust Lena**	**Pacific**	**Indian**
	1987-1998	**1990-1998**	**1985-1998**	**1985-1998**	**1988-1993**	**1988-1997**	**1984-1998**	**1993-1994**	**1987, 1990**	**1987**
Sulfur dioxide	0.3-0.8	0.2-0.8	0.08-0.15	0.5-2.4	1.8-3	1.2-3.5	0.1-0.5	(0.05-0.1)*	(0.02-0.1)*	(0.01-0.06)*
Sulphate aerosols	4.0-8.1	4-7	1.8-4.0	2.2-5.6	4-7	1.2-2.1	1.0-1.6	(0.1-0.2)*	(0.3-3)*	(0.1-5)*

*) – *monthly - mean values.*

Coordinates of observational stations and sampling points.

Astrakhan BR	45°45'N, 47°55'E	Woronezh BR	51°54'N, 39°36'E	Caucasus BR	43°42'N ,40°12'E
Oka-Terrace BR	54°54'N 37°48'E	Central Forest BR	56°36'N 32°48'E	Janiskosky	68°56'N 47°55'E
Barguzin BR	54°12'N 109°30'E	Ust Lena (Dunai island)	73°56'N 124°30'E		

Changes in Land Use and Land Cover in Arid Southern Mediterranean on the Long Term Using Remote Sensing and GIS - A Case Study of Northwest Egypt

Gad, A.

National Research Center, Giza, Egypt

Abstract

The study area is located along the northern Mediterranean coast of Egypt between El-Hammam town, to the east and El-Alamean town, to the west. The area, from the seacoast to the Libyan plateau is composed of calcareous formation of Pliocene and Pleistocene age, covered by recent sediments. The wind and fluvial activities rework the sediments. Adverse human activities (i.e. grazing, collection of plants for fuel and erection of coastal resort villas) have undoubtedly hindered the natural development of the vegetation and the soils. The current research aims to detect the land use / land cover changes as indicators of land degradation on the long term.

A number of SPOT XS satellite images dated 1987, 1988, 1993, 1996 and 1997 were digitally enhanced and geometrically corrected. Different remote sensing techniques were applied for the whole scenes, including visual and unsupervised classification. Field investigation was carried out on bases of the preliminary image analysis. Hybrid classification technique was carried out for the images of 1988 and 1997 using both ground truth and spectral classes of a window area having long term data record.

It was possible to construct two reliable land use / land cover maps for the years 1988 and 1997. Comparing the two sets of data showed the saline and hyper saline areas, which were changed during this period. It was found that area of short individual plants has increased on expense of tall individual ones. Resort houses and artificial limestone quarries as well as fig orchards that are maintained by medium to high level agricultural practices, have replaced coastal dunes and other ridge components.

The European Commission (DGXII INCO DC) funded the CAMELEO project, in the context of which the current work was performed.

Introduction

The problem of aridity extends mostly to more than half of the Globe. Moreover, related environmental problems arise in response to variable physical, biological, socio-economic and political situations in different countries. The world is classified into six climatic zones: hyper-arid, arid, semi-arid, dry-sub-humid, humid and cold. The countries of the southern Mediterranean basin are the most severely affected by desertification because of their inherent ecological fragility and high population growth. The common problem to all arid zones is the fragility of the arid ecosystem and the consequent threat of desertification (Abdel-Razik, 1976; UNESCO, 1977; El-Raey, 1998).

Egypt is generally classified as hyper-arid area. However, the northern coast of Egypt represents the arid zone of the southern Mediterranean (UNEP, 1992). Large parts of this zone have been desertified during the past few centuries and the process is still converting many vulnerable areas into deserts (UNESCO, 1979; Ayyad and Le Floch, 1983). Remote sensing, with advantages of synoptic view, multi-temporality and multi-wavelength bands could be

I. Barnes (ed.), Global Atmospheric Change and its Impact on Regional Air Quality, 195–201.

used effectively to monitor the environmental variables and to manage the natural resources (Gad and Abdel-Samie, 2000).

The purpose of this article, in the context of the CAMELEO project, is to use the potentiality of merging ground and satellite observations to assess changes in land use and land cover on the long term. This may develop into an operational system for monitoring and assessment of desertification, through the integration and coordination of data collection, exchange, analysis and strengthening of scientific capabilities among CAMELEO partners.

Location of study area

The investigated area was chosen along the northern coast of Egypt delimited by latitudes 30° 45′ N at the south and the Mediterranean Sea coast at the north. It is bordered from the east by El-Hammam town and from the west by El-Alamein town (Figure 1). Khashm El-Eish ridge is the southern extent of the study area. El-Omayed observatory region is the core of the proposed area, where the UNESCO biosphere reserve is a reference. El-Omayed observatory is a part of the international ROSELT * network, having an area of about 5,000 hectares.

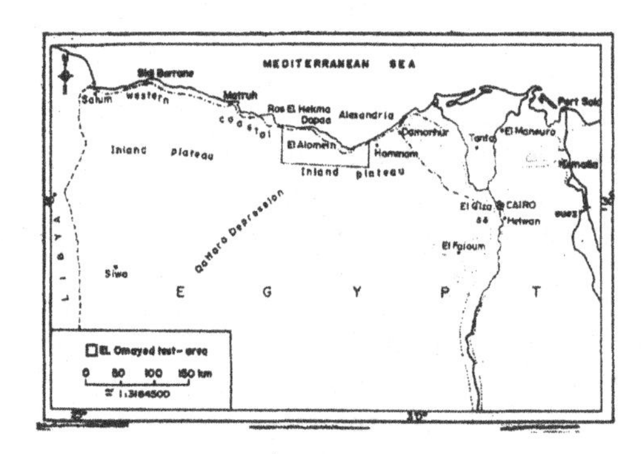

Figure 1. Location map of the study area.

Materials and methods

In this study five sets of SPOT images, covering the study area were used (Table 1). The images were digitally enhanced and geometrically corrected by applying several stretching processes. False Color Composites (FCC's) of five enhanced sub-scenes were created using the combination of bands 3, 2, 1 rendered in Red, Green and Blue, respectively. Unsupervised classifications were applied on the images (Schowengerdt, 1983). Different trials were carried out and eventually a number of 28 classes were chosen for the preliminarily interpretation.

Field investigation missions were carried out with the purpose of collecting ground truth information concerning landscape, soil, and vegetation. A number of 65 observation sites were studied, where different environmental parameters were described. Representative soil and water samples were also collected from different horizons for laboratory analysis.

The signature listing of different classes were studied carefully. Also, each class was displayed separately and a certain theme was assigned on bases of collected ground truth information and old thematic maps (Lillesand and Kiefer, 1987). Combining different classes resulted in the output classification having 11 classes for the images of 1988 through 1997. In order to figure out the changes that occurred in the land use/land cover of study area, a sub-set of El-Omayed observatory area, which has a long-term data record by ROSELLET network, was cut from both spring SPOT images of 1988 and 1997. Hybrid classification technique was generated, using both ground truth and spectral classes (Singh, 1989). A GIS model was developed, using ARC/INFO software to include both produced land use/ land cover maps of 1988 and 1997. Cross tabulation was elaborated, including area coverage of different classes to detect changes that occurred in a ten years period.

Table 1. Satellite Data Specification

No.	Type	Path	Row	Date	Bands
1	SPOT-XS	107	288	11/7/1987	1, 2, 3
2	SPOT-XS	107	288	03/4/1988	1, 2, 3
3	SPOT-XS	107	288	09/4/1993	1, 2, 3
4	SPOT-XS	107	288	10/8/1996	1, 2, 3
5	SPOT-XS	107	288	28/5/1997	1, 2, 3

Results and Discussion

Visual analysis of original SPOT images

The original false color composite images of the multi-temporal set reflect obvious changes in land use/land cover. Along the sea coast, construction of resort houses is going on as they did not appear in the image of 1988 and show the maximum coverage in 1997. Variations in the reflection of near infrared band are noticed and can be related to the seasonal variation in vegetation cover. The images of April 1988 and 1993 show the shade of red color, which is assigned to band 3, however image classification and NDVI elaboration are recommended for better understanding. Road network and channel courses are progressing from 1987 to 1997. Also, the coverage of limestone quarries and digging outputs increase with the time.

Unsupervised classification of SPOT images

The unsupervised classification could provide a preliminary legend for different land cover classes present in the study area. Various classes of land, water bodies and vegetation conditions are identified and displayed in the unsupervised classification images. Combining different classes resulted in the output classification, having 11 classes for the different images (Figure 2). Signatures listing of each produced image were studied by separating each class on the screen. It was found that class 1 is mostly related to "Deep water" whereas class 2 indicates the shallow water, as it is an extension of the previous class on the shoreline. The signature characteristics of class 3 are close to the signature of proposed "shallow water" and is located inland near the coast. Thus, "very shallow water" is then assigned. Sabkha areas are located around the proposed very shallow water, however it was possible to separate "wet sabkha" areas related to class 4 from "dry sabkha" areas, related to class 5. Class 6 could easily identify roads and settlements on the basis of its pattern and distribution. Class 7 indicates both sand beaches and limestone quarries as they have identical nature. Settlements are replacing calcareous sand beaches in recent images, whereas limestone quarries are increasing along Alexandria-Matrouh road. Classes 8, 9, and 10 reflect subclasses of vegetation cover. Combining the signature of band 3 and ground truth observation made it possible to assign "dense vegetation" to class 8, "sparse vegetation" to class 9 and "short trees" to class 10. Class 11 could be related to "bare soils", as it is characterized by less reflection in the near infrared band.

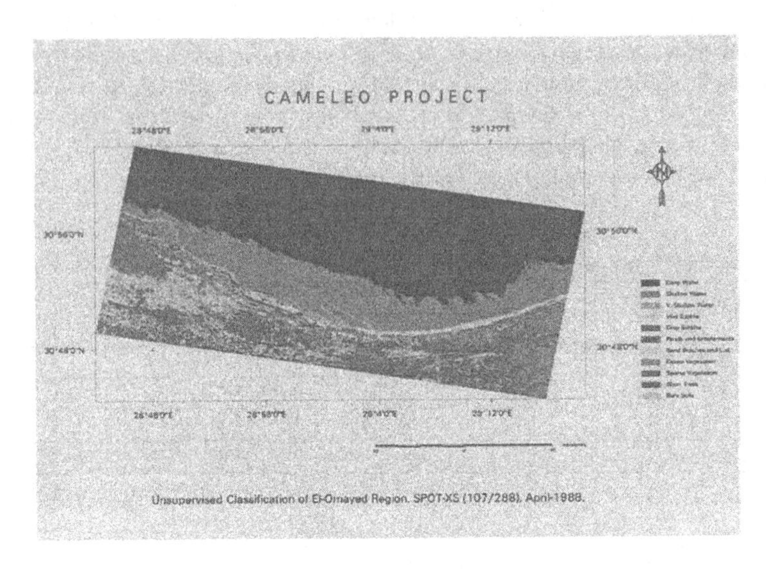

Figure 2. Unsupervised classification of SPOT XS image of 1997

Supervised classification of SPOT images

Unsupervised classification was carried out first with a choice of 60 classes, maximum number of iterations was 30 and the convergence was 0.99. The signatures corresponding to the 60 classes were compared with signature of ground truth sites using both histograms and mean plots in addition to image alarm. The supervised classification of the defined signatures was performed to produce the final classified image of El-Omayed 1988 and 1997 (Figure 3). The resulted supervised classes of both SPOT images 1988 and 1997 were carefully checked and evaluated on basis of recent field verification and data records of ROSELLET network. The following tuning actions were considered for the adjustment of the supervised classification output.

1- Beach sand should be included with Limestone as (land filling).
2- Wet Sabkha and dry Sabkha should be included with Halophytes as salt marshes and saline soil proper respectively.
3- Limestone class was found to include two different land cover types related to the calcareous structures along the coastal plain (i.e. consolidated dune rocky surfaces and coastal ridge). Both stabilized and consolidated dunes hold unique vegetation comprising special assemblage of plant species.
4- Shrubs 1 represent the type of vegetation characterized by plant species that grow as short individuals (average height is less than 50 cm) while having a relatively high land cover percentage (average of 25%).
5- Shrubs 2 represent the type of vegetation characterized by plant species that grow as tall individuals (average height exceeds 50 cm) while having a relatively low land cover percentage (less than 20%). It includes overgrazed grassland as indicated by numerous features (excessive trampling, dung, and poor growth of herbs and browsing of woody plants).
6-Halophytes (miss-interpreted as Fig trees 1): This class was originally intended to represent areas that include fig cultivation with the least agricultural practices. This type of plantation is

confined, in most cases, to the relatively high terrain within the elongated limestone ridges with contours of 10-20 m, which holds sparse vegetation type.

Conversely, this class should be further interpreted as one component of the class designated as "Degraded land" described later. This image class (renamed as halophytes) represents a special type of vegetation that inhabit saline depressions and salt marshes proper and has one of the highest plant cover in the area which exceeds 50%. Also included is the vegetation that tolerates saline soil types. It includes plants in the form of relatively large shrubs forming sand mounds and hammocks (reflectance may approximate that of fig plantations) and/or halophytic plants growing on the saline and hyper-saline soil.

7- Fig trees 2 refer to land areas occupied by appreciable fig orchards that are maintained by medium to high levels of agricultural practices. This type of plantation is common in the vicinity of local villages proximate to Bedouin houses and around provided by logistic facilities. Such fig orchards are more obvious as strips on the calcareous sand slopes of the coastal ridges bordering on the saline depression.

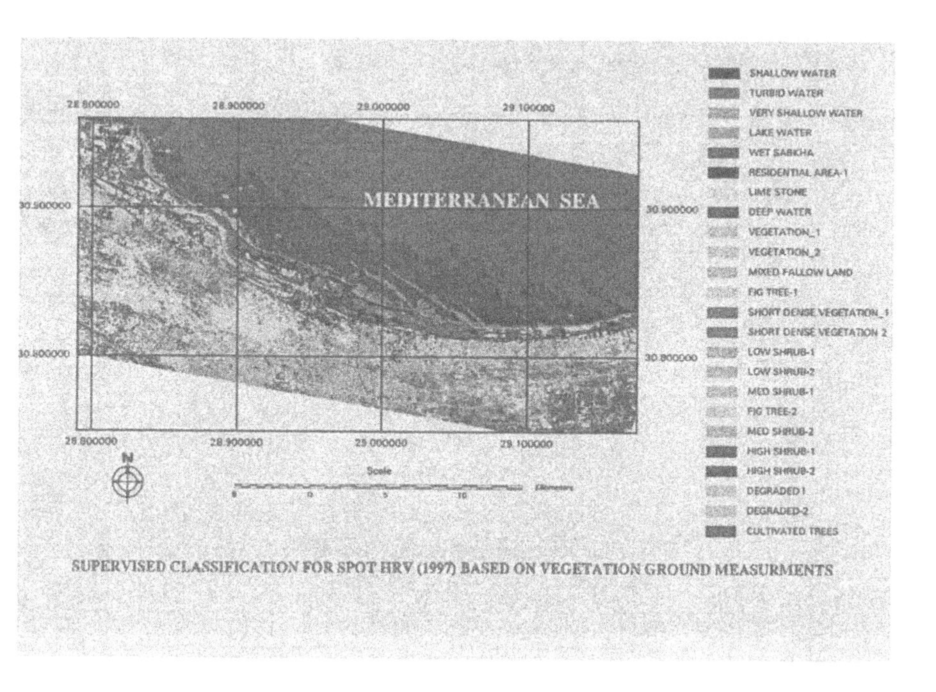

Figure 3. Supervised classification of SPOT image, 1997.

Change detection output maps and cross-tabulation 88/97

Cross tabulation technique resulted in detecting the changes during ten years difference. Table (2) presents the outcomes of area measurements in km^2. The following conclusions can be outlined;

1. The total area classified as "Halophytes" in 1988 (including Wet Sabkha, Dry Sabkha and Fig tree 1 in the pre-tuned classes) covers about 10.47 km^2 of which 2.46 km^2 represents saline soil proper. The coverage of this class increased in 1997 to about 14.10 km^2, one third

of which is devoted to saline soil proper (saline and hyper-saline soils) whereas the remainder represents less saline areas.

2. The coverage of the class "Shrub 1" increased from 3.962 km^2 in 1988 to 16.18 km^2 in 1997 on the expense of mainly the classes of "Shrub 2" (8 km^2), "Degraded" (4 km^2) and "Halophytes" (3 km^2).

3. The coverage of the class " Shrub 2" decreased from 62.98 km^2 in 1988 to 42.89 km^2 in 1997 mainly into the classes of "Degraded" (9 km^2), "Shrub 1" (6 km^2) and "Halophytes" (3 km^2).

4. The coverage of the class " Limestone" decreased from 12.96 km^2 in 1988 to 5.26 km^2 in 1997 mainly into the classes of "Settlements" (2.7 km^2), "Halophytes" (1.6 km^2) and "Burrowing" (1.1 km^2).

5. The coverage of the class "Settlements" increased from 1.51 km^2 in 1988 to 3.32 km^2 in 1997 mainly in the class of "Limestone" (1.31 km^2).

Table 2. Change detection in Land Use/Land Cover during the period 1988 -997

Classes	Unclassified	Deep Water	Shallow Water	Wet Sabkha	Dry SabkhaS	Shrubs (1)	Shrubs (2)	Fig trees (1)	Fig Trees (2)	Degraded Lands	Settlements	Beach Sands	Limestone	Total
Unclassified	41.28	0.00	0.00	0.00	0.00	0.00	0.00	0.00	0.00	0.00	0.00	0.00	0.00	41.28
Deep Water	0.04	32.49	10.59	0.00	0.00	0.00	0.00	0.00	0.00	0.00	0.00	0.00	0.00	43.12
Shallow Water	0.00	0.78	8.52	0.08	0.10	0.00	0.00	0.00	0.00	0.00	0.02	0.05	0.09	9.64
Wet Sabkhas	0.00	0.00	0.11	0.15	0.07	0.06	0.37	0.23	0.09	0.35	0.27	0.02	0.92	2.63
Dry Sabkhas	0.00	0.00	0.01	0.01	0.01	0.03	0.27	0.16	0.04	0.32	0.09	0.00	0.72	1.68
Shrubs(1)	0.03	0.00	0.00	0.00	0.00	0.80	8.01	2.96	0.12	4.08	0.01	0.00	0.17	16.18
Shrubs(2)	0.17	0.00	0.00	0.00	0.00	1.92	34.56	0.66	0.12	5.34	0.00	0.00	0.11	42.89
Fig Trees(1)	0.02	0.00	0.00	0.01	0.03	0.32	3.39	2.76	0.43	2.10	0.07	0.00	0.68	9.79
Fig Trees(2)	0.00	0.00	0.00	0.00	0.00	0.07	0.63	0.29	0.01	0.61	0.00	0.00	0.14	1.76
Degraded Lands	0.07	0.00	0.00	0.00	0.00	0.65	14.69	0.28	0.02	3.99	0.01	0.00	0.58	20.28
Settlements	0.00	0.00	0.07	0.08	0.09	0.04	0.37	0.24	0.11	0.38	0.56	0.02	1.40	3.32
Beach Sands	0.00	0.00	0.27	0.18	0.13	0.01	0.05	0.06	0.05	0.10	0.09	0.03	1.08	2.05
Limestone	0.01	0.00	0.03	0.04	0.09	0.01	0.06	0.06	0.04	0.37	0.05	0.03	4.47	5.26
Lake Water	0.00	0.14	1.77	1.08	0.19	0.01	0.00	0.01	0.04	0.00	0.04	0.03	0.19	3.51
Gardens	0.00	0.00	0.02	0.03	0.06	0.01	0.21	0.20	0.01	0.28	0.26	0.01	1.31	2.40
Digging Output	0.00	0.00	0.01	0.01	0.01	0.02	0.38	0.10	0.01	0.36	0.04	0.00	1.11	2.05
Total	41.63	33.41	21.38	1.68	0.78	3.96	62.99	8.01	1.07	18.27	1.51	0.19	12.96	207.85

Conclusions

The previous findings make it possible to conclude that the halophytic vegetation is expanding in the study area. This expansion can be attributed to the increase in water table level in soil especially those of the low land. Also, tall shrubby vegetation is in a process of replacement by shorter, but denser, type of vegetation (smaller individuals or species). The coastal dunes and ridge components are severely diminishing, having, in its turn, negative impact on the ecological system.

Acknowledgement

The authors would like to thank CAMELEO project team for helping with image processing, field surveys and interpretation.

References

Abdel Razik, M.S., 1976. A study on the vegetation composition, productivity and phenology in a Mediterranean desert ecosystem (Egypt). M.Sc. Thesis, Alex. Univ. pp. 89.

Ayyad, M.A. and E. Le Floch. 1983. An Ecological Assessment of Renewable Resources for Rural Agricultural Development in the Western Mediterranean Coastal Region of Egypt, Case Study: El Omayed Test-Area. Published with the financial assistance of the French Ministry of Foreign Affairs and the Staff and Facilities of the Mapping Workshop of C.N.R.S./C.E.P.E.L. Emberger, Montpellier.

El-Raey (1998). Strategic Environmental Assessment for Fuka - Matrouh, Egypt. PAP/RAC37-1995

Gad, A. and Abdel-Samie, A.G. (2000). Monitoring and evaluation of desertification processes in the Fayyoum depression, Egypt, Proceedings of "The third conference on desertification and environmental studies beyond the year 2000, Riyadh, Saudi Arabia, November 30 t0 December 4, 1999 ", pp. 83, King Saud University and UNEP: Sponsor.

Lillesand, T.M. and R.W. Kiefer, 1987. Remote Sensing and Image Interpretation (Second edition). John Wiley and Sons, New York.

Schowengerdt, R. A., 1983. Techniques for image processing and classification in remote sensing. Academic Press, INC, Orlando San Diego New York London Toronto Montreal Sydney Tokyo, pp. 249

Singh, A., 1989. Digital change detection techniques using remotely sensed data, International Journal of Remote Sensing, Vol. 10, No. 6, pp. 989-1003. Teillet, P.M., 1986. Image correction for radiometric effects in remote sensing, International Journal of Remote Sensing, Vol. 7,No. 12, pp. 1637-1651.

UNEP, 1992. World Atlas of desertification. Edward Arnold. A Division of Hodder Stoughton: London, New York, Melbourne, Auckland, pp. 69.

UNESCO, 1977. Development of Arid and semi-Arid Lands: Obstacles and Prospects. *MAP Technical Notes 6*. UNECSCO, Paris.

UNESCO, 1979. Trends in research and in the application of science and technology for arid zone development. *MAB Technical Notes 10*, UNESCO, Paris.

Investigation of PAH in Atmospheric Aerosols and Precipitation's of East Siberia

Alexander Gorshkov and Irene Marinaite

Limnological Institute, SB RAS, Ulan-Butorskaya st.3, Irkutsk, 664033 (Russia)
E-mail: agg@lin.irk.ru

Abstract -The evaluation of possible levels of total concentrations of polycyclic aromatic hydrocarbons (PAH) and of individual combinations of this compound class in aerosol, snow cover in industrial centres, unpopulated and background areas of Pre-Baikal and Lake Baikal Southern shore has been performed. A strong seasonal dependence of PAH concentration levels in the chemical composition of the aerosol is found. The local character of the pollution of snow cover with PAH is evident. Analysis of PAH was carried out by High Performance Liquid Chromatography (HPLC) with multiwave photometric detection.

Keyword – PAH, aerosol, snow cover, pollution, Eastern regions of Siberia, HPLC

Introduction

Eastern regions of Siberia play an important role in the formation of global aerosol. However, the experimental data on aerosol and precipitation chemical composition from this region are inadequate. The organic fraction, in particular, PAH (polycyclic aromatic hydrocarbons) has practically not been studied. For Siberia and Siberian cities, besides traditional PAH sources – metallurgy, processes of thermal and electric energy production, transport – other sources of emission of these compounds are characteristic. These are energy sources of small capacity (< 5 GW) using inefficient coal burning technologies (in Winter) and forest fires (in Spring and Summer). Taking into account the source intensities, the contribution of PAH into the organic aerosol fraction may be significant both at local level and regional scales. In order to determine the role of PAH in the aerosol chemical composition, we have carried out the following studies: i) determination of the concentration levels of PAH in aerosol in the areas of the emission sources; ii) seasonal variations in concentration levels of PAH in aerosol; iii) levels and rates of accumulation of PAH in the snow cover of industrial centres, unpopulated and background regions. The results obtained are published in the following original papers (Gorshkov *et al.*, 1996; Gorshkov *et al.*, 1998; Gorshkov and Marinayte, 2000; Khodger *et al.*, 1999; Koroleva *et al.*, 1998; Marinayte and Gorshkov, 2000; Raputa *et al.*, 1998). The present paper is an overview of the papers cited, it allows conclusions to be made on the role of PAH in the chemical composition of Eastern Siberia aerosol.

Experimental

Sampling. Aerosol and snow samples were collected in Irkutsk, Shelelhov, Slyudyanka and Baikalsk cites, Listvyanka, Tankhoi, Babushkin and Boyarsky villages, monitoring station background monitoring in Mondy (Figure 1) during the periods December 1993 – August 1999 (aerosol samples), in February- March 1994-1999 (snow samples). Aerosol was sampled for 4-24 h onto Whatman-41 paper filters (Great Britain), Schllicher & Schnell glass wool filters (Gegmany) and filters with Al_2O_3 (Russia), as well as the combination of filter and Octadecyl, Baker cartridges. A pump (Germany) with a rate of 4 m^3/h was used for sampling. The volume of the pumped air was measured with a Schlumberger, Gallus 2000 controller (Germany). A probe with a filter was set at 2 to 10 m above the ground. Snow samples were

I. Barnes (ed.), Global Atmospheric Change and its Impact on Regional Air Quality, 203–208.

collected within the precincts of the towns, settlements and outside them along the coast, on ice of Lake Baikal and background regions. Snow samples were taken as cores with a base area of 200x200 mm, with no less than 2-3 cores from each point of sampling. The snow samples were placed in polyethylene pockets and kept at -15 ÷ - 20 ^{0}C.

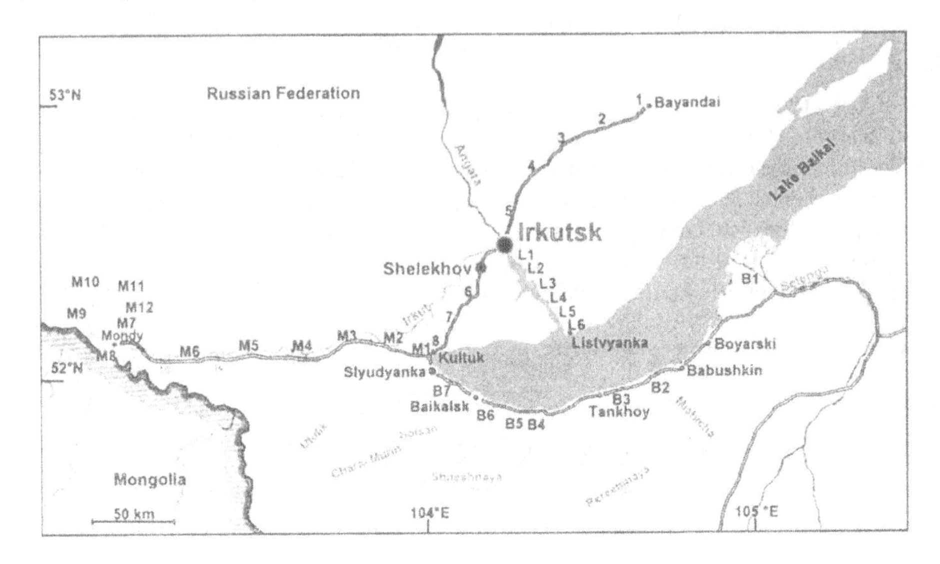

Figure 1. Sampling sites

PAH quantification. The method of quantification was based on extraction of PAH from the filters (aerosol samples and snow samples) and the filtered snow water by *n*-hexane. Extracts with PAH were evaporated on a rotary vacuum evaporator, the residues were dissolved in methanol and the solutions were analysed by HPLC (Gorshkov *et al.*, 1998; Gorshkov and Marinayte, 2000). Identification and quantitative determination was done using a HPLC method for 12 PAHs including in the list of priority pollutants: phenanthrene, anthracene, fluoranthene, pyrene, benz[a]anthracene, chrysene, benzo[b]fluoranthene, benzo[k]fluoranthene, benzo[a]pyrene, dibenzo[a,h]anthracene, benzo[g,h,i]perylene, indeno[1,2,3-c,d]pyrene.

Application of a micro-bore column (2x75 mm; Nucleosil 100-5, C18, PAH; Macherey-Nagel, Germany) and the chromatograph "Milichrom A-02" (EcoNova, Russia) allows significant increase of sensitivity and efficiency of the analysis with the total error of measurements 10 % within concentration range of individual PAH in the snow cover (0.4 – 80 μg/m^2) and the aerosol (0.3-450 ng/m^3). Simultaneous multi-wave detection (at 250, 260 and 290 nm) provides the correctness of the peak identification and the precision of quantitative measurements.

Results and Discussion

PAH in the atmospheric aerosol. Strongly marked seasonal variations were found in both the solid phase of aerosol and the relative content of PAH in it. The maximum concentration of the solid phase occurs in spring, whereas the maximum content of PAH in it is observed in winter. At the concentration of the solid phase of aerosol from 40 to 260 μg/m^3, the content of PAH in the examined samples does not exceed 0.3 %. The determined levels of the PAH concentration in the samples of atmospheric aerosol collected in different districts of the southern coast of Lake Baikal and Irkutsk, as an example of an industrial centre in eastern regions of Siberia, for three years are given in the Table 1.

Table 1. The PAH concentration levels in the solid phase of the aerosol

#	Snow-sampling points	Total PAH, ng/m^3	Benz[a]pyrene, ng/m^3
1	Irkutsk city	<0.3-300	0.3-30
2	Slyudyanka city	5.6-110	0.3-20
3	Baikalsk city	<0.3-3.8	<0.3
4	Lystvyanka village	1.6-56	<0.3-5
5	Tankhoi village	1.0-5.9	<0.3-0.5

The amounts of PAH from 25 to 300 ng/m^3 are typical of the winter period of Irkutsk city (total concentration of all identified compounds). In summer, the total PAH concentration in the atmospheric aerosol is 20 to 100 times lower. During forest fire periods, which result in drastic emissions of PAH into the atmosphere, the concentrations of these combinations in the atmospheric air can reach 15- 20 ng/m^3 (Irkutsk city, may 1996).

As is seen from Figure 2, the PAH concentrations in areas of the southern coast of Lake Baikal were appreciably different. The amounts of PAH and relations between them depend significantly on the sampling site, i.e., on the local sources of pollution. Relatively low concentrations of PAH in the aerosol (and snow cover) near Baikalsk are particularly worth noting. The Baikalsk Paper and Pulp Plant burns tremendous amounts of different fuels (coal, lignuine, petroleum products and bark) in pulp production. The observed low level of PAH may be caused by a combination of efficient purification of gaseous emissions, emissions of products into high atmosphere layers and dispersion over large areas.

Three compounds prevail in the total mass of PAH identified in the solid phase of the aerosol sampled in the winter period. They are phenanthrene, pyrene and fluoranthene. Their total amount reaches a half of the total mass of detected PAH, whereas the relative concentration of benz[a]pyrene is 4-10 % (4-22 ng/m^3, Irkutsk town). Among the recovered PAH in the aerosol sample collected in summer time (Irkutsk town, Listvyanka village) the major compounds were benz[a]pyrene, benzo[g,h,i]perylene and indeno[1,2,3-c,d]pyrene.

Significant difference in the PAH concentrations in winter and summer is connected with the following: i) heat sources operate more intensely in the cold season; ii) the anticyclonic weather is established in the region during November until March; this weather is characterised by weak winds, strong near-ground temperature inversions and frequent fogs that favour retention of the high level of atmospheric pollution; iii) the phase equilibration between the content of PAH in the solid and gas phase of aerosol shifts toward the latter as the ambient temperature increases from –20 to 25 ^{0}C; as result, PAH transfers fully (phenanthrene and anthracene) or partially (fluoranthene and pyrene) into the gaseous component of the atmospheric air.

PAH in the snow cover. Determination of the PAH was carried out in liquid and solid phases of snow cover. In snow water, a small number of PAH were found: phenanthrene, fluoranthene pyrene and sometimes anthracene and chrysene with concentrations varying from <5 to 300 ng/l. Concentrations of these compounds in the snow cover samples collected in the urban zones are significantly higher and are (in ng/l): phenanthrene – 400-8500, fluoranthene – 500-3700, pyrene – 350-2800, anthracene – 5-100. In the solid phase of snow water, i.e., in solid particles with sizes ≤ 6 nm (up to 98 % of their number), the maximum number and amount of accumulation PAH were determined. The total content of phenanthrene, fluoranthene and pyrene in all samples was more than 50-80 % of the amount of detected compounds.

Table 2. Concentration levels and rates of accumulation of PAH

#	Snow sampling points	Total PAH, mg/m^2	Benz[a]pyren, mg/m^2	Benz[a]pyren, mg/m^2 week
1	Irkutsk city	89-840	<0.4-21	<0.02-1.31
2	Shelekhov city	1400-15000	2.8-770	0.17-48
3	Slyudyanka city	120-160	1.3-5.0	0.08-0.31
4	Baikalsk city	19-21	0.4-0.8	0.02-0.05
5	Lystvyanka village	60-120	4.6-7.4	0.29-0.46
6	Tankhoi village	32-75	0.7-2.4	0.04-0.15
7	Babushkin village	15-28	0.8-1.7	0.05-0.11
8	Boyarsky village	22-28	3.0-3.6	0.19-0.22
9	Irkutsk-Bayanday profile	4.1-21	<0.4-1.0	0.02-0.06
10	Irkutsk- Slyudyanka	42-52	1.5-2.4	0.09-0.15
11	Irkutsk-Lystvyanka profile	13-98	<0.4-3.0	0.02-0.19
12	Southern shore of lake Baikal	11-98	0.5-3.0	0.03-0.19
13	Kultuk-Mondy profile	0.7-17	<0.4-0.4	<0.02-0.02

An extremely high accumulation value of PAH (1400-14500 μg/m^2) was found near Shelekhov sity (Irkutsk Aluminium Plant). On the Southern shore of lake Baikal local zones of relatively high accumulation of PAH were identified in the territory of the cities Slyudyanka, Baikalsk, Babushkin and the villages Kultuk, Tankhoy, Boyarsky (Table 2). Far from settlements, the PAH accumulation sharply decreased. In the profile points of unpopulated areas these values did not exceed 0.7-20 μg/m^2 (Irkutsk-Bayanday profile, Kultuk-Mondy profile).

Estimation of accumulation levels and spread areas of PAH. Mathematical models that were developed for determining the spread of anthropogenic suspended solid pollutants in the surface layer of the atmosphere can be used for forecasting the regional dispersion of PAH from their emission sources based on quantitative detection of PAH on solid particles of the snow cover (up to 70-90 %). Such models demonstrated that in winter, under conditions of the Siberian anticyclone, solid particles of gas emission settle mainly near their sources. The north-west winds transfer of the air mass along the Angara river can contribute to the air pollution at the coast of lake bringing pollution from the industrial areas of the Angara region as the air pollution there is rather high. Winds with the monsoon component in the Southern Baikal region can transport pollutants from the emission sources towards Lake Baikal.

Experimentally established levels of PAH accumulation agree convincingly with the results of the mathematical simulation. The PAH accumulation is, in fact, bounded by the precincts of the town. As seen from Figure 2, there is a sufficiently reliable correlation (r = 0.9879) between the mean PAH concentration in the aerosols of the ambient air in winter and their accumulation level in the snow cover. So the level of PAH accumulation in the snow cover, as an integral characteristic, permits us to estimate the mean value of the surface aerosol pollution. Taking into account the fact that in winter PAH are almost completely associated with hard particles aerosol (up to 95 %), determination of the PAH accumulation in the snow cover yields an estimation of pollution of the ambient air by these compounds.

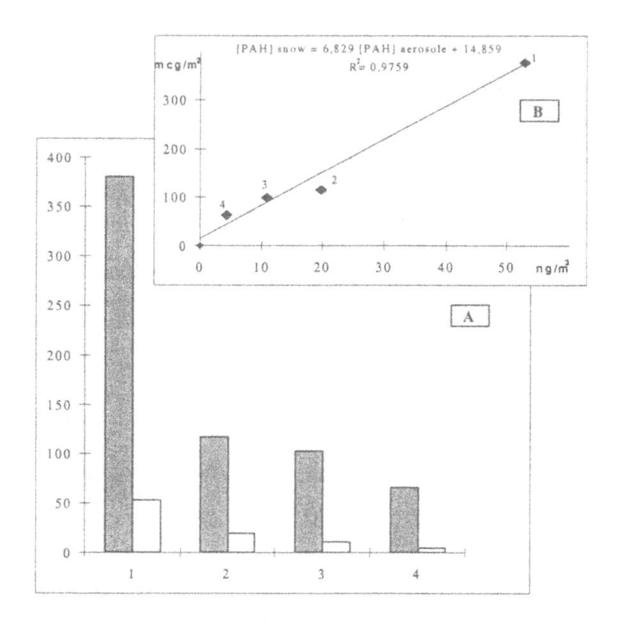

Figure 2. Average concentrations of PAH the aerosol and in the snow cover (A) and their interdependency (B)

Profile points in unpopulated areas at the Southern shore have relatively high PAH accumulation levels compared to the Irkutsk–Bayanday and Kultuk–Mondy profiles (Table 2) probably as a result of local transfer from industrial centres and (or) spraying out of the emissions from the Baikalsk Puip and Paper Plant. The small value of PAH concentrations in snow samples collected on ice of the lake near Slyudyanka and Baikalsk is more likely caused by shorter accumulation time and instability of snow cover on ice as compared with coast than by the absence of the PAH dispersion in this direction (PAH concentrations on ice of the lake is 0.02-2 μg/l, in the territory of Tankhoi – 0.5-1.9 μg/l).

A comparative estimate of pollution with PAH in Eastern regions of Siberia is possible with the use of two parameters: i) concentration of benz[a]pyrene in the ambient air as a "carcenogenity" index; and ii) the PAH accumulation rate in snow cover. The concentration levels of benz[a]pyrene in Irkutsk (maximum concentration at 30 ng/m^3) correspond to atmospheric pollution in large industrial centres. In the Southern coast of Lake Baikal, near Slyudyanka, where the level of PAH accumulation in the snow cover is maximum it varied widely in aerosol. Minimum values correspond to the level of concentration detected in background regions, maximum values - concentration of benz[a]pyrene in industrial centres. This indicated a certain periodicity of high PAH concentration in the ambient air and the considerable purification capacity of the atmosphere. Maximum accumulation rates of benz[a]pyrene in snow cover near Slyudyanka, Lystvyanka Tankhoi, Babushkin and Byarsky are almost 25-450 times smaller than for the industrial regions of Baikal (e.g., Irkutsk and Shelekhov) and industrial centers of Western Europe. These date are indicative of smaller

PAH flow from the atmosphere to the underlying surface and, consequently, lower pollution of the atmospheric air. In the territory of the background monitoring in Mondy the PAH accumulation rates in the snow cover are comparable with those for background regions.

Conclusions

PAH are characteristic micro-compounds in the chemical composition of the aerosol in the eastern regions of Siberia. PAH concentration levels and their ratio have a large seasonal dependence. During cold season, the pollution of the aerosol and of snow cover is local. Due to the small area of dispersion of PAH emissions into the atmosphere and to the duration of the winter period (150-200 days), PAH accumulation levels in the snow cover in the areas with sources of their emission are characterised by extreme values.

References

Gorshkov, A.G., I.I. Marinayte, V.A.Ovolkin, G.I.Baram and T.V. Khodger; Polynuclear aromatic hydrocarbons in the surface aerosol of the southern coast lake Baikal; in: M.Kilmala and P.E.Wagner (eds), Nucleation and Atmospheric Aerosol. *Proc. Conference on Nucleation and Atmospheric Aerosol,* Helsinki (1996) 597-600.

Gorshkov, A.G., I.I. Marinayte, V.A.Ovolkin, G.I.Baram and T.V. Khodger; Polycyclic aromatic hydrocarbons in the snow cover of the southern coast lake Baikal, *Atmos. Oceanic Opt.* **11** (1998) 780-784.

Gorshkov, A.G. and I.I. Marinayte; Monitoring of the ecological toxicants in the environmental objects of Baikal region. Part 1. Determination of polycyclic aromatic hydrocarbons in aerosol of industrial centres (using Irkutsk as an example), *Atmos. Oceanic Opt.* **13** (2000) 889-902.

Khodger T.V., A.G. Gorshkov and V.A.Ovolkin; The composition of aerosol over the East Siberia, *J. Aerosol Sci.* **30** (1999) S271-S272.

Koroleva G.P., A.G. Gorshkov, T.P. Vinogradova, E.V. Butakov, I.I. Marinayte and T.V. Khodger; Study of snow cover pollution as accumulating medium (Southern Pre-Baikal), *Khimia v interesakh ustoïtchivogo razvitiïa* **6** (1998) 327-337 (in Russian).

Marinayte I.I. and A.G. Gorshkov; Monitoring of polycyclic aromatic hydrocarbons in the snow cover of Southern Pribaikalye (results of a 5-year survey), *Abst. Third Vereshchagin Baikal Comference*, Irkutsk (2000) 150.

Raputa V.F., T.V. Khodzher, A.G. Gorshkov and K.P. Koutzengii; Aerosol falls on snow cover on the outskirts of siberian towns, *J. Aerosol Sci.,* **29** (1988) S807-S808.

Regional Atmospheric Changes Caused by Natural and Antropogeneous Factors

Kazimir Karimov and Razia Gainutdinova

Institute of Physics of National Academy of Sciences
Chui Prosp., 265-A, Bishkek, 720071 Kyrgyz Republic
E-mail: kazimir@academy.aknet.kg

Introduction

Any increase in the concentrations of greenhouse gases in the atmosphere entails changes in the radiation balance of the Earth and results in changes of temperature of the atmosphere, ocean, dynamics of the atmosphere and weather processes. The influence of greenhouse gases on climate can be shown directly by radiation influence and indirectly by means of change of some chemical processes in the atmosphere.

The temperature of air in the lower layers of the atmosphere is a very sensitive measure of climatic changes. In the opinion of many scientists, if the concentration of carbonic gas or the appropriate concentration other greenhouse gases reaches a level twice that of its concentration before the industrial epoch, then an increase in the temperature of air of some degrees can be expected (Karimov and Gainutdinova, 1999). The model calculations show that the average temperatures all over the world will be raised by 1,5-4,5 degrees depending the choice of model of circulation (Trend'90, 1990).

1.1 Variations of carbon dioxide concentration in atmosphere within Kyrgyzstan

For the realization of an estimation of the rate of temperature rise above the region of Kyrgyzstan caused by natural and anthropogenic factors, the data of a long-term series of measurements of the relative concentration of CO_2 in the atmosphere at station Issyk Kul were used (Arefiev *et al.*, 1996).

The relative concentration of CO_2 in an atmosphere was defined by a spectrometric method based on the registration of direct solar radiation in an infrared area of a spectrum and the definition of the function of the absorption band of CO_2 centered at 2,06 microns.

Figure 1 presents the data of measurements of the relative CO_2 concentration as monthly average values (C) for the period between 1980 and 1998 (thin line). A thick line designates the trend. Apart from the seasonal variations in the CO_2 concentration a continuous increase is observed over the whole analysis period. The seasonal oscillations prevail throughout the measurement period: the maximum CO_2 concentration was observed in April - May, the minimum in August - September. The amplitude of the seasonal changes in the CO_2 concentration is determined by a phase shift in an annual cycle "photosynthesis-destruction of continental biota", by a degree of seasons change and is dependent on circulation and dynamic processes in the atmosphere (Anthrogeneous changes of a climate, 1987).

The method of average sliding was chosen for the allocation of a trend component from a temporary number of monthly average values of the CO_2 concentration and an analysis of lower-frequency changes. This method allows fast fluctuations to considerably extinguished with the periods less than the width of the filter.

The analysis of the received data shows, that the contents of CO_2 in the atmosphere above Kyrgyzstan in the period between 1980-1998 continuously grew from a value of 338 ppmv at the end of 1980 up to a value of 378 ppmv at the end of 1996, i.e. the concentration of CO_2 has

I. Barnes (ed.), Global Atmospheric Change and its Impact on Regional Air Quality, 209–214.

increased on an average of 0,7 % per year, which is approximately twice as fast as, the average planetary gain presented in "Anthrogeneous changes of a climate" (1987).

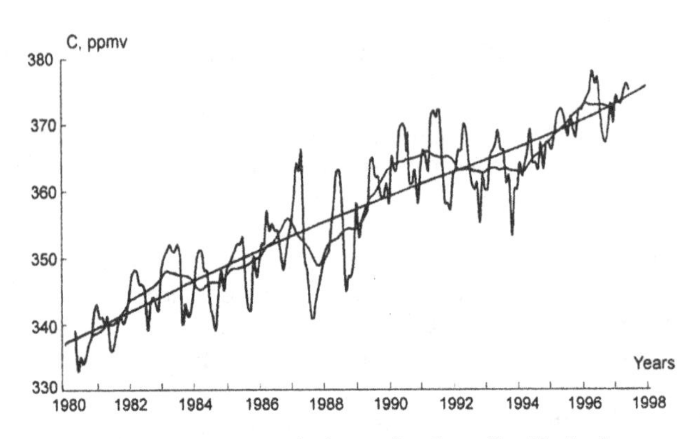

Figure 1. Long-term variations of carbon dioxide in the atmosphere above Kyrgyzstan

The fluctuations within the 44 month period are allocated in slow variations of the CO_2 concentration, which is possible to interpret as change of speed trend of CO_2 (V_T) for an analyzed interval of time not only on size, but also on a mark. So, in 1983-1984 and 1987 the mark of the trend speed of CO_2 was negative.

The significant speeds 0,8-1,6 ppmv/year of CO_2 trends are observed at Arctic and middle latitudinal stations and in the European region, in particular, in Germany from the data of work for the 1980ies this speed reached on average 2,7 ppmv/year (Greenhouse effect, 1989).

The analysis of the received results shows as in the area of northern Tien Shan the trend speed of the CO_2 concentration in the atmosphere changes. In 1989 it reached the maximum value equal to 4,3 ppmv/year. Similar values of V_t were observed in 1982. In 1985 the trend speed of CO_2 was equal to 2,6 ppmv/year. In 1983 and 1984 inter-annual changes of CO_2 were practically not observed and in 1987 the trend speed of CO_2 was negative: V_t = -3,1 ppmv for one year.

The observable considerable changes of the trend speed in the CO_2concentration in the area of Northern Tien Shan, on the one hand, are the response to a variation of the quantity of burnt fuel in the Northern hemisphere in the last decades, and on the other - reflect changes in the efficiency of the biosphere in this region. In the work of Khrgian and Mohov (1991) it was noted, that between 1980-1983 a of reduction of the global emission of carbon to the atmosphere from 5,37 billion tons in 1979 to 5,08 billion tons in 1983 was observed. This was linked to a reduction in the burning of liquid fuel during an economic crisis. From the data of work Khrgian and Mohov (1991) in 1978-1979 was an increase of carbon of 5,6%. It is possible to connect the increase of carbon in 1988 with the increase in the growth rate of the CO_2 concentration. It is reflected in the low-frequency trend of the CO_2concentration, beginning from the second half of 1988. In 1996 in the seasonal variations of the CO_2 concentration two maximums were observed, and in 1994-1995 two minimums took place. Thus, the seasonal variability of the CO_2

concentration above Northern Tien Shan has a complex character caused, first of all, by features of a dynamic mode of the atmosphere and also the geographical position of the region.

1.2 Seasonal variations of temperature deviations in the lower layer of atmosphere

According to the estimations in (Final Application, 1991) within one-decade emissions of CO_2 in atmosphere make 55% of the contributions in radiation balance caused by anthropogeneous greenhouse gases. Therefore approximately one half in the expected increase of temperature will be connected with growth in concentration of CO_2 in the atmosphere.

In the work of Budyko (1980) the formula to account for the increase in the average temperature in a lower layer of the atmosphere caused by greenhouse effect was given as:

$$\Delta T_p = \gamma \Delta T_c[\ln(C_2/C_1 + \beta)]\ln 2,$$

ΔT_p – increase of temperature caused by amplification of greenhouse effect;
ΔT_c – increase of temperature at doubling concentration of CO_2;
γ – coefficient describing inertia of climatic system;
C_1, C_2 – concentration of atmospheric CO_2 at the beginning and end of the analyzed period;
β – coefficient describing additional amplification of greenhouse effect caused by increase of small greenhouse gas components.

The estimation of temperature change in Kyrgyzstan was carried out using the formula given above and the results of the measurement of the CO_2 concentration at the station Isyk Kul. In 1981 the content of CO_2 was equal C_1 = 340 ppmv, in 1990 C_2 = 364 ppmv. This increase has made 7%. Supposing as well as in (Budyko, 1980), $\Delta T_c = 3^0C$, $\gamma = 0{,}6$, $\beta = 0{,}04$ we calcualte that the temperature in the lower layer of atmosphere should increase by $\Delta T = 0{,}28^0C$ from 1981 till 1990.

In order to compare the results of the calculation of ΔT_p with the data of temperature change, we shall allocate a long-term average seasonal component from a number of temperatures, which we shall name as "norm". In many modern studies an average of five years of values of temperature is used as "norm". The analysis of the climatic data in Kyrgyzstan shows, that in seasonal variations of temperature some anomalies were observed in 1984-1985 and in 1987-1989. Therefore, as the basic period the interval between 1980 -1983 was used for definition of "norm".

The analysis of variations of deviations of smoothed values of temperature from average long-term variations was carried out on the basis of the data from the meteorological stations Frunze and Cholpon Ata. In a summer season at the Frunze station the average temperature for the 10-year period was $11{,}5^0C$ with deviations of $\pm 1{,}0^0C$. The maximum values of temperature were observed in 1984, 1990-1991, minimum in 1981 and 1987-1988. The large decrease of the average background temperature no more than $0{,}2^0C$ for 10 years is marked. The error in the determination of the monthly average temperature is $\sigma T \sim 0{,}1^0C$.

For the Cholpon Ata station the average temperature for the 10-year period was $+8{,}5^0C$ with a maximal deviation of $\pm 0{,}9^0C$. These deviations also, as well as at the Frunze station, were observed in 1984, 1990-1991 and minimums - in 1981, 1987-1988.

In the winter period at the Frunze station a constant growth of deviations of temperature from $12{,}0^0C$ (1980) up to $7{,}5^0C$ (1991) was observed. The deviation ΔT from the background temperature is $\pm 4{,}5^0C$. The maximum deviation occurred in 1982 and 1987, minimum - in 1985 and 1990.

At the Cholpon Ata station the increase of the background temperature was from -7,4^{0}C in 1980 up to 7,0^{0}C in 1991 with a maximum deviation $\Delta T = \pm 1,2^0$C. The maximum deviations of temperature also occurred in 1984, 1986-1987, minimums in 1982, 1984, and 1981.

From a comparison of the seasonal variations in winter temperatures between the Frunze and Cholpon Ata stations it is visible, that at Frunze the temperature trend is much larger, and there are also rather large fluctuations in the temperature deviations compared to the background. This fact, obviously, is connected with the influence in the variability of the anthropogeneous factor in the industrial zone of the Chui Valley. The atmosphere above Cholpon Ata (Issyk Kul region) is less subject to this factor.

In changes of temperature in a winter season in 1980-1994 the variability of temperature with the periods about 5 years is visible. To a first approximation it is possible to represent these changes of temperature for the 10 years as a linear dependence. A dashed line in a figure 2 (a) gives the linear interpolation of variations of temperature. This dependence of the experimental data for the Frunze station is represented by the expression:

$$T^0C(t) = -1,1^0C + 0,15^0C(t)$$

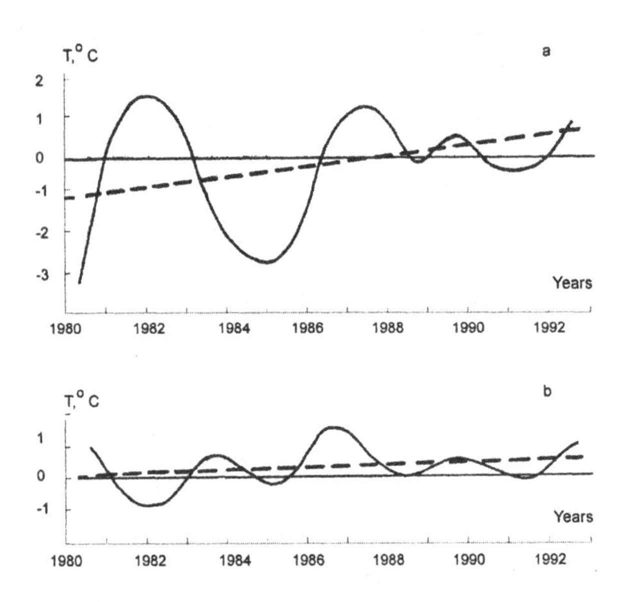

Figure 2. Long-term variations of temperature in lower atmosphere in winter period: a - Chui Valley, b - Isyl Kul region

The positive inclination of the linear trend is visible from this expression. It means, that the average temperatures in the winter period grow at +0,15^{0}C per one year.

Figure 2 (b) presents the similar change of temperature for the winter period in the Issyl Kul area (Cholpon Ata). The small periodic changes of temperature with fluctuations of temperature within the limit ±1,0^{0}C with the period about three years are visible. The periodic changes of temperature at the Cholpon Ata station will not be correlated to the periodic change of temperature on the Frunze station. Probably these fluctuations of temperature are connected with a periodicity of ~2,8 years, which is caused by processes in low and subtropical latitudes and rendering influence on a free atmosphere of average latitudes (Karimov and Gainutdinova, 2000). The constant increase of temperature can be interpolated as a linear dependence. The linear changes of temperature are presented in a figure 2 (b) by a dashed line with a positive inclination of the linear trend. This linear dependence is described by the equation:

$$T^0C\ (t) = -0{,}4^0C + 0{,}05^0C\ (t).$$

A positive linear trend specifies an increase of temperature in the lower atmosphere within the Issyk Kul region of +0,05^{0}C per year. Accordingly for ten years the growth of the background temperature will total + 0,5^{0}C. On the basis of the data above mentioned, the increase of temperature is caused by a real greenhouse effect. So, at the Cholpon Ata station the temperature should be increased at +0,28^{0}C for 10 years.

From a comparison of these two figures it is visible, that 50% of the real increase of temperature is connected with a greenhouse effect. The rest can be caused by other factors, and, first of all, will be connected with the dynamic factors - transfer of heat from low latitudes (Karimov and Gainutdinova, 1997). The data on the change of temperature at the Frunze station in three - five times is exceeded by the change of temperature at the Cholpon Ata station. It specifies the rather large variability of the anthropogeneous factor in Chui Valley and Issyk Kul region (Karimov and Gainutdinova, 2000).

Conclusions

1. The basic regularities of change of atmospheric temperature in Kyrgyzstan have been investigated in connection with variations in the concentration of carbonic gas in the atmosphere.
2. At periodic variations of temperature in summer period there are fluctuations with the periods per 5,5 years. In the winter period the fluctuations with the period about 2,8 years obviously connected with two-years periodicity of atmospheric processes in low and subtropical latitudes and rendering influence on an atmosphere of average latitudes are revealed.
3. The connection of amplitudes of interannual variations of temperature and concentration of carbonic gas is obtained. The maximal and minimal values in variations of temperature are late for 0,5-1,0 years concerning maximal and minimal in variations of carbon dioxide.
4. The contribution of greenhouse effect in long-term growth of atmosphere temperature in Kyrgyzstan is estimated. The greenhouse effect caused by growth of the concentration CO_2 on 7% for 10 years, should result in increase of temperature of air of 0,28 degrees, that makes 56% from its general actual increase. Others 44% can be caused by other factors and, first of all, dynamic - transfer of heat from low latitudes.

References

Anthropogeneous changes of a climate. Under edition of M.I. Budyko and Yu.A. Israel, L: Hydrometeoizdat, (1987) 4-7.

Arefiev, V.N., Kamenogradsky, N.E., Kanin, F.V., Semeonov V.K. Changes of speed of accumulation of carbonic gas in an atmosphere on measurements in Issyk Kul, *Izvestia RAN, Atmospheric and Ocean Physics,* **4** (1996) 437-439.

Budyko, M.I. A climate in last and future., L: Hydrometeoizdat, (1980) 4-6.

Final Application of scientific and technical session II World Climatic conference, *Meteorology and Hydrology,* **4** (1991) 8-20.

Greenhouse effect, climate change and ecosystems. Under edition of B. Bolino, B.R. Dess, B.R. Yager, R. Warrin, M.: Hydrometeoizdat (1989).

Karimov, K.A. and Gainutdinova, R.D. Some results of environmental monitoring in Kyrgyzstan: Atmospheric transfer of contaminants. Proceedings of *the NATO ARW on Integrated Approach to Environmental Data Management* Systems, Dordrecht, The Netherlands, Kluwer Academic Publishers, NATO ASI Series 2: Environment, **31** (1997) 465-472.

Karimov, K.A. And Gainutdinova, R.D. Environmental changes within Kyrgyzstan. Proceedings of *the NATO ARW on Environmental Change, Adaptation and Security.* Dordrecht, The Netherlands: Kluwer Academic Publishers, NATO ASI Series 2: Environment, **65** (1999) 201-204.

Karimov, K.A. and Gainutdinova, R.D. Changes of regional climate caused by natural and antropogeneous factors, Ecology of Kyrgyzstan: Problems, Forecasts, Recommendations, Bishkek, Ilim, (2000) 66-81.

Khrgian, A,Kh. And Mohov, I.I. The review of the book: Trends' 90, *Izvestia AN SSSR, Atmospheric and Ocean Physics,* **8** (1991) 892-895.

Trend'90. Oak Ridge, (1990) 260.

Environmental problems connected with air pollution in the industrial regions of Ukraine

Mykola Kharytonov[1], Natalia Gritsan[2], Larisa Anisimova

[1]State Agrarian University, Voroshilov st.25, Dnipropetrovsk, 49600,Ukraine
[2]Nature Management & Ecology Institute of Ukrainian Academy of Sciences, Moscovskaya 6,Dnipropetrovsk, 49600,Ukraine

Introduction

There is global, regional and local contamination of the atmosphere. The level of industrial pollutants and the circulation of air streams influence the extent of contamination. Industrial pollution in Ukraine has many negative socio-economic consequences. The level of pollution is related to the geographical situation, prior patterns of natural-resource use, the structure of the national economy, and ecological conditions. The pollution contributes to the global greenhouse effect and to the depletion of the ozone layer.

When Ukraine was a part of the former Soviet Union, heavy industries were founded and emphasis was put on the mining, metallurgical and chemical industries. These types of industries use enormous quantities of resources and energy and pollute the environment as a result of obsolete production technologies and lack of relevant waste-treatment facilities. The high concentration of the industries, especially in large cities, aggravated the very negative impact on natural ecosystems and human health. Industrial emissions contribute to climate change by enhancing the greenhouse effect and by depleting the ozone layer. The enhanced greenhouse effect is caused directly by the emissions of carbon dioxide, methane, nitrous oxide and fluorine-containing compounds, and indirectly by substances such as nitrogen oxides, carbon monoxide and volatile organic compounds. Thus human activities have resulted in concentrations of the various greenhouse gases in the atmosphere far in excess of their natural values. Acidification is one consequence of atmospheric pollution by acidifying components and substances that directly or indirectly decrease the rate of ozone formation. In Ukraine, total ozone content (TOC) is monitored by six stations in Kiev, Borispol, Boguslav, Odessa, Lviv and Karadag observatory. TOC was calculated as the average for each year. For instance, three days in 1995 (1 & 2 Feb. and 23 Aug) had TOC levels 2.6 times the climate norm. Six cases with ozone depletion were recorded too [1]. Year by year the total ozone concentration in Ukraine has declined.

It is known that emissions for the three main gases in Ukraine in 1990 total 657 million tons (CO_2-648 million tons, N_2O-25.000 tons, CH_4-9,5 million tons). It is important to emphasize that gas emissions have been decreased from 1980 to 1996. Emissions of SO_2 decreased by 57%, and NO_x decreased by 54%. From 1985 to 1996, there were decreases in emissions of 50% for organic substance emissions and 70% for CO.

Contamination of the atmosphere by heat occurs also. It is known that emissions of over-heated industrial gases leads to higher air temperatures near the earth's surface due to heat entrapment and reflection. Hothouse-gas emissions from the heat-power complex of Ukraine are responsible for 65-70% of the total emissions of over-heated gases.

Within Ukraine, 79% of all atmospheric emissions occur in the Donetsk-Pridneprovsky Region. Major stationary emissions sources are: centralized heat and –power production (33%), metallurgical industries (25%), coal mining (23%), and chemical and petroleum industries (2%). The spectrum analyze of environmental pollution showed that

I. Barnes (ed.), Global Atmospheric Change and its Impact on Regional Air Quality, 215–222.

benzo-pyrene, hydrogen sulfide, ammonium, formaldehyde, and nitrogen dioxide are the primary pollutants here [2].

Materials and Methods

The investigations were conducted in the areas of industrial air and soil pollution. The managed and non-managed emission sources of hazardous substances in Dnipropetrovsk region were monitored with stationary stations.

There are several industrial zones in the Pridneprovsky Region.

A) The Dnepropetrovsky zone including Dneprodzerzhinsk, Dnepropetrovsk and Novomoskovsk. This industrial zone area is about 1164 km^2.

B) The Krivoyrozhsky zone including Shirokovsky, Krivorozhsky, Pyatikhatsky districts with area about 2660 km^2.

C) A manganese ore mining zone including Ordzhonikidze, Marganets, Nikopol. This industrial zone area is about 312 km^2.

D) The Western Donbass industrial Zone including Pavlograd and Yurievsky Districts. This industrial zone area is about 1195 km^2.

Soil and quarry dust samples took away on the distance 0,5-1,5,3-5,5-7 and 10-12 km from source of contamination. Investigations of some pollutants toxicity were carried out in the model laboratory experiments. Aliquots of the untreated black soil (1 kg soil DW) were treated with salts of heavy metal, gases of acids and rehydrated to 60 % field capacity with distilled water.

The salts of heavy metals (Mn,Zn,Cu, Pb,Cd) were included in the soil separately in doses of 50 mg/kg and together in doses of 50 mg/kg. Enzyme activity of soil samples was determined after one week at 28 °C.

Dust and soil samples were prepared for heavy metal analyses by extraction with 1 N HCl. The content of heavy metals in the samples was determined by flame atomic-absorption spectrophotometer. Agrochemical and biochemical analyses of the soil samples were made by the standard methods.

Results and Discussion

The managed and non-managed emission sources of hazardous substances in the Dnipropetrovsk region are numerous. The main pollution sources are facilities of metallurgy, power generation, mining, and chemical and petrochemical industries. These managed emission sources are controlled better than the non-managed sources including quarries, waste depositories, and ash heaps.

Up to 80% of the total industrial emissions in the southeast part of Ukraine are connected with enterprises of mining-metallurgical complex. For example, in the Dniepropetrovsk region the concentration of mining and metallurgical production is 7-10 times the national average. Production and processing of manganese ore is 100% of the national total, iron ore is 86%, cast iron production is 50%, and the production of steel is 47% of the national total. There are 57 metallurgical companies in the Prydneprovsky Region. The metallurgical sector accounts for 64% of total regional emissions (about 530.000 tons per year) and is especially a source of the following substances (in thousand tons per year): dust – 90, SO_2 -- 50, CO -- 370, NO_2 - 20, VOC (volatile organic compounds) -- 2.

Eleven thermal power stations cause tremendous air pollution (total amount was about 250 thousand tons per year, including dust – 80, SO_2 --120, CO -- 15, NO2 -- 50). The wastes of all kinds from chemical companies are very different and are toxic for the environment, but the total amount of emissions to the atmosphere is about 2000 tons per year. The greatest emissions took place in the cities of Krivoy Rog, Dnipropetrovsk, and Dniprodzerzhinsk because of the high concentration of environmentally dangerous industries. A decreasing trend emission for some industrial gases was observed over the last years for these three cities (Table1).

Table 1. Dynamics of pollutant concentrations in the atmosphere of the industrial cities, relative to Maximum permissible Concentration (MPC).

Pollutant	MPC, mg/m^3	1985	'86	'87	'88	'89	'90	'91	'92	'93	'94	'95	'96	'97	'98
Dnepropetrovsk															
Dust	0.15	2.0	2.0	1.3	1.3	1.3	1.3	1.3	1.3	1.3	1.3	1.3	1.3	1.9	1.9
Sulfur dioxide	0.05	0.6	0.6	0.4	0.4	0.2	0.2	0.1	0.2	0.2	0.2	0.3	0.4	0.4	0.5
Carbon monooxide	3.00	0.3	0.7	0.7	0.7	0.7	0.7	0.7	0.7	0.7	0.7	0.7	0.7	0.5	0.7
Nitrogen dioxide	0.04	1.2	1.2	1.2	1.0	1.0	1.0	1.0	1.0	1.0	1.0	1.3	1.5	1.6	1.6
Nitrous oxide	0.06	0.5	0.3	0.3	0.5	0.7	0.7	1.0	0.7	0.7	0.5	0.7	0.7	0.7	0.7
Ammonia	0.04	1.7	2.0	1.7	1.5	1.5	0.8	0.8	1.0	1.3	2.0	2.3	2.5	2.4	2.6
Krivoy Rog															
Dust	0.15	2.7	2.0	2.7	2.7	2.7	2.7	3.3	2.7	1.3	2.0	2.0	2.0	2.4	2.2
Sulfur dioxide	0.05	4.5	3.6	1.3	0.7	1.2	1.1	1.4	1.5	1.2	0.6	1.5	1.1	0.8	1.3
Carbon monooxide	3.00	0.7	0.7	0.7	0.7	0.7	0.7	1.0	0.7	0.7	1.0	1.0	1.0	0.8	0.8
Nitrogen dioxide	0.04	2.8	2.0	2.0	2.0	2.0	2.0	3.0	2.3	1.8	1.5	1.5	1.5	1.4	1.4
Nitrous oxide	0.06	0.6	0.4	0.4	0.6	0.7	0.6	0.6	0.6	0.4	0.3	0.4	0.4	0.5	0.5
Ammonia	0.04	1.5	1.3	2.5	2.3	3.5	3.0	3.5	2.8	2.5	2.0	2.2	2.2	2.6	2.5
Dneprodzerzhinsk															
Dust	0.15	3.3	1.3	2.0	2.0	2.7	2.0	1.3	2.0	0.7	0.7	0.7	1.3	0.8	1.1
Sulfur dioxide	0.05	6.8	4.4	0.2	0.2	0.2	0.2	0.2	0.2	0.2	0.1	0.3	No data	0.3	0.2
Carbon monoxide	3.00	0.7	0.7	0.7	0.3	0.7	0.3	0.2	0.2	0.2	0.3	0.7	0.7	0.6	0.6
Nitrogen dioxide	0.04	1.5	1.0	1.0	1.3	1.3	1.3	1.5	1.8	1.8	1.2	1.8	1.8	1.8	1.6
Nitrous oxide	0.06	0.5	0.5	0.5	0.8	0.8	0.8	0.8	0.5	0.8	0.7	0.7	0.7	0.6	0.8
Ammonia	0.04	2.3	2.3	2.8	5.3	5.3	2.0	3.0	1.8	2.8	1.8	2.8	2.2	2.8	2.5

Table 2. Emissions to the atmosphere from industrial enterprises, thousand tons.

Emission	Total	Pb	Benzo-pyrene	Other	Gas and Liquids	SO_2	CO	NO	CH_4
Dnepropetrovsk	296 357	2.23	0.014	72 589	233 766	110 759	61 013	40 997	9 227
Dneprodzerzhinsk	315 699	0.03	0.073	52 847	262 851	46 500	184 633	22 346	3 926
Krivoy Rog	1 252 712	0.10	0.034	207 940	104 478	98 119	902 291	36 755	3 096
Nikopol	77 057	0.11	0.002	13 405	63 651	2 078	56 448	4 015	433
Novomoskovsk	2 140	--	--	593	1 547	381	230	408	416
Marganets	5 295	0.01	--	4 000	1 295	603	215	350	107

	H_2SO_4	F	CS	H_2S	Cl	Other	Kg/human
	121	9.5	44	517	0.1	1078	241
	290	1.0	501	1240	1.895	3404	1087
	33	0.4	--	2551	0.3	1925	1617
	163	188.0	--	22	--	304	484
	0	46.2	--	--	--	66	28
	0	0.0	--	--	--	20	96

Obviously, the reduction in SO_2, CO, and N0 is partly the result of the reduction in industrial production during the period. At the same time, this is not a desirable solution for the future. The reduction is also partly the result of mitigation measures. However, in spite of considerable decreases in industrial emissions in recent years, the indexes of human health related to the environment continue to be bad in these industrial regions [1]. In Table 2 the average annual concentrations of measured pollutants in the atmosphere are shown for the cities of Dnepropetrovsk, Krivoy Rog, Dneprodzerzhinsk, Nikopol, Novomoskovsk.

For the last decade the annual average emission of industrial dust was 1.25 million tons for Krivoy Rog, and 1 million tons each for Dnipropetrovsk, Dniprodzerzhinsk, Nikopol and Novomoskovsk. The amount and type of emissions in different places of the Dnipropetrovsk region depends on the kind of industrial enterprises. In particular, iron ore mining-metallurgy in Krivoy Rog results in high levels of industrial dust, oxides of sulphur, carbon and nitrogen. Meantime the iron ore mining in the Krivoy Rog quarries involves blasting. The iron ore dust ascends to 600 m. The volume of dust - gas cloud is within 10-20 mln m^3 and dispersed to distances of 12-15 km [3]. Table 3 shows the concentrations of heavy metals in soil different distances from the place of blasting (the quarry of the Krivoy Rog iron ore deposit).

Table 3. Soil contamination in the top 10 cm by heavy metals near a quarry in the north part of the Krivoy Rog iron ore deposit, mg/kg (numerator - limits, denominator - average value).

Distance, km	Fe	Mn	Zn	Cu	Pb
0.5-1.5	11100-14500 11439	190-614 443	26-46.7 30.8	10-18 12.0	6.2-12.0 9.6
3-5	11300-12100 11814	338-514 401	26-34 31	10-18 12.6	6.0-12.0 9.6
5-7	8800-13300 11667	300-395 347	24-47.5 32.3	6-12 10.2	4.3-21.5 12.5
10-12	7500-11250 9025	319-491 390	14-38 24	10-18.5 12.4	10.4-23.2 14.6

Table 4 shows the heavy metal concentrations in the quarry dust in the primary cloud (which settles within 1 km) and the secondary cloud (which spreads as much as 10 km or more).

Table 4. Maximum content of heavy metal composition in quarry dust mg/kg.

Metal	Fe	Mn	Zn	Cu	Ni	Co	Pb	Cr	Cd
Quarry dust cloud									
Primary	15000	800	80	32	25	25	20	8	5
Secondary	385	210	259	67	14	63	35	37	14

Table 5 shows the variability in soil enzyme activity depending on the distance to the source of dust cloud after blasting.

Table 5. Variability in soil enzyme activity per 100 g soil depending on the distance to the source of dust cloud after blasting.

Distance from the source of contamination, km	Urease, mg NH_3 in 24 hours	Phosphatase, mg P_2O_5 in 1 hour	Invertase, mg glucose in 24 hours
1.5	14.4-22.6 19.2	6.2-9.5 7.8	28.2-31.0 29.5
3.0	20.4-26.4 23.7	8.1-10.0 9.1	33.2-35.9 34.2
4.0	5.8-11.0 9.1	6.2-11.1 8.4	29.8-31.8 30.9
5.0	2.7-8.4 5.6	0.92-4.36 2.64	27.5-32.8 30.2

It appears that the strong reduction of the activity of phosphatase and urease that occurred between 4 and 5 km from the quarry is a result of the specific pattern of the aerial spreading of the primary and secondary dust cloud after blasting. It is possible to minimize the spread of the blast-cloud by use of a water canon at the instant of the blast.

It was shown also that in the vicinity of the manganese-ore concentrating plants and waste-sites, soils are highly contaminated for 2-5 km from the source. Heavy metal concentrations at 3 km were (natural control in brackets): manganese 535-1038 (538), zinc 8,5-33,5 (4,6), copper 7,6-16,0 (7,7), cobalt 6,4-7,4 (4,0), nickel 9,6-10,5 (6,1), lead 7,2-14,5 (7,0) mg/kg. Oxides and hydroxides of Fe and Mn are ordinary components of black soil, but high concentrations can lead to significant changes in the local geochemical balance because these compounds have the ability to absorb other elements [4]. Heavy metals and quarry dust from the primary cloud were added to natural black soil (chernozeom) and enzyme activity was measured one week later. Copper, and to a lesser extent, cadmium and zinc caused a large reduction in urease and phosphotase activity (Figure 1). Invertase activity was not strongly reduced by any one metal, but was strongly reduced by the various metals combined.

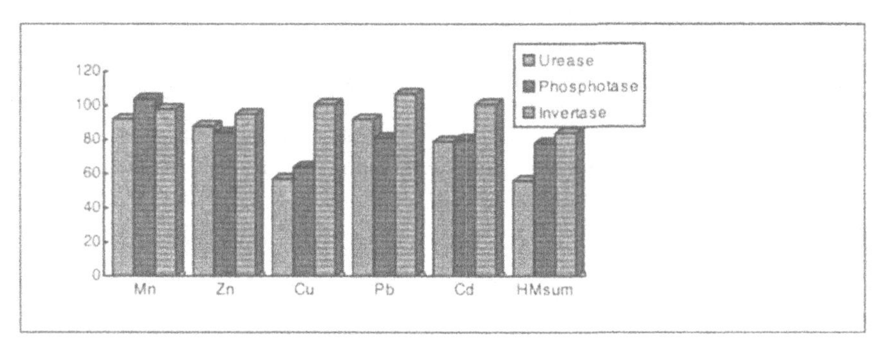

Figure 1: Influence of heavy metals on soil biological activity (control 100%).

It is well known that pollutants (heavy metals, etc) have considerable influence on the quantity, specific composition and vital functions of the soil microbiota [5,6] They inhibit the activity of the soil enzymes, and the general physiological activity of the soil deteriorated. In this connection peroxidase was selected because of its part in molecular condensation reactions during humic acid formation. In a laboratory experiment, H_2SO_4 and HNO_3 gases increased peroxidase activity in the soil, while HCl decreased it (Table 6).

Table 6. The relation between gas pollutants (H_2SO_4, HNO_3, HCl) and soil peroxidase activity (relative to chemical reaction) in a laboratory experiment.

Pollutant	Concentration (expressed as MPC)	Peroxidase activity, (in relative units)
Control		0.40 ± 0.038
H_2SO_4	1	1.04 ± 0.40
H_2SO_4	10	1.30 ± 0.045
HNO_3	1	0.16 ± 0.012
HNO_3	10	0.50 ± 0.041
HCl	1	0.22 ± 0.018
HCl	10	0.15 ± 0.013

Peroxidase activity for soil samples taken in different places of Dnepropetrovsk was very depressed (3 relative units compared to 7.8 ±0.09 relative units for the control). The regression between SO_2 and N2O emissions (Table 1) and peroxidase activity in soil samples from 1997 to 1999 is shown in Table 7.

Table 7. Relationship between SO_2 and N_2O emissions and peroxidase activity in soil

X	Y	Regression
SO_2 mg/m3	Peroxidase activity, (in relative units)	Y= - 321.957x+7.725
N_2O mg/m3	Peroxidase activity, (in relative units)	Y= -114.956x + 7.7112

It may be possible that soil microorganisms contribute to the greenhouse effect by release of CO_2, CH_4, N_2O. Therefore, any atmospheric changes, including pollutants that change the enzymatic activity of soil microorganisms may influence the intensity of the greenhouse effect [7]. It is known that N_2O production leads to a high level of heat entrapment by the atmosphere (approximately 150 times the effect of CO_2) [8].

The control of emissions of pollutants is based on their comparison with their MPC (maximum permitted concentration). It is helpful to try to decrease the emissions to the MPC. However, this is often not enough because so many ingredients of industrial emissions are not typical for the natural atmosphere, and even small quantities of some pollutants can create additional environmental problems.

The data on the forests and agricultural lands distribution within Dnepropetrovsk province can be taken into account to assess the biological potential for some gas pollutants reutilization including SO_2 accumulation (Table 8).

There are several reasons to make changes in forests and agricultural land distribution. The first reason is connected with biological conservation for mine land, washed soils, etc. On the other hand a high industrial load requires improving forest planting for toxic ashes reutilization as well.

Conclusions

The air basin of large industrial cities is extremely polluted in spite of the permanent convection and inter-boundary transfer. The atmosphere contains undesirable quantities of some substances (eg. SO_2, NO_x, NH_4). This has negative effects on the global environment and on local health. Also, the reduction of biological activity of industrially contaminated soils is one important cause of soil degradation. To limit the pollutant emissions Ukraine needs to develop environment legislation, environmental impact assessment procedures, and networks of monitoring locations to provide good environment monitoring and auditing for enterprises. It is necessary to implement new mitigation measures: to decrease the general levels of industrial air pollutants, to limit ozone depletion, to burn fuel completely to ashes, etc. Thus, changes in forest and agricultural land distribution in Dnepropetrovsk province are needed urgently.

Acknowledgements

Credit is due to people who were involved in this work including, Anatoly Drizhenko, Alexander Zberovsky, Natalia Torkhova, Nikolay Lukashenko and Paul Gibson.

References

1.National report on environment state in Ukraine, 1995. (1997). Raevsky Press. Kiev, 96p.(Ukrainian)

2.N..Gritsan,Babiy A.P. Case study. Hazardous materials in the environment of Dnepropetrovsk Region (Ukraine). (2000). Journal of Hazardous Materials, A76: 59-70

3.N. Kharytonov. Ecotoxicological problems under mining at the ukrainian steppe, 30th International Geological Congress, Beijing,China,(1996).

4.A.Kabata-Pendias, Pendias H. Trace Elements in Soils and Plants,2nd edn., Levis.Publ.Boca Raton.Fl. 365 p.

5. H.Babitch , Stotzky G.1985. Heavy metal toxicity to microbe - mediated ecologic processes: a review and potential application to regulatory policies. // Environ. Res. (1992). 36(1): 111-137.
6. T.Duxbury . Ecological aspects of heavy metal responses in microorganisms.
// Adv. Microb. Ecol. (N.Y.; London).(1985) 8: 185-235.
7. L Machta . The ozone depletion (an example of harm commitment)//MARC. A general report.(London) (1976). 33 p.
8. M.M,Umarov. Soil microbiology and problem of "greenhouse" effect. 4th NIS Conference: Microorganisms in Agriculture. Pushchino. (1992) p.202 -203.(Russian)

Table 8. Agricultural land and forest distributions in the districts of the Dnepropetrovsk province (numerator – area, ha; denominator - % from total district square)

DISTRICTS	Volume of SO_2 emission thous.ton	Area covered with forests, thous.ha	Ability of areas covered with forests to SO_2 Accumulation, thous.ton	Arable Land area, thous.ha	Ability of areas covered with arable lands to SO_2 accumulation thous.ton	Total potential ability to SO_2 accumulation,thous.ton
1	2	3	4	5	6	7
Apostolovsky	65,153	2,2	0,880	104,428	4,177	5,057
Vasilkovsky	0,006	2,2	0,880	119,002	4,760	5,640
Verkhnedneprovsky	0,152	12,0	4,800	86,330	3,453	8,253
Dnepropetrovsky	72,817	5,9	2,360	110,864	4,435	6,795
Krivorozhsky	27,126	3,8	1,520	102,185	4,087	5,607
Krinichansky	0,002	1,8	0,720	149,414	5,977	6,697
Magdalinovsky	0,003	2,3	0,920	137,842	5,514	6,434
Mezhevskoy	0,066	3,1	1,240	112,104	4,484	5,724
Nikopolsky	0,812	3,1	1,240	133,492	5,340	6,580
Novomoskovsky	0,178	9,3	3,720	147,182	5,887	9,607
Pavlogradsky	2,746	8,9	3,560	113,411	4,536	8,096
Petropavlovsky	2,977	2,7	1,080	108,697	4,348	5,428
Petrikovsky	0,033	2,2	0,880	56,017	2,241	3,121
Pokrovsky	0,021	3,4	1,360	106,358	4,254	5,614
Pyatikhatsky	2,582	6,9	2,760	141,792	5,672	8,432
Sinelnikovsky	0,079	2,7	1,080	144,472	5,779	6,859
Solonyansky	0,043	3,3	1,320	152,905	6,116	7,436
Sofievsky	0,025	1,8	0,720	121,224	4,849	5,569
Tomakovsky	0,004	1,9	0,760	94,839	3,794	4,554
Tsarichansky	0,002	6,4	2,560	73,409	2,936	5,496
Shirokovsky	0,003	1,5	0,600	96,729	3,869	4,469
Yurievsky	0,001	2,3	0,920	79,502	3,180	4,100
Total for province	174,831	89,7	35,880	2492,198	99,688	135,568

Comparison of Experimental and Calculated Data on Ion Composition of Precipitation in the South of East Siberia

Ekaterina V. Kuchmenko, Yelena V. Moloznikova, Olga G. Netsvetaeva, Natalia A. Kobeleva, Tamara V. Khodzher

Limnological Institute of Siberian Branch of Russian Academy of Sciences, Irkutsk 664033, Box 4199, Russia
thermo@isem.sei.irk.ru

Permanent observations of the chemical composition of precipitation carried out by the Limnological Institute of SB, RAS, at the monitoring stations in the south of East Siberia (Irkutsk, Listvyanka, Mondy) make it possible to estimate the specific features of the ion composition of the precipitation in the region and to determine to some extent the anthropogenic impact on the chemical composition of the precipitation. However, based on the observational data alone it is difficult to separate the respective contributions of natural and anthropogenic sources of aerosols to the concentration of ions in the drops. In addition, in the process of fog and precipitation formation gaseous impurities are absorbed from the air (*Stepanov*, et al., 1997). Absorption of gaseous impurities by the drops of precipitation and fogs need to be estimated to evaluate the impact of abnormal weather conditions on the pollution level in the cities (*Bezuglaya*, 1985).

The object of the paper was to compare experimental data with the results of numerical thermodynamic modeling of the equilibrium composition of precipitation in clean and polluted atmospheres. Thermodynamic models are suggested for the description of the complex processes in the atmosphere including phase changes and chemical inter-relations with incompletely studied reaction mechanisms. An undeniable advantage of the approach is that there are practically no constraints on the number of the substances that can be considered. Unlike the methods of chemical kinetics that are traditionally used when modeling transformations of substances in the polluted atmosphere the thermodynamic statement does not require information on a concrete mechanism of reactions, their speed and intermediate products.

The model of extreme intermediate states (MEIS), given in the monograph (*Gorban et al.*, 2001), makes it possible to describe both the state of the final equilibrium of a reacting system and a number of intermediate (incomplete) equilibria that are likely as the system relaxes towards the final equilibrium state.

Currently the Energy Systems Institute is developing MEIS versions that detail sequentially the description of hetero-phase processes in the atmosphere. The first elementary model represents a conventional block of final equilibrium of the model of extreme intermediate states (MEIS) [4]:

find:

$$\min\left[G = \sum_{j\in J_g}\left(G_j^0 + RT\ln\frac{x_j}{\sigma}\right)x_j + \sum_{j\in J_c} G_j^0 x_j\right] \quad (1)$$

subject to

$$\boldsymbol{Ax=b}, \quad (2)$$

$$x_j \geq 0, \quad (3)$$

where G and G_j^0 are the Gibbs energy of the system and the standard Gibbs energy of a mole of the j-th component, respectively; J_g and J_c are sets of components of gaseous and

I. Barnes (ed.), Global Atmospheric Change and its Impact on Regional Air Quality, 223–228.

condensed phases, T is the temperature, R is the universal gas constant, $\boldsymbol{x}$ is an n-dimensional vector of component amounts of the reaction mixture, σ is a total number of substances in the gaseous phase; A $[m \times n]$ is a matrix of chemical element content in the system components; m is a number of material balances, and $\boldsymbol{b}$ is a vector of amounts of elements.

As is known water droplets in the atmosphere are not a pure substance since watering of solid particles (condensation nuclei) requires less energy than formation of clean water drop-nuclei. Therefore, the next model modification took into account the possible presence of a dissolved electrolyte solution in the system. MEIS version describes dissolved solutions of strong electrolytes based on the statistic theory of Duebai-Hukkel (*Kaganovich and Fillipov*, 1995). The Gibbs energy of one mole of the j-th component of the solution can be represented by the form:

$$G_j = G_j^0 + RT \ln \frac{x_j}{\sigma_s} + RT \ln \gamma_j, \tag{4}$$

where σ_s is a total number of moles of solvent and dissolved substances and γ_j is a rational activity coefficient of the j-th component. For water solutions the value of γ_j is found from the equations:

$$\lg \gamma_{\pm j} = -\frac{\varphi \left| Z_{1j} Z_{2j} \right| I(x_e)^{0,5}}{1 + I(x_e)^{0,5}} + \Psi(x_e), \tag{5}$$

$$\varphi = \frac{1.825 \cdot 10^6}{\varepsilon^{1,5} T^{1,5}}, \tag{6}$$

$$I \approx \frac{27.778}{x_{H_2O}} \sum_{j \in J_\pm} \left(Z_{1j}^2 \nu_{1j} + Z_{2j}^2 \nu_{2j} \right), \tag{7}$$

$$\Psi = 0{,}1 \cdot \left| Z_{1j} Z_{2j} \right|, \quad \gamma_{\pm j}^{\nu_j} = \gamma_{+j}^{\nu_j} \cdot \gamma_{-j}^{\nu_j}, \quad \nu_j = \nu_{1j} + \nu_{2j},$$

where ν_{1j} and ν_{2j} are stoihiometric coefficients of the cation and anion in the dissociation reaction of the j-th electrolyte; φ and Ψ are coefficients; Z_{1j} and Z_{2j} are charges of cation and anion, respectively, into which the j-th electrolyte dissociates; I is the ionic force of solution: $\boldsymbol{x}_e$ is a vector of quantities of electrolytes; ε is the dielectric permeability of water; and $J_\pm$ is a set of pairs of ions in the solution. Further unfolding of the model supposes consideration for Gibbs energy components, determined by the surface tension and electric charge of droplets.

In the present paper the mathematical modeling methods were used to analyze the impact of the chemical composition of atmospheric air and condensation nuclei on the composition of the atmospheric precipitation in the south of East Siberia. Aerosols were represented in the system in the form of solid salts of metals (Table 1).

The system considered was heterogeneous but spatially homogeneous and the particle sizes of aerosol and drops were not taken into account. The calculation simulated watering of a solid nucleus that is accompanied by its dissolution and absorption of gaseous pollutants. The case in point is composition of the polluted atmosphere: the sulfur oxide content exceeds 0.6 MPC and the aerosol concentration is almost 2 MPC for particulates (the concentration of particulates in Irkutsk is: 1.3 MPC - average; 3 MPC - maximum; that of sulfur oxide: 0.3 MPC - average, 1 MPC - maximum) (*Malevsky*, 2001; *Bezuglaya*, 2000).

Table 1. Example of an atmospheric precipitation composition calculation at T=298 K and P = 1 atm

Substance	G_0, Dj/mole	Concentration, mole/kg	
		Initial	Equilibrium
Gas phase			
N_2	-57072	26.7	26.7
O_2	-61110	7.15	7.15
CO_2	-457182	$1.19 \cdot 10^{-2}$	$1.19 \cdot 10^{-2}$
H_2O	-298051	1.22	1.06
O_3	70614	$1.42 \cdot 10^{-6}$	0.0
NO_2	-37345	$4.26 \cdot 10^{-7}$	$3.40 \cdot 10^{-11}$
NO	28487	$6.04 \cdot 10^{-12}$	0.0
HNO_3	-213410	$5.09 \cdot 10^{-8}$	$1.81 \cdot 10^{-11}$
CH_3Cl	-151824	$3.50 \cdot 10^{-7}$	0.0
SO_2	-370743	$3.49 \cdot 10^{-7}$	$2.48 \cdot 10^{-11}$
Solid phase			
C: $CaSO_4$	-1466387	$3.49 \cdot 10^{-7}$	0
C: Na_2CO_3	-1169418	$3.49 \cdot 10^{-7}$	0
Liquid phase			
H_2O	-306714	$3.49 \cdot 10^{-4}$	0.165
$H^+ \cdot NO_3^-$	-243873	0	$1.96 \cdot 10^{-7}$
$Ca^{2+} \cdot 2NO_3^-$	-1027536	0	$6.35 \cdot 10^{-8}$
$Na^+ \cdot NO_3^-$	-508125	0	$1.54 \cdot 10^{-7}$
CO_2	-713359	0	$8.47 \cdot 10^{-7}$
$Ca^{2+} \cdot 2HCO_3^-$	-1966506	0	$8.94 \cdot 10^{-9}$
$Na^+ \cdot HCO_3^-$	-977610	0	$3.53 \cdot 10^{-8}$
$2H^+ \cdot SO_4^{2-}$	-900192	0	$2.68 \cdot 10^{-7}$
$Ca^{2+} \cdot SO_4^{2-}$	-1439981	0	$2.33 \cdot 10^{-7}$
$2Na^+ \cdot SO_4^{2-}$	-1428694	0	$1.96 \cdot 10^{-7}$
$H^+ \cdot Cl^-$	-177537	0	$1.46 \cdot 10^{-7}$
$Ca^{2+} \cdot 2Cl^-$	-894863	0	$4.32 \cdot 10^{-8}$
$Na^+ \cdot Cl^-$	-441788	0	$1.16 \cdot 10^{-7}$
Weight of system, kg		0.0286	0.0286
G, Dj		-68335.8	-68336.6

In Table 2 the results of calculations for the polluted atmosphere (Table 1) are compared with the data of measurements in the region under consideration.

Table 2. Chemical composition of atmospheric precipitation (anions) in experimental and calculated data (%)

	HCO_3^-	SO_4^{2-}	NO_3^-	Cl^-
Mondy, summer	12.75	33.14	18.90	35.50
Mondy, winter	35.97	24.11	14.39	25.52
Irkutsk, summer	26.72	46.79	13.24	13.25
Irkutsk, winter	24.70	47.40	12.40	15.50
Calculation	24.63	47.28	15.49	12.59

It should be noted that the relationship of anions in the model example corresponds to that observed in the atmospheric precipitation in Irkutsk in winter. A calculated equilibrium concentration of solution (total of dissolved ions) appears to be 5 times higher than the average and 2 times higher than the maximum measured in atmospheric precipitation at the station of Irkutsk. Analysis of the results requires accounting for the following circumstances:

- In the initial data of the calculated example the concentrations of the main pollutants were set essentially higher than the average values for Irkutsk;
- The chemical composition of the solid phase (aerosol) was taken arbitrarily without consideration of the fact that real particles contain insoluble compounds of silicon, aluminum, iron, etc.

Thus, the agreement between measured and calculated values of precipitation concentration at the given stage can be considered quiet satisfactory.

In the calculation for the "clean atmosphere" (the amount of solid phase corresponds to the natural aerosol concentration), the equilibrium concentration of salts in precipitation decreases 10 times compared to the polluted atmosphere and corresponds to sufficient accuracy to the precipitation mineralization at the station of Mondy.

The experimental data presented in Table 2 were obtained based on the primary statistic processing of the chemical analysis of the results. The 1999 data only were considered (three points - Irkutsk, Mondy and Listvyanka). It is obvious that the amount of sulfate ion in the precipitation increases sharply compared to the background areas in the south of East Siberia (for instance - Mondy), (*Netsvetaeva et al.*, 2000).

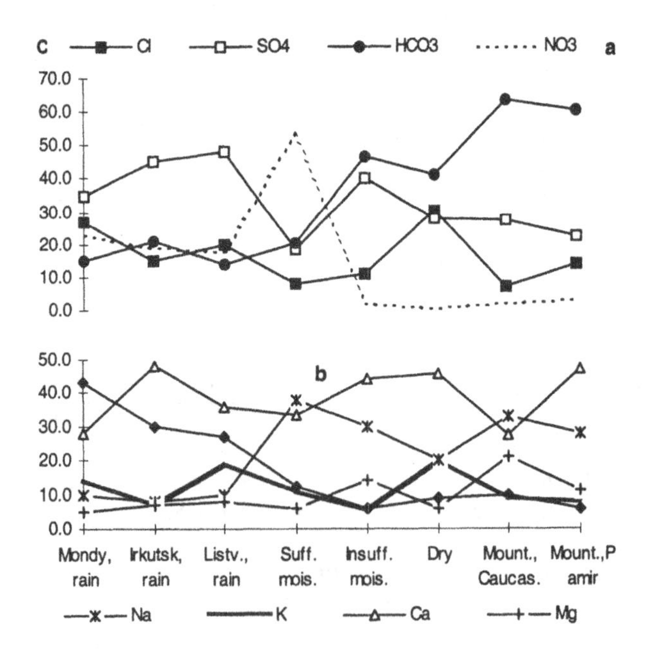

Figure 1. Precipitation composition in different zones of Russia

Figure 1 shows the observational data averaged for a year (chemical composition of atmospheric precipitation) along with average annual characteristics averaged for natural zones (*Glazovskaya*, 1988). The figure shows that in the studied area the concentration of sodium ions is lower and that of calcium and ammonium is higher compared to the average zonal. As to the anions, a similar analysis showed that at the stations of Irkutsk and Listvyanka SO_4^{2-}. Their noticeable increase, particularly in winter, is caused by worsening dispersion conditions of anthropogenic impurities.

An indicator of anthropogenic impacts at the station of Irkutsk in winter is the increasing concentrations of hydrocarbonate ions, calcium ions, sodium, etc. observed in the emissions from thermal energy enterprises. At the station of Mondy an increase in the total of main ions is observed in winter but on the average it is 5 times lower than in the precipitation at the station of Irkutsk. Probably the insignificant amount of snow (20-25 mm) plays an important role here as well as an increase in wind velocity in winter and the transfer of soil aerosols.

Apparently, the chemical composition and ion concentrations in atmospheric precipitation depend on the composition of atmosphere in the area of precipitation and on the change in the composition on the transported air mass. For the days with precipitation in 1999 the reverse trajectories were analyzed with the use of trajectory model and information presented in the Internet (http://www.arl.noaa.gov). Seven different types of trajectories for air masses at the heights of 3000 and 5000 m were distinguished. Averaging the composition of atmospheric precipitation by types of trajectories resulted in a rather characteristic picture (Figure 2):

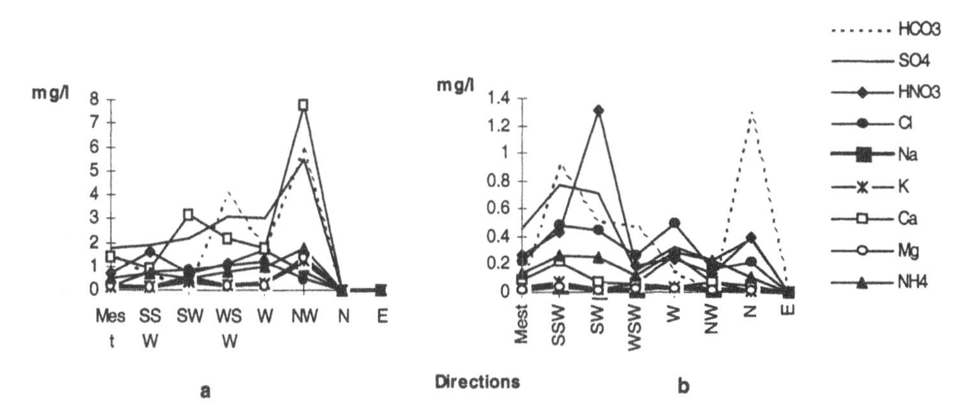

Figure 2. Precipitation composition in Irkutsk versus direction of air mass transport

At the station of Irkutsk maximum concentrations of ions of Ca^{2+}, SO_4^{2-}, HCO_3^- were obtained for northwest types of trajectories, i.e. when precipitation formed above the industrial areas of the Irkutsk region and Krasnoyarsk krai. Ion content in the precipitation coming from the South and Southwest is minimum. At the station of Mondy the highest content of main ions is observed in the atmospheric precipitation brought from the Southwest - from Mongolia and Kazakhstan. It is known from the literature (*Glazovskaya*, 1988; *Pivovarova*, 1983) that the particulate content in the air mass and natural mineralization of the precipitation increases from the North to South.

Thus, comparison of the results showed that thermodynamic modeling allows simulation of watering and dissolution of solid aerosol along with absorption of gaseous impurities from atmosphere by the droplets. The resulting concentrations of main ions in precipitation are,

- firstly in good agreement with the measured data in terms of per cent relationship between main cations and anions;
- secondly, to a first approximation the calculated characteristics of precipitation mineralization can be considered quite satisfactory.

Data on the chemical composition of air and nuclei condensation in the cities and background areas of East Siberia are required to specify the calculated characteristics.

Thus, thermodynamic modeling with a long-term series of observations can be used to estimate the contributions of anthropogenic and natural mechanisms of precipitation mineralization. Further unfolding of the thermodynamic approach is expected to provide an insight into heterogeneous phenomena on the surface of liquid and solid aerosols.

Acknowledgement

This research is partially supported by Russian Foundation of Basic Research (grant no. 01-02-16643). We acknowledge Prof. B.M.Kaganovich for his help and consulting in the preparation of this work.

References

Stepanov A.S. Zakharova I.M., Novikova L.D. Modeling the processes of pollutants accumulation in the fog drops// *Meteorology and hydrology*. **4**(1997) 25-36.

Bezuglaya E.Yu. Meterological potential and climatic peculiarities of air pollution in the cities. *Hydrometeoizdat*. Leningrad.(1985) 352 .

Gorban A.N., Koganovich B.M., Filippov S.P Thermodynamic equilibria and extrema. Analysis of attainability and partial equilibria in physico-chemical and technical systems. *Nauka*. Novosibirsk, (1995) 236.

Kaganovich B.M., Filippov S.P. Equilibrium thermodynamics and mathematical programming. *Nauka. Sib.branch*. Novosibirsk.(1995) 233 p.

On the environment situation in Irkutsk region in 1999. Edited by Malevsky A.L., Irkutsk (2001) 295.

Year-book on the atmosphere pollution in the cities on the territory of Russia. Edited by Bezuglaya E.Yu. St.Petersburg (2000) 280.

Netsvetaeva O.G., Khodzher T.V. Obolkin V.A., Kobeleva N.A., Golobokova L.P., Korovyajiva I.V., Chubarov M.P. Chemical composition and acidity of atmospheric precipitation in Pribaikalie. *Optika atmosfery i okeana*. **v.13, 6-7** (2000) 612-617.

Glazovskaya M.A. Geochemistry of natural and technogenic landscapes of the USSR. *Vysshaya shkola*.Moskva. (1988) 325.

Pivovarova Z.I. Radiation characteristics of climate in the USSR. *Hydrometeoizdat*. Leningrad. (1977) 335 .

Monitoring of the Spatial and Temporal Variation of the Particle Size Distribution and Chemical Composition of the Atmospheric Aerosol in Siberia

Koutsenogii K.[1], Koutsenogii P.[1,2], Smolyakov B.[3], Makarov V.[1], Khodgher T.[4]

[1] *Institute of Chemical Kinetics and Combustion SB RAS, 630090 Novosibirsk, Institutskaya ., 3*
[2] *Institute of Cytology and Genetics SB RAS, Novosibirsk*
[3] *The Inorganic Chemistry Institute SB RAS, Novosibirsk*
[4] *Limnological Institute SB RAS, Irkutsk*

Introduction

The presented study is accumulating the main results obtained within the framework of the project "Siberian Aerosol", which commenced in 1991. The project was stimulated by the large international project "Arctic Haze" (AA), which is connected with a study of long-range transport of industrial pollution in the Northern Hemisphere in the Arctic region.

At the present time a lot of attention is being paid to the effect of long-range transport of continental AA to the polar region. Studies in Norway and Alaska have shown, that Western and Central Siberia may considerably contaminate the atmosphere in the Arctic. The cities and many regions of Southern Siberia are strongly contaminated by heavy metals. The level of such contamination is considerably higher, than in industrially developed countries of Europe and North America. Thus, the local pollution by AA may represent a health hazard for people living in these regions. In many cases such powerful point sources of polluting industrial emissions are rather simple with respect to the environment chemistry and one can expect to obtain important scientific results with a minimum of effort and cost. It is necessary to outline some specific aspects of studies of Siberian AA. In Siberia we have remote areas, which are located at huge distances from industrial and highly polluted centers. According to generally accepted opinion, the characteristics of aerosols in these regions are considered to be "background", i.e. aerosols, which are formed due to natural processes and little influenced by polluting substances.

Many years of studies of the characteristics of atmospheric aerosols in different regions of the Earth show that a considerable fraction of the aerosol mass is comprised of particles produced due to wind erosion from the surface of soil and oceans. These are so-called dust particles and sea salt particles, respectively. The content of particles of natural and anthropogenic origin is rather low. Central Siberia is distanced a few thousand km from the sources of erosion particles. In winter, the ground in Siberia is covered by snow, and water surfaces, including the ocean, are covered by ice. Therefore, during winter, the conditions are very favorable for the study of the long-range transport of industrial pollutants. Thus, the targets of the project "Siberian Aerosol" are:

- Investigations of the laws of formation, transformation and transport of aerosols in the Siberian region on local, regional and global scales for the determination of their sources and sinks.
- Estimation of the influence of AA on the quality of the atmospheric air, levels of contamination of vegetation, soil and water, fate of different substances and elements in objects of the environment.
- Estimation of the impact of AA of different nature on the health of people and animals.
- Investigation the influence of AA on atmospheric processes and climate.

I. Barnes (ed.), Global Atmospheric Change and its Impact on Regional Air Quality, 229–235.

The whole complex of studies may be divided into three steps:

1. Probe sampling and AA monitoring in Siberia. It foresees probe sampling in highly polluted areas, regions affected by technogenic emissions, and remote areas with minimum influence from anthropogenic pollution.
2. Measurement of aerosol particle size distribution and chemical composition for aerosol particles in different geographical zones and during different seasons.
3. Statistical analysis of experimental results, collection and evaluation of existing data.

Figure 1a shows a map of the monitoring station network in the Northern Hemisphere for the study of "Arctic Haze" in early 1991. Of the some 22 stations, not one was located in the territory of Siberia. Figure 1b shows the AA probe sampling station network, existing in Siberia, at the time of writing.

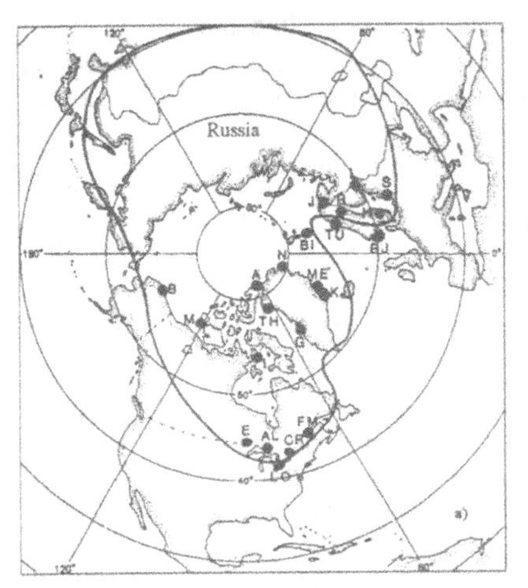

Fig.1a.The mean position of the Arctic front in January (solid line) from Barrie and Hoff (1984) and the location of air monitoring stations providing observations.

Fig.1b. Network of atmospheric aerosol monitoring stations in Siberia.

Results of studies of atmospheric aerosols

Particle size distribution. The distribution has three modes and may be described by a function, which according to the Whitby classification characterizes remote continental AA. The parameters of this function, obtained experimentally, are shown in Table 1.

Table 1. Microphysical characteristics AA in Siberia.

Characteristic AA	**Mode 1 Nucleation**	**Mode 2 Accumulation**	**Mode 3 Coarse**
Number concentration (N_i) d_{i50}, μm	0,024	0,17	1,6
Specific surface (S_i) d_{i50}, μm	0,030	0,21	3,4
Mass concentration (M_i) d_{i50}, μm	0,033	0,24	5,0
σ_{ig}	1,6	1,6	2,4
$N_i/\Sigma N_i$, %	80	20	-
$S_i/\Sigma S_i$, %	7,6	90,7	1,7
$M_i/\Sigma M_i$, %	0,1	15,2	84,6

The coarse fraction of Siberian atmospheric aerosols is mainly determined by soil erosion processes. In summer pollen and large forest fires exert a significant influence on the chemical composition of this fraction (Golovko, 2001; Koutsenogii K.P.et al., 1996).

Chemical Composition of AA. Studies of the multi-element composition of AA allow significant information about the type of aerosol particles and possible sources of their formation to be derived. It was found, that the elements which may be determined in aerosol particles could be separated into three classes according to their relative content in particles of different size fractions. The first class includes aerosols, where elements like Ca, Ti, Fe, Rb, Sr, Y are situated mainly (up to 98%) in the coarse size fraction (d > 1 μm). These particles are formed due to wind soil erosion. The second class includes aerosols, where elements like V, Cu, Br are mainly (up to 64%) found in the sub-micron fraction. These are technogenic particles. To the third class belong particles, where elements like Mn, Zn, As, Pb are homogeneously distributed though all the size fractions. These are particles of mixed type (Koutsenogii et al., 1998).

An analysis of the space-time change of the multi-element composition of AA allows the determination of the types of aerosol sources and scales of their influence on the contamination level of the environment. Figure 2 illustrates the laws of change of the relative concentration ($\langle X_{iFe} \rangle$) of different elements in AA, experimentally observed in the South (a, b, c) and in the North (d) of Western Siberia. The dark points in Figure 2a, b, c represent geometric mean values, measured at the Klyuchi sampling site (forest-steppe zone), the open rectangles represent the values from the Karasuk sampling site (steppe zone). In Figure 2 d the dark points represent values for measurements in Tarko-Sale (forest-tundra zone), the open rectangles – values from Samburg (tundra zone). The values $\langle X_{iFe} \rangle$ were calculated, using the following equation:

$$\langle X_{iFe} \rangle = \left\langle \frac{C_i}{C_{Fe}} \right\rangle, EF_i = \frac{X_{iFe}}{X_{FeCrust}}$$

where: C_i –mass concentration of i-element in aerosol particles; C_{Fe} – the same value for ferrous concentration, < > - geometric mean. From the presented data, it is evident, that the dark points and the open rectangles are positioned rather close to each other. As the distance between Klyuchi and Karasuk is about 450 km and the distance between Tarko-Sale and Samburg is about 250 km, this observation shows that the multi-element composition of the aerosol particles in Western Siberia depends only weakly on the geographical position of the sampling site for the chosen season.

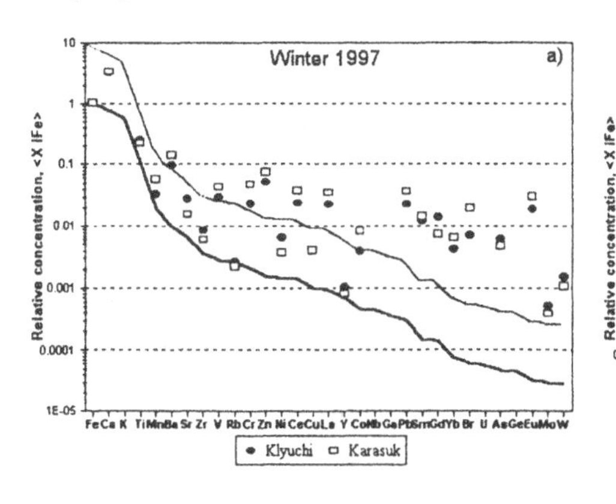

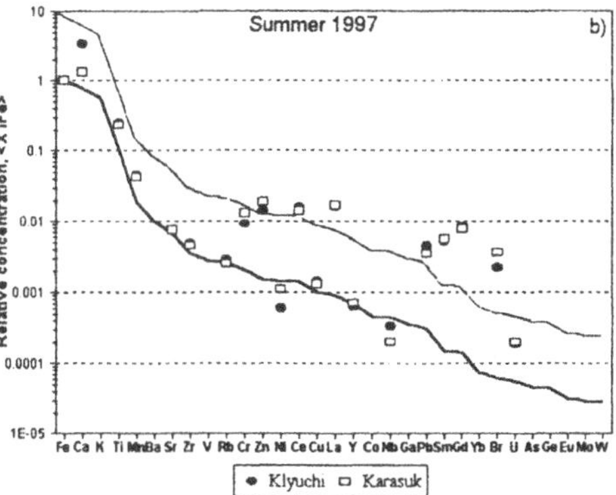

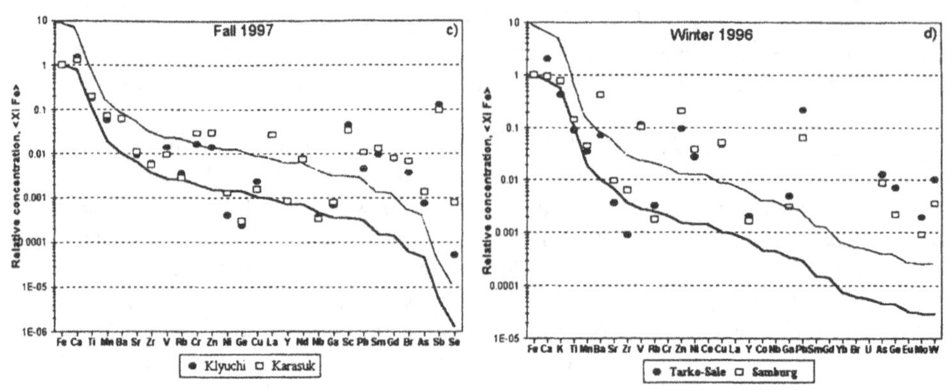

Fig.2. The temporal-spatial variation of the chemical composition of the atmospheric aerosol in West Siberia.

The bold curves represent the relative content in the earth's crust (so-called Clark of elements). The thin lines correspond to EFi=10. These lines divide the elements into two groups. To the first group (EF_i <10) belong the elements, which are formed due to soil erosion. To the second group (EF_i >10) belong the elements, which mainly originate from anthropogenic sources.

The quantity relations between the levels of contamination of the atmosphere of the northern and southern regions of West Siberia is illustrated by the data in Table 2.

Table 2. Mass concentration of the different elements in the atmospheric aerosol from the North and South West Siberia regions in winter.

Region	**Site**	**Anthropogenic sources, ng/m^3**							
		V	Zn	Ni	Cu	Pb	As	Ge	
North	Tarko-Sale	16	18	5	9	40	6	1	
	Samburg	13	27	4	7	9	1	0,3	
South	Klyuchi	21	38	2	7	15	1	0,7	
		Soil-erosion sources, ng/m^3							
		Ca	Ti	Mn	Sr	Zr	Rb	Y	Fe
North	Tarko-Sale	380	17	8	0,7	0,2	0,7	0,2	190
	Samburg	120	18	6	1,2	0,8	0,3	0,2	130
South	Klyuchi	2400	300	90	21	12	6	1,2	1800

This table demonstrates the most important feature of the space variation in concentrations from technogenic and soil erosion sources. The concentration of the latter (soil erosion) is much higher in the south, compared with the north. This implies that the characteristics of the soil erosion component of AA has a regional character, therefore, its elemental composition is determined by the regional peculiarities of the soil composition, its erosion properties and the climatic peculiarities of the region. An absolutely different picture is observed in the space-time deviation of AA composition, which is defining by influence. From Table 2 data it can be seen, that values for the concentrations of the elements of technogenic origin for the whole territory of West Siberia are within a factor of 2-3 the same. This indicates that technogenic contamination appears on a global scale.

The global character of the contamination in Siberia is supported by results from the multi-element analysis of individual aerosol particles obtained by an electron probe X-ray microanalysis (EPXMA) (Van Malderen et al., 1994; Van Malderen, 1996; Koutsenogii P.K.

et al., 1996). This method allows the determination of the elemental composition of individual aerosol particles larger than 0.3 μm. In the analysis particles in the size range 0.3-2 μm are considered. Table 3 presents the results of the multi-element analysis of individual aerosol particles in different regions of Siberia.

Table 3. Multi-elemental composition of the individual aerosol particles.

Baikal Lake, summer					
Upper zone		**Middle zone**		**Lower zone**	
Content	Abud.,%	Content	Abud.,%	Content	Abud., %
Al, Si, Fe	45.0	Al, Si, Fe	71.7	Al, Si, Fe	35
Org	25.2	Ca, S	14.3	quartz	24.5
Ca, S	13.3	Fe-rich	5.3	Al-rich	7.1
P, K, Cl, S	5.8	Cl, K, P, S	2.4	P, K, Cl	6.0
Fe-rich	4.4	Al-rich	0.6	Pb-rich	0.2
S-rich	3.9	S-rich	3.7	S-rich	6.4
Zn-rich	1.2	Zn-rich	1.1	Ca, S	4.7
Al-rich	0.9	Ti-rich	1.0	Ti-rich	0.4

Novosibirsk region							
Karasuk, winter		**Karasuk, summer**		**Klyuchi, winter**		**Klyuchi, summer**	
Content	Abud.,%	Content	Abud.,%	Content	Abud.,%	Content	Abud.,%
Al, Si, Fe	57.4	Al, Si, Fe	64	Al, Si, Fe	57.3	Al, Si, Fe	53.5
Quartz	11.5	Fe-rich	10.4	Si-rich	24.3	Ca, S	19.6
Ca, S	13.3	Ca, S	10.3	Organic	9.9	Zn, Fe, Ti	8.6
Fe-rich	9.1	Cl, K, P, S	7.2	Ca, S	5.1	Fe-rich	8.0
Pb-rich	5.2	Ca, Si	6.8	Fe-rich	3.4	P, S, K,Cl	6
S-rich	3.5	Pb-rich	1.3			S-rich	3.6
Zn-rich	1.2						
Ti-rich	0.8					Pb-rich	0.9

There is no significant difference in the element composition of individual particles, collected in different regions of Siberia in the same season. This supports the suggestion, that AA (diameter from 0.3 to 2 μm) in Siberia are well mixed and form the system of global scale – "Siberian Haze"(Koutsenogii P.K. et al., 1996).

In cases, when the type of source is known, or the source may be positively identified, it is not complicated to determine the multi-element composition from the specific source and estimate its share in the total mass concentration of AA.

In particular, the multi-element compositions from large forest fires and the pollen component of AA in Siberia are well established (Golovko, 2001; Koutsenogii K.P.et al., 1996).

Ionic composition. Another important component of Siberian AA is a water-soluble part, which may be characterized by its ionic composition. In the winter period the soluble fraction share in the total aerosol mass reaches 47%, in summer it decreases down to 3,7%. The significant increase of the insoluble share in the total aerosol mass in summer is caused by enhanced soil erosion in the period, when the ground is not covered by snow. This supports once more the increase of importance in soil erosion in summer compared to winter for the chemical composition of AA.

Spatial and seasonal variations of AA ionic composition have been presented and discussed in Smolyakov et al. (2001). The average ratio of ion equivalents (% in the sum of cations and anions, respectively) is shown in Table 4. The main anion in AA total composition is sulfate. The ratio of equivalents $[NH_4^+]/[SO_4^{2-}]$ in any case are less than 0.5.

This fact indicates that for the neutralization of sulfate and other anions, the presence of significant amounts of soil typical cations in the water-soluble fraction of AA is required. Cations like Ca^{2+}, Mg^{2+}, Na^+, K^+ have no volatile form and should be introduced through soil erosion or sea-salt aerosols. The true effects of these sources result from the different contributions of (Ca^{2+} + Mg^{2+}) and Na^+ in the northern and southern sites of observation. The contribution of Cl^- is higher in the southern sites wherever the contribution of NO_3^- is higher in the southern sites. However the ratio of equivalents [Na^+]/[Cl^-] in the south of West Siberia may reach 1 in summer. This fact may be explained by the transport of salt particles from the Aral-sea or salty soils in the Kulunda steppe. In Samburg the sum of cations (without H^+) is small in relation to sum of anions, and a higher acidity AA occurs.

Table 4. Mean ionic composition (% equivalent) in the water-soluble fraction of atmospheric aerosols in the West Siberia.

Sites	NH_4^+	(Ca+Mg)	Na^+	K^+	H^+	HCO_3^-	F^- + $HCOO^-$	Cl^-	NO_3^-	SO_4^{2-}
Samburg	29.6	22.6	27.8	5.7	14.3	11.3	5.9	8.0	3.3	71.5
Tarko-Sale	36.8	31.4	21.7	6.3	3.8	19.2	3.7	5.1	4.8	67.2
Klyuchi	30.2	49.2	9.4	6.7	4.6	20.8	3.0	2.2	12.2	61.2
Karasuk	43.2	34.4	10.4	8.5	3.5	7.7	2.5	2.3	13.6	73.9

These conclusions about the variable action of different sources on the stoichiometric ionic composition of AA are supported by statistical (factor) analysis of the concentration variation of different ions in each series of observations. In Samburg, the separate group is often defined by the combination NH_4^+, SO_4^{2-}, H^+, which reflects gas-phase conversion processes and an acid factor as well as the combination Na^+, Cl^- (sea factor). In Klyuchi and Karasuk the combination Ca^{2+}, Mg^{2+}, HCO_3^- (soil-erosion factor) is abundant.

The changes in the chemical composition of AA in the south and north of West Siberia, taking into account not only the multi-elemental and ionic compositions, but also the content of organic and inorganic carbon, are shown in Figure 3.

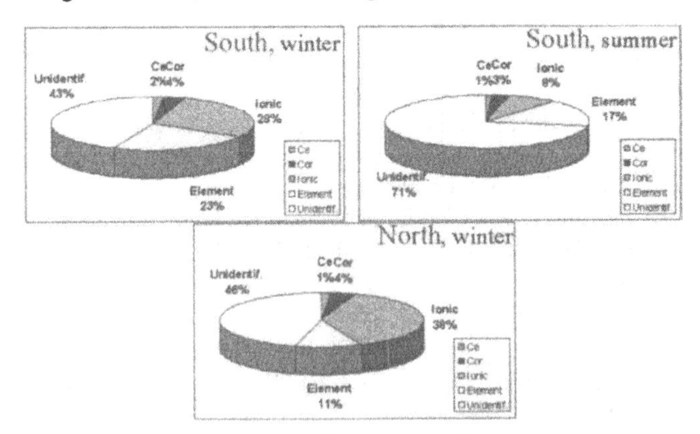

Fig. 3. Chemical composition of atmospheric aerosol in West Siberia.

The concentration of organic carbon is 2-4 times higher than inorganic, the total content of carbon in the aerosol mass is rather small, on average between 4% and 6%. The shares of mineral and water-soluble fractions of AA change significantly from summer to winter in the northern and southern regions of West Siberia. The presented data on space-time changes in the chemical composition of Siberian AA, show, how complicated the process of formation is of one of the most important feature of AA, deciding considerably their properties.

Conclusions

The system of stationary monitoring, created in Siberia, includes main soil-climatic zones of the vast region and allows studies of changes in chemical composition and the particle size distribution of AA of this region on local, regional and global scales to be performed. From complex studies of the main characteristics of AA in different soil-climatic zones and during different seasons, the natural and anthropogenic factors were determined, which influence the particle size distribution and chemical composition of the aerosols.

The investigations was supported by grants from RFBR, INTAS, Department Science of USA, and SB RAS.

References

Golovko V.V., The study of the atmospheric aerosol pollen component in south of Western Siberia. Thesis of Ph.D., 2001, Novosibirsk, 16 p.

Koutsenogii K.P., Valendik E.N., Bufetov N.S., Baryshev V.B. Emission from large forest fire in Siberia, *Siberian ecological journal* **1** (1996).

Koutsenogii P.K., Boufetov N.S., Smirnova A.I., Koutsenogii K.P.; Elemental composition of atmospheric aerosols in Siberia, *Nucl. Instr. Meth. Phys. Res.* **A405** (1998) 546-549.

Koutsenogii P.K., Van Malderen H., Hoornaert S., Van Grieken R., Koutsenogii K.P., Bufetov N.S., Makarov V.I., Smolyakov B.S., Nemirovski A.M., Osipova L.P., Krukov Yu. A., Ivakin E.A., Posukh O.L., Bronshtein Smolyakov B.S., Koutsenogii K.P., Pavluik L.A., Filimonova S.N. and Smirnova A.I.; Monitoring of ion composition of atmospheric aerosol in the West Siberia, *Atmospheric and oceanic optics* ? (2001) ?.

E.L.; Siberian Haze. Complex study of aerosols in Siberia, *Atmospheric and Oceanic Optics* **9** (1996) 451-454.

Van Malderen H., Van Grieken R., Bufetov N., Koutsenogii K.; Chemical characterization of individual particles in Central Siberia, *Envir. Sci. Technol.* **30** (1996) 312-321.

Van Malderen H., Van Grieken R., Khodzher T.V., Bufetov N.S., Kutsenogii K.P.; Analysis of individual aerosol particles in Siberian region. Provisional results, *Atmospheric and Oceanic Optics* 7 (1994).

Whitby K.T.; The physical characteristics of sulfur aerosols, *Atmospher.Environ.* **12** (1978) 135-159.

Model Accumulation Mode of Atmospheric Aerosol Including Photochemical Aerosol Formation, Coagulation, Dynamics of the Atmospheric Boundary Layer and Relative Humidity

P. Koutsenogii

Institute of Chemical Kinetics and Combustion SB RAS, 630090 Novosibirsk, Institutskaya, 3

Introduction

The results of experimental and theoretical studies on the seasonal variations of the diurnal cycle of the sub-micron atmospheric aerosol (AA) characteristic in Siberia are presented. Measurements of the particle size distribution of AA in two geographical locations, i) in the Novosibirsk region and ii) in the region of Lake Baikal, in the size range 10^{-3} to 10^{3} μm have shown that the distributions in both regions are practically identical (Koutsenogii, 1992). As the distance between regions of observation is about 1300 km, this supports that global processes influence on the particle size distribution of the observed AA. This distribution has three modes and may be described by a function, which according to the Whitby (Whitby, 1978) classification characterizes remote continental AA. The results are illustrated in Figure1a.

The sub-micron mode of AA in remote continental regions in summer is formed as a result of photochemical conversion of chemically active gas precursors (Seinfeld, 1986). Studies of the dynamics of diurnal variations of particle size distribution and concentration confirms the results of experimental curve interpretation (Koutsenogii, 1992; Koutsenogii and Jaenicke, 1995). The specific feature of the diurnal variation of the number particle concentration (see curve 2 on Figure1b) is the presence of two peaks, i.e. one in the morning and one in the evening. Since the rate of photochemical gas - to - particle conversion is proportional to the intensity of solar light and concentration of chemically active gas precursors a satisfactory theoretical description of experimental data can be given when it is taken into account that the concentrations of chemically active gas impurities is also depend on the height of the boundary layer. The results of calculations of the diurnal particle concentration according to the model proposed in Koutsenogii (1992), Koutsenogii *et al.* (1995) and Koutsenogii *et al.* (1998) are shown in Figure 1b, curve 1. The mechanism of photochemical formation of the sub-micron AA fraction from gas precursors is also confirmed by experimental data on the diurnal size distribution variations for sub-micron aerosols (see Figure1c) and by the data obtained in smog chamber measurements (Koutsenogii, 1992; Koutsenogii and Jaenicke, 1995). In particular, during the morning the maximum of the particle number concentration (see Figure1b, curve 2) is bimodal with a high concentration of very small particles (with a radius of ~ 10 nm) and the second maximum in the range of ~ 100 nm (see Figure 1c, curve 1). In the daytime the size distribution exhibits a rather wide mono-modal distribution (Figure 1c, curve 2), while at night a more narrow mono-modal distribution is observed (see Figure 1c, curve 3). These regularities highlight the important influence of coagulation in the dynamics of the particle size distribution in the sub-micron range of AA.

The accumulation mode (0,1 - 1 μm) is the longest lived fraction of AA and its characteristics are not so variable with time. It is convenient to study the dynamics of AA mass concentration changes in long-term (seasonal) cycles of atmospheric processes using the methods most sensitive to the properties of the particles in this size range. Nephelometry is one of the simplest methods to study the dynamics of accumulation mode behavior.

I. Barnes (ed.), Global Atmospheric Change and its Impact on Regional Air Quality, 237–242.

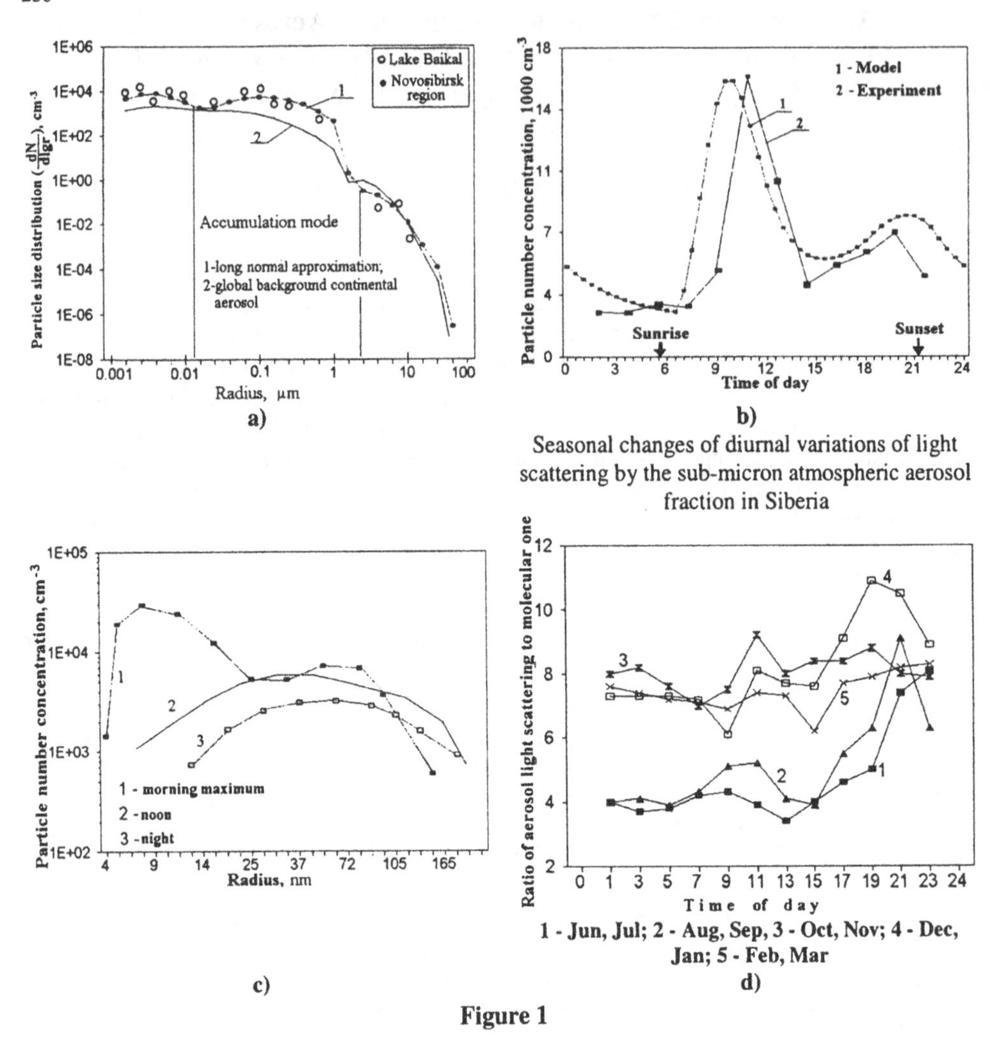

Seasonal changes of diurnal variations of light scattering by the sub-micron atmospheric aerosol fraction in Siberia

Figure 1

Experimental results

As has been shown (Koutsenogii et al., 1995) the main light scattering is attributable to aerosol particles in the size range 0,1 - 0,3 µm. In this size range the sensitivity the of nephelometer FAN-A has a mean dependence on particle size of r^3 (Koutsenogii et al., 1995). This may be shown by calculation of aerosol light scattering in the viewing volume of FAN-A according to Mie-theory (Bohren et al., 1983). This means that the total aerosol light scattering is proportional to the total mass of the nucleation mode of the aerosol size distribution, or the total mass of sub-micron particles.

For aerosol light scattering (S-1) it is possible to calculate the coefficient of proportionality to the aerosol mass of the sub-micron particles (M_{sub}) (Koutsenogii, 1997).

$$M_{sub} = 2.6 * (S - 1) \qquad (1)$$

Figure 1d shows the seasonal changes in the diurnal variations of light scattering by the AA sub-micron fraction in the Novosibirsk region (Academ-town site).
In winter (Figure1d, curve 4) when the solar radiation is minimum both in intensity and in duration, the diurnal peaks of light scattering, which is proportional to the mass concentration of the accumulation mode becomes smooth and the time interval between the appearance of the peaks decreases. In addition, compared to the warm period (Figure 1d, curve 1) in winter the overall light scattering is several times higher. This can be explained by the decrease in the height of the atmospheric boundary layer in winter and the increase in the intensity of anthropogenic pollution with chemically active gases.

Data from simultaneous measurements of the light scattering in different sites of the Novosibirsk region in winter and summer are shown in Figure2a and Figure 2b, respectively. From these figures two principal peculiarities are apparent. The firstly, is the very close similarity of the diurnal light scattering cycles at the Klyuchi and Karasuk sites. The second is the very different diurnal cycle forms between the summer and winter seasons.

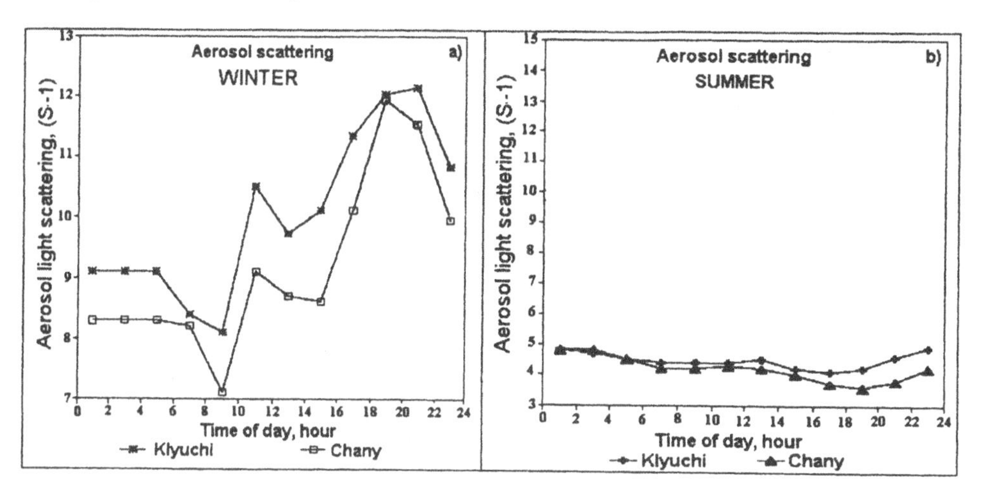

Figure 2

Theoretical Model

Let us assume that the sub-micron fraction of the atmospheric aerosol is a mixture of two mono-disperse aerosol particles. In this case the total number concentration (N) is sum,

$$N=N_1+N_2 \qquad (2)$$

Where N_1 is the number concentration of the nuclear mode and N_2 is the number concentration of the accumulative mode.
The lifetime of the second mode is 10 days and the particles homogeneously distribute in the atmospheric boundary layer. The lifetime the small particles from the nuclear mode depend on the coagulation with the large particles and transport of the aerosol from the atmospheric boundary layer (ABL) (Koutsenogii and Jaenicke, 1995).
During the coagulation

$$\frac{dN_1}{dt} = -k_{coag} N_1 N_2 = -\frac{1}{\tau_1} N_1 \tag{3}$$

According to Seinfeld (1986) the coagulation constant (k_{coag}) for the radius 23 nm particle and 0,16 μm is $5 \cdot 10^{-2}$ cm^3/s. From experimental data (Koutsenogii, 1992; Koutsenogii and Jaenicke, 1995) the number concentration of the accumulation mode is 2000 cm^{-3}. τ_1 is approximately 10 hours. In summer the aerosol particle transport time from the ABL is roughly 3 hours. Because the mass concentration of the sub-micron fraction AA is proportional to the total scattering light (σ) we can write

$$\sigma = \sigma_1 + (\sigma_0 + \sigma_2) \tag{4}$$

whereσ_1 is the light scattering of the nuclear mode, σ_0 is the stable part of the light scattering of the accumulation mode, σ_2 is the additional part resulting from the coagulation small and large sub-micron particles. The temporal change of two modes are

$$\frac{d\sigma_1}{dt} = -\frac{1}{\tau_1}\sigma_1; \quad \frac{d\sigma_2}{dt} = +K_M \frac{1}{\tau_1}\sigma_1 \tag{5}$$

$$K_M = -\frac{d\sigma_2}{d\sigma_1} = +\left(\frac{d\sigma_2}{dM_2}\right) / \left(\frac{d\sigma_1}{dM_1}\right) \tag{6}$$

because $-\frac{dM_2}{dM_1} = 1$

Theoretical calculations (Koutsenogii et al., 1995) have shown that the function σ (r) can be approximated by

$$\sigma = \sigma_0 \frac{1 + (r/R)^{1,5}}{1 + (R/r)^6} \tag{7}$$

In (7) R=0,16μm. Using (6) and (7) results in K_M=210 if r=23 nm. This result implies that the increase of the light scattering as a consequence of the coagulation between two aerosol modes is important. Thus for $\frac{d\sigma}{dt}$ may be represented by

$$\frac{d\sigma}{dt} = C_o f(t+T) - \frac{\sigma - \sigma_0}{\tau_2} Z(t) \tag{8}$$

$1/\tau_2$ is the amplitude of the vertical transport rate. This value is 0.3 $hour^{-1}$. Z(t) is the temporal rate of the vertical diurnal exchange. All the parameters to solve equation (8) may be obtained from experimental data (Laikhtman, 1961).

WINTER

$$T \approx 1 \text{ hour}, \quad f(t) = \begin{cases} 0, \ if: \ 0 \le t \le 9; \ 17 \le t \le 24 \\ \sin\left(\frac{\pi}{8}(t-6)\right), \ if: \ 9 \le t \le t \end{cases} \tag{9}$$

$$Z(t)=1 \tag{10}$$

because the turbulent diffusion coefficient for the whole day is constant in this season.

SUMMER

$$T \approx 2 \text{ hours}, \ f(t) = \begin{cases} 0, \ if: \ 0 \le t \le 6; \ 22 \le t \le 24 \\ \sin\left(\dfrac{\pi}{16}(t-6)\right), \ if: \ 6 \le t \le 22 \end{cases} \tag{11}$$

$$Z(t) = 1{,}2 + \sin\left(\frac{\pi}{2}(t-1) + \frac{3}{2}\pi\right) \quad \text{(see Figure 32 from (Laikhtman, 1961))} \tag{12}$$

In summer it is necessary to take in account the dependence on the relative humidity

$$\sigma = \sigma_0\left(1 - f_{RH}(t)\right)^{-0{,}75} \tag{13}$$

$$f_{RH}(t) = \begin{cases} 0{,}75 \ if: \ 0 \le t \le 6 \\ 0{,}75 - 0{,}25\sin\left(\dfrac{\pi}{18}(t-6)\right) \ if: \ 6 \le t \le 24 \end{cases} \tag{14}$$

In winter f_{RH} is constant during the whole day.

Figure 3a and Figure 3b show the results of the theoretical calculations obtained with the proposal model. These calculations are represented by the curves marked "model calculation". The experimental data are also shown in the figures.

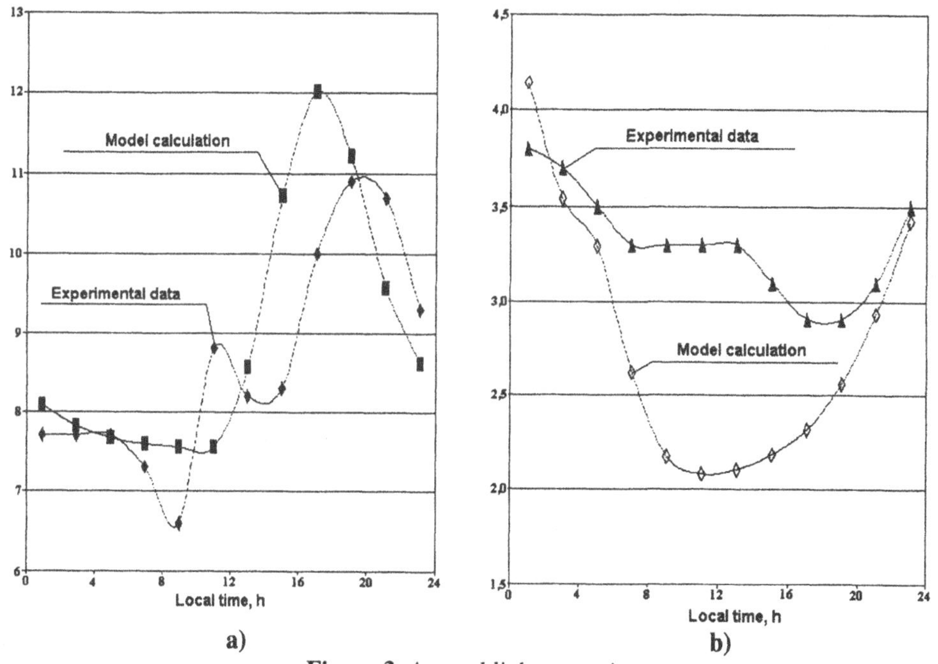

a) b)

Figure 3. Aerosol light scattering.

The above comparison of the experimental data and theoretical calculations show that the latter describes the main peculiarities of the experiment measurement.

Conclusion

A theoretical model has been developed to describe the seasonal variation of the diurnal cycle of the accumulation mode of the atmospheric aerosol including the photochemical aerosol formation, coagulation, dynamics of the atmospheric boundary layer and relative humidity. The calculations using this model show that one can correctly reproduce the experiment data.

Acknowledgements

These investigations were supported by a grant from SB RAS for the joint research project "Siberian Aerosol".

References

Bohren C.F., Hufman D.R.; Absorption and Scattering of Light by Small Particles, in: John Wiley (ed.) N.Y. (1983)

Koutsenogii P.; Aerosol measurements in Siberia, *Atmosph. Res.* **44** (1997) 167-173.

Koutsenogii P.; Measurements of remote continental aerosol in Siberia. Ph.D. Dissertation Johannes-Gutenberg University, Mainz, Germany (1992).

Koutsenogii P.K., Bufetov N.S., Kirov E.I., Shuysky S.I.; Dynamics of 24-hours and Seasonable Cycles of Aerosol Formation in Atmosphere from Measurements in Novosibirsk Region, *Atmospheric and Oceanic Optics* **8** (1995) 1355-1365.

Koutsenogii P., Jaenicke R.; Number concentration and size distribution of atmospheric aerosol in Siberia, *J. Aeros. Sci.*, **25** (1995) 377-383.

Koutsenogii P.K., Levykin A.I., Sabelfeld K.K.; Numerical modeling of diurnal cycle of particle size distribution of submicron atmospheric aerosol, *Atmospheric and Oceanic Optics* **11** (1998) 540-542.

Laikhtman D.L.; The physics of the atmospheric boundary layer, in: GIMIZ, Leningrad (1961) 253 (In Russian).

Seinfeld J.H.; Atmospheric Chemistry and Physics of Air Pollution, in: J. Willey & Sons (eds.), (1986) 738.

Walter H.; Coagulation and size distribution of condensation aerosols, *J.Aeros.Sci.* **4** (1973) 1-15.

Whitby K.T.; The physical characteristics of sulfur aerosols, *Atmospher.Environ.* **12** (1978) 135-159.

Volatile Organic Compounds (VOCs) in the Atmospheric Air of Moscow: The Impact of Industry and Transport

Laritcheva Olga O., Semenov Sergei Yu., Smirnov Valentin N., Tananikyn Nikolai I.

Ministry of Health of Russian Federation Russian Research Center of Emergency Situations,
46,4 Shabolovka str., Moscow 117419, Russija Phone: (095) 954-70-45,
Fax:(095) 113-22-50, E-mail sysemenov@mtu-net.ru

Introduction

The Russian Research Center of Emergency Situations (RRCES) conducted a survey of Moscow air in the presence of organic compounds in the period from May to October. It is known, that the pollution of the environment by organic pollutants is one of the major causes of adverse health effects on people. Earlier RRCES surveys studied the contamination of Moscow atmospheric air by dioxins and furans (1, 2).

The purpose of this research was the identification and quantitative determination of organic compounds in Moscow air, as well as an estimation of the pollution degree, and contributions of the different sources. City roads, city background territories (park zones) and other areas near industrial zones were examined. The influence of motor-vehicle transport, large plants, including MSWI, power plants, petroleum refineries etc. was studied.

Methods

A total of 46 sites were inspected and more than 100 samples were taken. Sampling sites (Table1) were selected so that it was possible to estimate the influence of motor-vehicle transport (Sadovoe Kolco, Prospect Mira), industrial zones (Kaloshino, Ochakovo, Kapotnya, ZIL, Serp i Molot), incinerators (Beryulevo, Altufevo) on the composition and quantity of organic compounds in the air.

Several tens of individual air pollutants related to the following classes were determined in each experiment:

- Saturated and non-saturated hydrocarbons
- Polyaromatic hydrocarbons
- Cyclic and aromatic hydrocarbons
- Ethers and esters
- Chlorinated organic compounds
- Oxygen- and sulfur-containing compounds

Sampling was executed by pumping a definite air volume (1-10 m^3) for 50 – 70 min on average through a trap with Tenax. Sampling was conducted at a height of 1.5 m at 9.00 a.m. For a better estimation of the organic compound concentrations repeat sampling was conducted the following day at the same place and time. The data were averaged if the results diverged by no more than 30%, otherwise, supplementary measurements were conducted.

Comparisons of air probes sampled at different times of the day were conducted. It was established that air probes sampled at 9 a.m. contained 1.5 times more pollutants than probes sampled at 11.30 p.m. at the same place.

The analysis of the air was performed by GC/MS according to "The method of determination of the organic compounds in atmospheric air No. 1-M/5.6" (3). Identification of the compounds was

I. Barnes (ed.), Global Atmospheric Change and its Impact on Regional Air Quality, 243–247.

by comparison of retention times and mass spectra with those of authentic samples. The standards of the Environmental Protection Agency (EPA), USA were used. The detection limit of the method was < 0.001 mg/m^3.

Table 1. List of sampling sites

No.	SAMPLING SITES	No.	SAMPLING SITES
1	Valovaya str., 6/ 8	24	Sukharevskaya square
2	Zacepskij val str., 14	25	Sukharevskaya square, near 70 m
3	Ochakovo, Zheleznodorozhnaya str.,1	26	Kaloshino, Montazdnaya str., 40
4	Verbnaya str.- TEC	27	Rublevo-Uspenskoe high road, 2 km
5	Kapotninskij 1 st. lane, 41	28	Osennyay str. , 4
6	Veshnyakovskaya str., 27	29	Osennyaya str., 4
7	Baykalskaya, str., 4	30	Altuvevskoe high road, UM-37
8	Stupinskij blind alley, 6	31	Profsoyuznaya str., at station Metro "Profsoyuznaya."
9	Kapotnja, Verkhnie polya	32	Zubovskij boulevard, 15
10	Kapotnya, 1-quarter, market	33	Varshavskoe high road, at station Metro "Varshavskaya"
11	Kapotnya, station bus №49	34	Batajskij lane, 33
12	Ochakovskoe high road 17, TEC near 800 m	35	Amurskaja str., 27
13	Dnepropetrovskaya str., Forestry	36	Bajkalskaya str., 16
14	Ochakovo, Generala Dorokhova str.	37	Amurskaya str., 30
15	Rizhskoe high road 1,5km from MKAD	38	Mytnaya str. 62
16	1-Gajvoronovskij lane, 5	39	Altufevskoe high road, 33a
17	Gajvoronovskaya str. with Lyublinskoj str.	40	Krylatskoe, row canal
18	Gajvoronovskaya str.	41	Avtozavodskata str., ZIL
19	Gruzinskij Val str. near square Beloruskij Vokzal	42	Tyufeleva Roshcha, ZIL
20	Biryulevo-West contra MWSI	43	Entuziastov high road "Serp i Molot."
21	Biryulevo-West, platform"Pokrovskaya"	44	Lane near "Serp i Molot"
22	Prospekt Mira, Square Rizhkij Vokzal	45	Ugreshskaya str.
23	Kashirskoe high road, 70	46	Ugreshskij passage

Results and Discussion

In all the samples more than 50 compounds were quantitatively identified. Aliphatic and aromatic compounds dominated in all samples; PAH (naphthalene) and oxy-chemicals (acetone) are present. Chlor-organic compounds were also detected in 12 samples. Figure 1 shows the data on the content of aliphatic and aromatic compounds an all the sampling sites. Total concentrations varied over a wide range, from 0.07 mg/m^3 (sample No. 29) up to 1.37 mg/m^3 (sample No. 1).

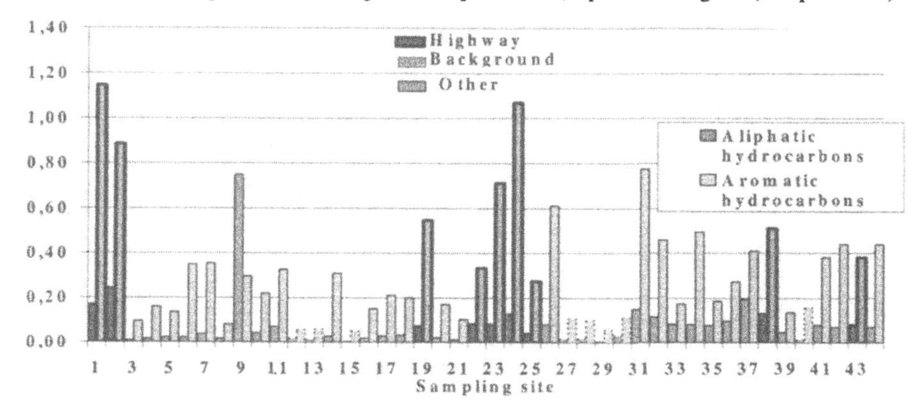

Figure 1. Aliphatic and aromatic compounds (mg/m^3) in Moscow air.

An estimation of the background air pollution was conducted by sampling 8 probes in the clean districts of Moscow, situated a long way from the sources of pollutant emission: No. 3, No. 5, No. 13; No. 15, No. 27, No. 28, No. 29, No. 40. Most contaminated are the sampling sites located directly on the large automobile roads (highways), where the concentration of organic compounds is 5 - 20 times in excess of the background values.

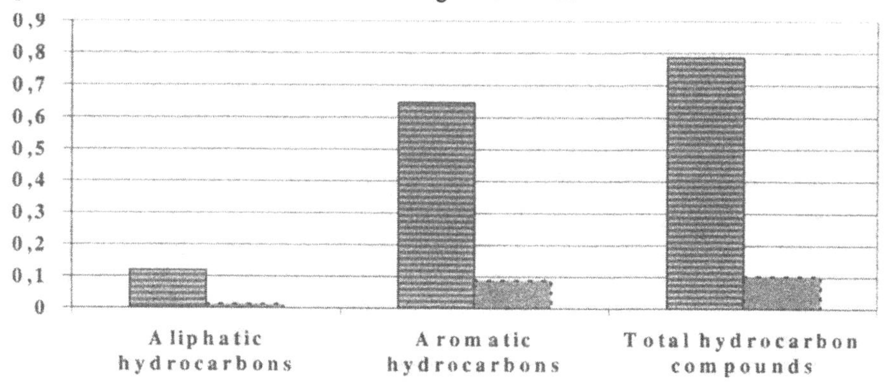

Figure 2. Highway/background average concentrations (mg/m^3) of aliphatic and aromatic hydrocarbon compounds in Moscow air.

Figure 2 shows the typical distribution by the main classes of compounds. The major contribution is made by aromatic hydrocarbons, which as a rule, account quantitatively for 80 to 90 % of the detected organic compounds.

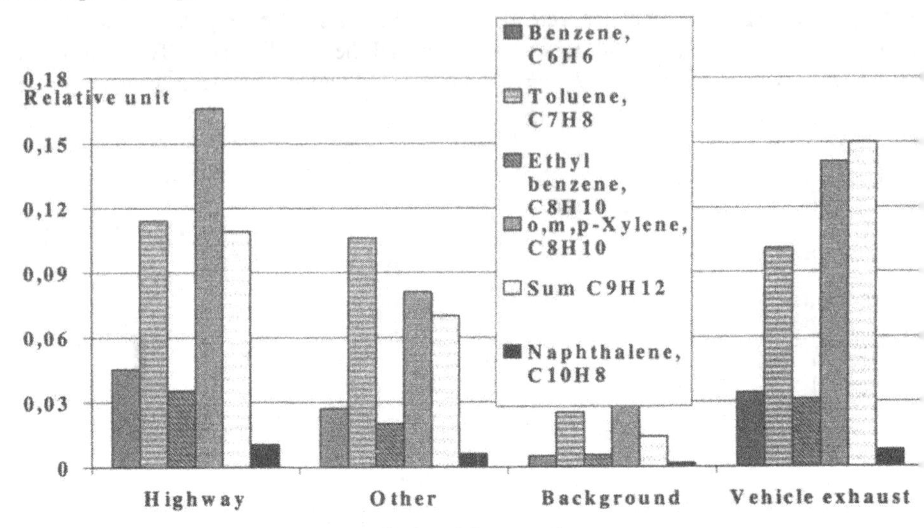

Figure 3. Ratios of major components polluting Moscow air.

The greatest contribution to the concentration of aromatic compounds is yielded by benzene, ethyl benzene, naphthalene, toluene, o-, m-, p- xylenes and C_9H_{12} isomers, which include isopropylbenzene, trimethylbenzene, methylethylbenzene. The relative content of these species to another in the background sites, in other sampling sites, as well as in vehicles exhausts shows only a small variation.

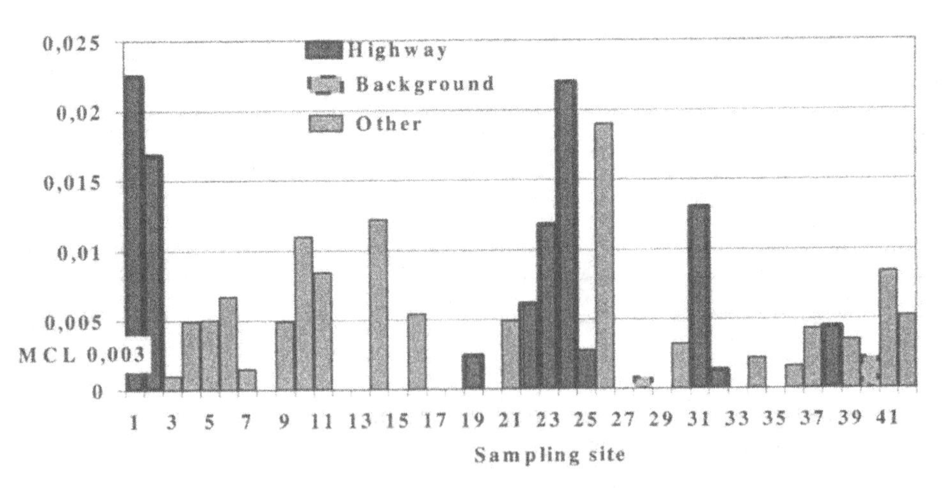

Figure 4. Naphthalene (mg/m^3) in Moscow air.

This demonstrates the same constant source of contamination of the air by organic compounds, i.e. from vehicles. The maximum contaminated level (MCL) is exceeded in many sampling sites by compounds such as naphthalene (Figure 4), trimethylbenzene (Figure 5), methylethylbenzene, ethylbenzene, isopropylbenzene, styrene.

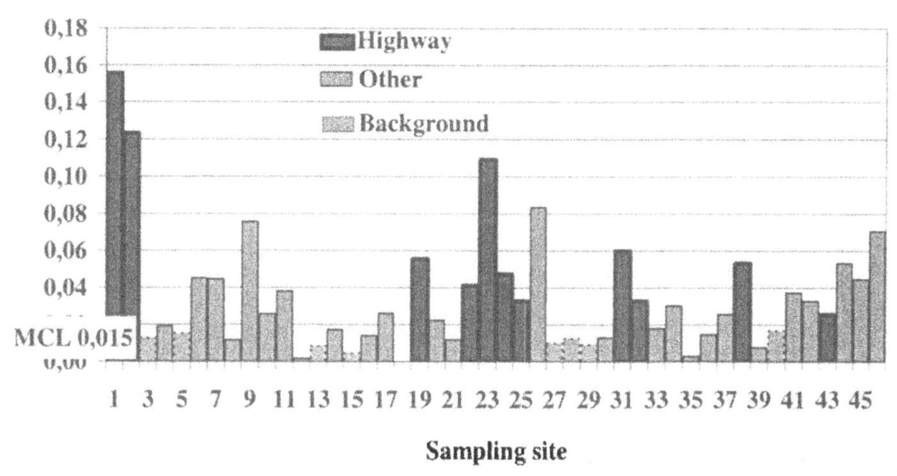

Figure 5. Trimethylbenzene (mg/m^3) in Moscow air.

The study has shown that Moscow air is contaminated by organic compounds, predominately, by aromatic hydrocarbons up to 90%. Vehicle exhausts are the main source of the organic compound emissions in the Moscow atmosphere. Influence of the transport is observed even in the background areas, i.e designated "clean" regions of the city. Concentrations of the organic compounds at the highway do not exceed those in the clean areas by more than 20 times. The study has revealed that the concentrations of some aromatic compounds at many sampling sites such as naphthalene, methylethylbenzene, isopropylbenzene, trimethylbenzene and ethyl benzene exceed MCL. The contamination has extremely adverse effects on the health of the Moscow population.

References

1. Semenov S.Y., Smirnov V.N., Zykova G.V., Finakov G.G. (1998). PCDD/PCDF emission from Moscow municipal solid waste incinerator. Organogalogen compounds, 1998, vol. 36, 301-305. 18th Symposium on Halogenated Environmental Organic Pollutants, "Dioxin '98".
2. Semenov S.Y., Zykova G.V., Smirnov V.N., Finakov G.G. Koverga A.V., Gorban O.M. (1999). Dioxin and Furan Concentrations in Snow of Moscow Region, Russia. Organogalogen compounds, 1999, vol. 43, 231-235. 19th Symposium on Halogenated Environmental Organic Pollutants, "Dioxin '99".
3. Method MKP "The method of determination of the organic compounds in atmospheric air No. 1-M/5.6", (1996).

Transcontinental Observations of the Surface Ozone and Nitrogen Oxide Concentrations by using the Carriage-Laboratory

T.A. Markova and N.F. Elansky

A.M. Oboukhov Institute of Atmospheric Physics, RAS
Pyzhevsky 3, 109017, Moscow, Russia
E-mail: markova@omega.ifaran.ru

Introduction

This work contains an analysis of a set of data on the O_3, NO, and NO_2 concentrations measured in the atmospheric surface layer by using the carriage-laboratory moving along the Trans-Siberian Railway. The measurements were performed in the course of the Russian-German TROICA (Trans-Siberian Observations of the Chemistry of the Atmosphere) experiments organized by the Oboukhov Institute of Atmospheric Physics (Russian Academy of Sciences) and the Max Planck Institute of Chemistry (Germany). The main goal of these experiments was monitoring of the gaseous pollutants in the atmospheric surface layer over Siberia. This region does not have stations of atmospheric monitoring. The periods covered by the TROICA experiments, the routes of the carriage-laboratory, and the parameters measured are listed in Table 1.

The carriage-laboratory was coupled to passenger trains. The ozone concentration in air was measured with Dasibi 1008-RS and Dasibiby 1008-AH instruments calibrated against a standard. The NO and NO_2 concentrations were measured with an AC-30M instrument with an accuracy of 1 ppb. Air samples were taken through Teflon tubes at an elevation of about 4 m above the ground.

Table 1. Periods, routes, and parameters measured.

Experiment	Period	Route	Parameters
TROICA – 1	17.11.95-02.12.95	N.Novgorod – Khabarovsk – Moscow	O_3, NO, NO_2, meteorological parameters
TROICA – 2	26.07.96-13.08.96	N.Novgorod – Vladivostok – Moscow	O_3, NO, NO_2, CO, VOC, black carbon aerosol, meteorological parameters
TROICA – 3	01.04.97-14.04.97	N.Novgorod – Khabarovsk – Moscow	O_3, NO, NO_2, CO, CH_4, VOC, black carbon aerosol, meteorological parameters
TROICA – 4	17.02.98-07.03.98	N.Novgorod – Khabarovsk – N.Novgorod	O_3, NO, NO_2, CO, CH_4, VOC, black carbon aerosol, meteorological parameters
TROICA – 5	26.06.99-13.07.99	N.Novgorod – Khabarovsk – Moscow	O_3, NO, NO_2,CO, CO_2, CH_4, ^{222}Rn, VOC, black carbon aerosol, meteorological parameters

Principle results

Figure 1 represents the 10-km mean concentrations of O_3 and NO_2 measured along the Trans-Siberian Railway in the course of the westward TROICA-1 - TROICA-5 expeditions. The $[O_3]$ variations are determined by atmospheric photochemical processes, by air transport

I. Barnes (ed.), Global Atmospheric Change and its Impact on Regional Air Quality, 249–254.

of different scales, and by wet and dry deposition. Over a large area covered with our measurements, the effect of these factors changed from one experiment to another and within the period of each of the experiments.

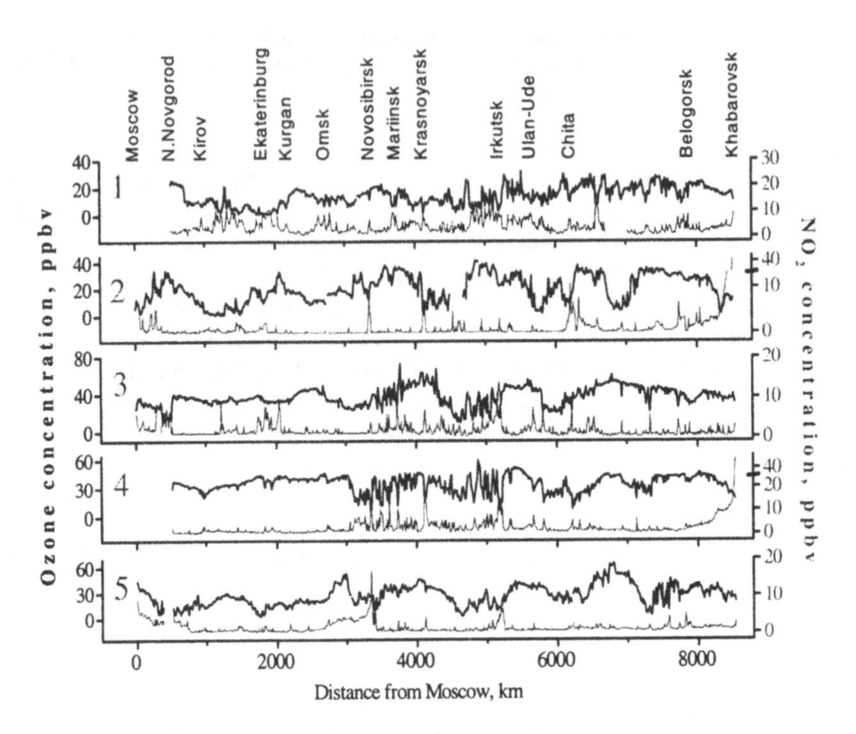

Figure 1. Behavior of the 10-km mean concentrations of O_3 and NO_2 along the westward sections of the TROICA-1 - TROICA-5 expeditions. The big figure at the beginning of each curve specifies the corresponding expedition.

A peculiarity in the ozone distribution over the continent is a commonly occurring increase of the surface ozone concentration in the eastward direction. The maximum and minimum gradients equal to 0.4 and 2.2 ppbv per 1000 km were measured during the eastward routes of the TROICA-5 and TROICA-3 expeditions, respectively. The gradient is most significant in fall and summer under near-uniform meteorological conditions over the entire area under observation.

The Moscow-Khabarovsk railway lies within a narrow latitude zone between 48.5° and 58.5°N. Therefore, this gradient can be associated with an effect of the transcontinental zonal transport of central-European pollutants supplemented with anthropogenic emissions from populous and industrial regions of European Russia and the Urals. Shakina et al., (2001) concluded that the zonal gradient can be partially associated with a rather intensive stratosphere-troposphere exchange over the eastern zone of the Continent influenced by the active subtropical high frontal zone. This stratosphere-troposphere exchange, most active in April, increased the gradient observed in the course of the eastward TROICA-3 expedition.

In the regions subjected to anthropogenic effects, the ozone behavior is peculiar. Under action of solar radiation, temperature and humidity of air, and NO_x, CO, and VOC (volatile organic compounds) concentrations, the ozone content in polluted air can either increase up to several MACs (maximum allowable concentration) or decrease significantly.

The former phenomenon is characteristic for western Europe, the USA, Japan (Oltmans and Levy, 1994; Kley et al.,1994; Appenzeller et al., 1996) and the latter is characteristic for Moscow (Elansky and Smirnova, 1997) and other localities. The measurements of pollutants within large industrial zones, towns, and inhabited localities around the Trans-Siberian Railway allow us to design a rather comprehensive pattern of the ozone seasonal behavior in the polluted atmosphere over a prolonged Russian Continent.

The major criterion for the industrial and transport pollution of the atmosphere is the NO_x concentration level. Figure 2 contains information on the NO_x concentration in the atmospheric surface layer along the railway. The NO_x concentrations measured by the carriage-laboratory coupled immediately behind a locomotive were averaged over warm seasons and over cold seasons. The area from Moscow to the Urals covered with a dense network of industrial enterprises and highways is most polluted. In addition, this area is influenced by pollutants transported from western and central Europe. This influence reveals itself in a CO-concentration zonal gradient (Bergamaschi et al., 1998). Carbon monoxide is less reactive than NO_x and, therefore, its behavior may provide comprehensive information on the long-range transport of pollutants and zonal gradient of some VOCs (Elansky and al., 2000). A great number of areas characterized by enhanced NO_x concentrations was revealed in southern regions of central Siberia. The NO_x concentrations were most high within cities, such as Novosibirsk, Krasnoyarsk, and Irkutsk. In the regions to the east of Lake Baikal, the effect of anthropogenic sources was very low; however, in the vicinity of Khabarovsk, it increased again.

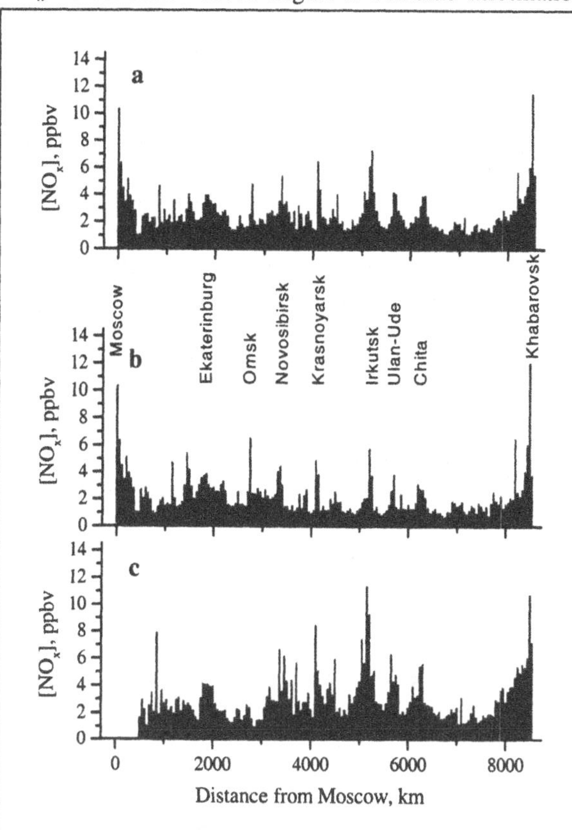

Figure 2. Behavior of the 50-km mean concentration of NOx along the Trans-Siberian railway: (a) data averaged over the TROICA-2 - TROICA-5 expeditions; (b) data averaged over the warm seasons, the TROICA-2 and TROICA-5 expeditions; (c) data averaged over the cold seasons, the TROICA-3 and TROICA-4 expeditions.

The NO_x and O_3 concentrations are in photochemical equilibrium and their changes are opposite in sign. Indeed, in Figure 1, almost all local maxima in the NO_x concentration are accompanied with minima in the O_3 concentration. We correlated the mean level of the ozone concentration within cities of more than 100-thousand population with that outside of these cities and their plumes. We considered 170 such events. In 139 events, the ozone contents in urban air were decreased; in 26 events, they were almost unchanged; and only in 5 events, they were increased relative to the surrounding rural areas.

In Figure 3, the data on the decrease in the urban ozone concentrations as compared to the corresponding rural valuess are presented. The data are given for most populated cities and are averaged over all expeditions. Within European cities, the decrease in the ozone concentration was rather significant and ranged from 30 to 40%. Within cities located in western Siberia on the plains ventilated well, the decrease was minimum and ranged from 13 to 25%. The decrease was most significant within cities located in mountainous central and eastern Siberia. This decrease is caused by the following factors. First, these cities include big industrial enterprises. Second, they are located in valleys where air exchange with the surroundings is weakened. Third, for these cities, a high frequency of surface inversions is characteristic, especially, in cold seasons when this area is influenced by vast Siberian anticyclones promoting accumulation of NO_x and of other pollutants.

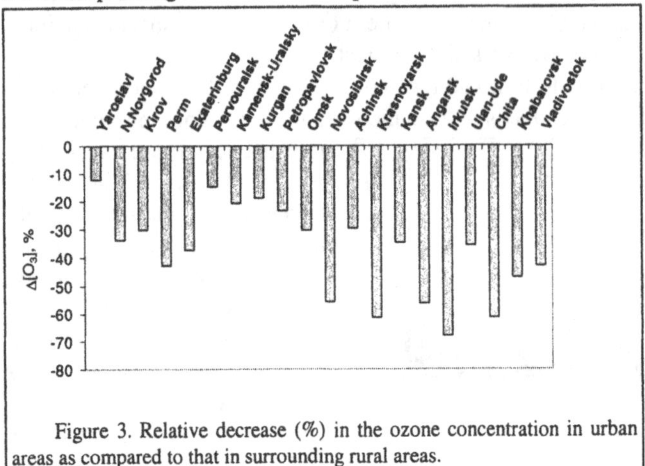

Figure 3. Relative decrease (%) in the ozone concentration in urban areas as compared to that in surrounding rural areas.

In plumes of cities, the ozone content was always decreased almost without exception. The city impart can be significant at a distance as long as several hundred kilometers. For example, on November 19, 1995, the TROICA-1 expedition found the Novosibirsk plume prolonged for 350 km from the city.

For areas of Russian cities, a decrease in the surface ozone concentrations is typical. However, sometimes, the ozone concentration within cities can be increased as a result of photochemical reactions. In the course of the TROICA-5 expedition, such an event was observed within polluted regions of the Khabarovsk Krai. On July 2, 1999, in the course of the eastward movement of the carriage-laboratory, an increase in the ozone concentration to 166 ppbv was observed within Birobidzhan (Figure 4). On July 3, 1999 (Saturday), in the course of the westward movement of the same

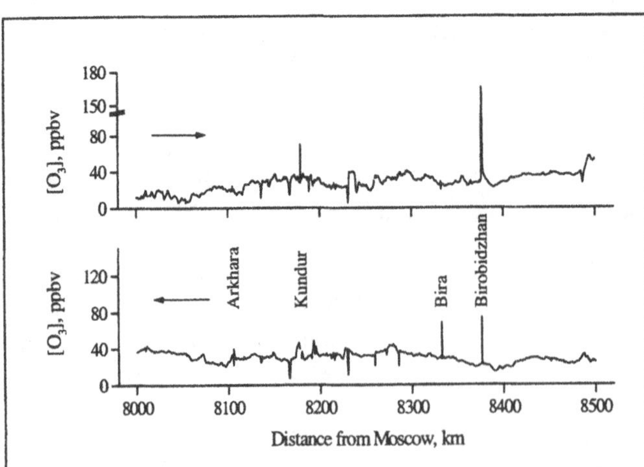

Figure 4. Ozone concentration increase due to atmospheric photochemical activity over the Khabarovsk Krai in the course of the TROICA-5 expedition.

expedition, a smaller increase in the ozone concentration was observed within Birobidzhan and other towns of this region. On July 2 and 3, 1999, the NO_x concentration in this region was nearly the same. The measurements were performed under hot cloudless conditions at a daytime temperature of 25-32°C. Analyses showed that air samples collected in this region were characterized by a high content of unsaturated hydrocarbons and their derivatives with a maximum in the vicinity of Birobidzhan (within the town of Birobidzhan, air was not sampled) (Elansky et al., 2000). A trajectory analysis revealed a prolonged transport of polluted humid air masses from Japan to the Khabarovsk Krai. A combination of intensive solar illumination, high temperature and humidity of air, and organic pollutants in the air masses led to intensive generation of ozone up to a concentration level constituting a hazard to human health. This observation shows that clearly pronounced smog events, similar to those in other regions of the globe, are sometimes observed in Russia.

In the course of summer and spring expeditions, we repeatedly registered extraordinary changes in the air composition due to the burning of grass and forests. The most intensive fires were registered in summer, 1996 and 1997, over eastern Siberia. Burning biomass emits reactive pollutants, such as CO, CO_2, CH_4, NO_x, and VOCs. As a result of photochemical reactions in the atmosphere, the ozone concentration in surface air can increase.

Measurements performed on the basis of the carriage-laboratory in the summer of 1999 revealed abnormally high concentrations of carbon monoxide, aerosol (including soot one), and ozone. Figure 5 shows the CO and O_3 concentrations measured along the railway section between 6400 and 7800 km from Moscow. This area was influenced by an air mass transported from China and Mongolia. The CO concentration was as high as 500-600 ppmv and exceeded significantly its mean level of 100-150 ppmv characteristic for unpolluted air. In correspondence with this level of the CO concentration, the daytime ozone concentration increased up to 65 ppbv. In the nighttime, the ozone concentration varied significantly. These variations were caused by a high ozone concentration level in the above-inversion polluted air and by quick destruction of ozone in the under-inversion layer. Along almost the entire railway section between 6400 and 7800 km from Moscow, the ozone concentration increased by about 20 ppbv. Several local spikes in the ozone concentration can be caused by individual sources of ozone precursors leading to ozone generation in the vicinities of these sources under definite meteorological conditions within mountain valleys poorly ventilated.

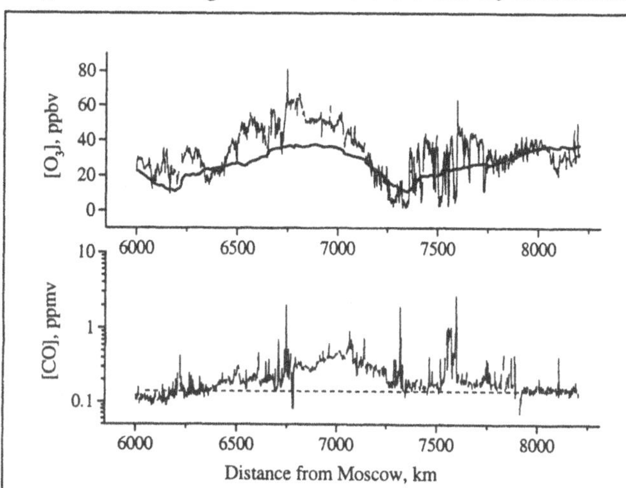

Figure 5. Ozone and carbon monoxide concentrations in the fire plume observed in the course the TROICA-5 expedition. The bold line at the top section of the figure corresponds to the daily mean behavior of the ozone concentration in this series of measurements; the dotted line at the top section of the figure corresponds to the background CO concentration characteristic for this region.

A like phenomenon was observed in the course of the TROICA-2 expedition. In the neighborhood of the above-mentioned region, between 7200 and 8100 km from Moscow, high CO and soot aerosol concentrations ranged from 600 to 700 ppbv and from 1.0 to 1.5 $\mu g/m^3$, respectively, were found in the atmosphere. These concentrations are of the same order of magnitude as the concentrations found in the summer of 1999. Radioactive tracer analyses of carbon and oxygen in air samples collected in this region showed uniquely that CO is a product of biomass burning (Bergamaschi et al., 1998). Analogously to the TROICA-5 expedition, it was found that, throughout the entire polluted region, ozone was generated actively in the daytime and its concentration increased episodically up to 45 ppbv in the nighttime. Ozone concentrations enhanced by about 10-20 ppbv were observed throughout this region almost without exception.

Conclusion

The carriage-laboratory provides a unique means for monitoring the chemical composition of the atmosphere in remote regions poorly studied. The comprehensive information on the concentrations of minor atmospheric gases gives a notion on their spatial and temporal distribution. The positive eastward gradient typical for the ozone-concentration can be associated with the zonal air-transport from central Europe or with the active stratosphere-troposphere exchange at the east of the Continent. In urban air, a decrease in the ozone concentration is usually observed. However, in the Khabarovsk Krai, clearly pronounced smog situations, similar to those in other regions of the globe, are sometimes observed.

Acknowledgement

We are grateful to G.S. Golitsyn and P.J. Crutzen for the concept of railway experiments and for their promotion and realization and to C.A.M. Brenninkmeijer and I.B. Belikov for contributing to the creation of the scientific laboratory and performance of the scientific measurements on the basis of the standard railway carriage.

References

Appenzeller C., Holton J.R. and Rosenlof K.H. Seasonal variation of mass transport across the tropopause. *J.Geophys. Res.* **101** (1996) 15071-25078.

Bergamaschi P., C.A.M. Brenninkmeijer, M. Hahn, T. Rockmann, D. Schaffe, P.J. Crutzen, N.F. Elansky, I.B.Belikov, N.B.A. Trivett, D.E.J. Worthy. Isotope analysis based on source identification for atmospheric CH_4 and CO sampled across Russia using the Trans-Siberian railroad. *J. Geophys. Res.* **103** No. D7 (1998) 8227-8235.

Elansky, N.F. and O.I. Smirnova, Ozone and Nitrogen Oxide Concentrations in the Atmospheric Surface Layer over Moscow. *Izv. Academ. of Sciences, USSR, Atmos. and Oceanic Physics.* **33** No 5 (1997) 551-565.

Elansky, N.F., G.S. Golitsyn, T.S. Vlasenko and A.A. Volokh, Concentrations of Volatile Organic Compounds in Surface Air along the Trans-Siberian Railway. *Dokl. Ross. Akad. Nauk* **373** No. 6 (2000) 816-821.

Kley D., Geiss H. and H.Mohnen V.A. Tropospheric ozone at elevated sites and precursor emissions in United States and Europe. *Atmos. Environ.* **28** (1994) 149-158.

Oltmans, S.J. and H. Levy II. Surface ozone measurements from a global network. *Atmos. Environ.* **28** (1994) 9-24.

Shakina N.P., A.R. Ivanova, N.F. Elansky, and T.A. Markova, Transcontinental surface-ozone measurements in the course of the TROICA experiments: 2. Effect of the stratosphere-troposphere exchange. *Izv. Academ. of Sciences, Atmos. and Oceanic Physics.* (2001) in print.

Evaluation of On-Road Vehicle Emissions: The Lundby Tunnel Study in Gothenburg, Sweden

Monica Petrea, Jutta Lörzer, Ralf Kurtenbach, Raluca Mocanu and Peter Wiesen

Physikalische Chemie, BUGH-Wuppertal, Germany
Gaußstr.20 D-42097 Wuppertal
E-mail: mpetrea@uni-wuppertal.de

Introduction

Transport activities contribute significantly to air pollution in Europe and, therefore, is a key element in the evaluation of any transport policy.

Gas analysis is an important tool in the vehicle engine and combustion research fields. In the vehicle engine field it is used to make the necessary measurements of pollutant species. Particular attention is given to the nitrogene oxides (NO_x) emissions of heavy duty vehicles (HDV) since several tunnel studies over the last few years showed a large underprediction of the NO_x emissions of HDV by the emission models (Sturm et al., 2000), in addition, only a few studies on volatile organic compounds (VOC) emissions from HDV exist (Becker et al., 2000; Lonneman et al., 1986). Preliminary results from this study will be compared and discussed with similar data from other tunnel studies (John et al., 1999).

The determination of the "real world emissions" in the tunnel study is based on statistical modelling. The measurement campaign carried out in the Lundby tunnel, Gothenburg, Sweden, in March 2001 was aimed at evaluating the existing emission model and also the development of an emission model for all transport modes, providing emission estimates at different regional and global levels.

Measurement Site

The Lundby tunnel has a length of 2 km, and consists of two independent tubes. The sample ports were located at 350m from the entrance and 400m from the exit.

Experimental equipment

The NMVOCs were monitored using a compact GC instrument (Airmovoc 2010) with enrichment system (Patyk and Hoepfner, 1995) and FID detector. C_2-C_{10} aliphatic and aromatic hydrocarbons (Kurtenbach et al., 2001) were measured with a time resolution of 20 min and detection limits in the ppt range, e.g. ethane 16pptV, isopentane 14 ppt and toluene 8 pptV, and approx. 50 different hydrocarbons were detectable. The instruments including the sampling systems were calibrated with a certified calibration gas mixture of 30 different C_2-C_{10} VOC from National Physical Laboratory (NPL).

The GC double system with ECD and TCD detector (Chromato-sud) was used to measure nitrous oxide and carbon dioxide. The time resolution was 6 min, with the following detection limits: nitrous oxide, 20 ppbV and carbon dioxide, 10 ppmV. Calibration of the GC system including the sampling system was made with different nitrous oxide and carbon dioxide gas

I. Barnes (ed.), Global Atmospheric Change and its Impact on Regional Air Quality, 255–259.

mixtures. In addition, measurements of NO_x, CO, CO_2, THC, N_2O, SF_6, PM1, PM2,5, PM10 were performed by other groups. Video recording and contact loops were used to monitor the traffic. Meteorological parameters such as wind speed, temperature, pressure and relative humidity were also measured.

Results and discussion

Prior to the measurements a quality assessement was carried out, in order to assure the scientific integrity of the data and harmonization of the instruments.

Figure 1 shows a typical chromatogram from tunnel air at the exit. More than 50 different compounds are detectable and the concentration has been quantified for 24 compounds.

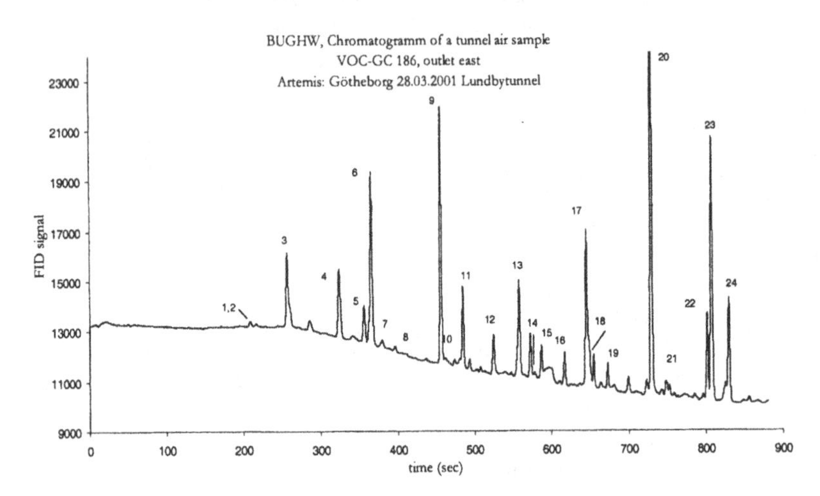

Figure 1. Chromatogram of tunnel air sample

1 ethane/ethyne; 2 ethane; 3 propene; 4 iso-butane; 5 iso-butene; 6 n-butane; 7 trans-butene; 8 cis-butene; 9 iso-pentane; 10 1-pentene; 11 n-pentane; 12 2,2-dimethylbutane; 13 cyclopentane; 14 1-hexene; 15 n-hexene; 16 methylcyclopentane; 17 benzene; 18 cyclohexane; 19 n-heptane; 20 toluene; 21 n-oktane; 22 ethylbenzene; 23 m,-p-xylene; 24 o-xylene;

The validation of the GC systems was performed by positioning the two Airmovoc instruments at the same location. The higher concentrations of the NMVOC, especially toluene and iso-pentane, which were observed at the exit location show, in comparison with the results of the entrance air measurements, that these compounds were emitted as primary pollutants from traffic (Lies et al., 1988, Staehelin et al., 1995)

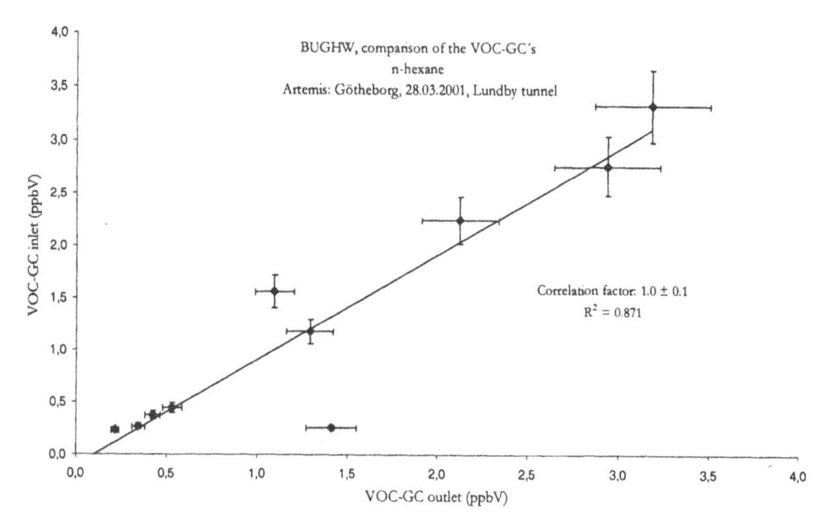

Figure 2. Comparison of the VOC-GCs for n-hexane

Figure 2 shows as an example, the correlation plot of n-hexane for the two instruments. For most compounds good agreement was observed between the two systems.

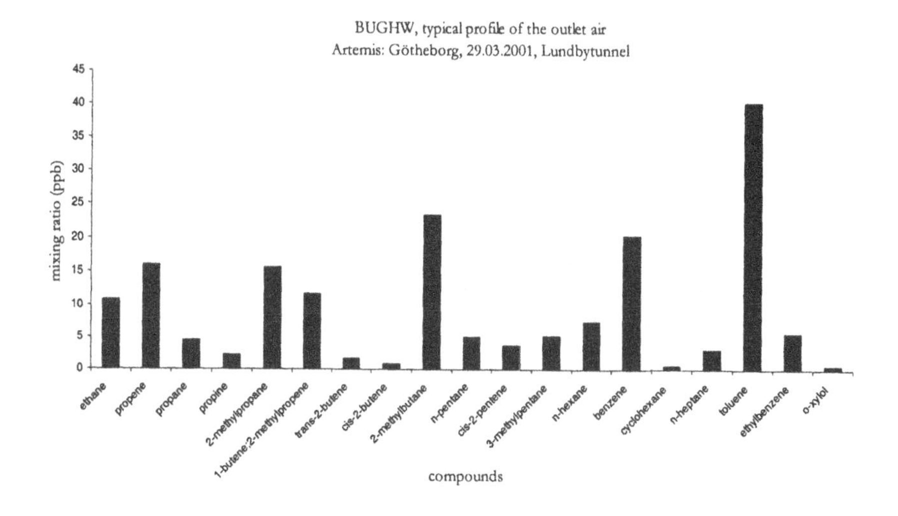

Figure 3. Typical NMVOC profile of the tunnel exit air

Figure 3 presents a typical NMVOC profile of the exit air in the Lundby tunnel. Among the aromatic hydrocarbons, toluene showed the largest mixing ratio, whereas iso-pentane and propene had the largest mixing ratios among the alkanes and alkenes, respectively.

This NMVOC profile is in good agreement with other studies,e.g. Kiesberg tunnel, Wuppertal, Germany (Gomes, 2001; Hassel et al., 1994; Kurtenbach et al., 2002; Schmidt et al., 1998).

From the measured data emissions ratios relative to CO_2 were calculated for the NMVOC. Figure 4 presents a typical correlation plot n-hexane/CO_2.

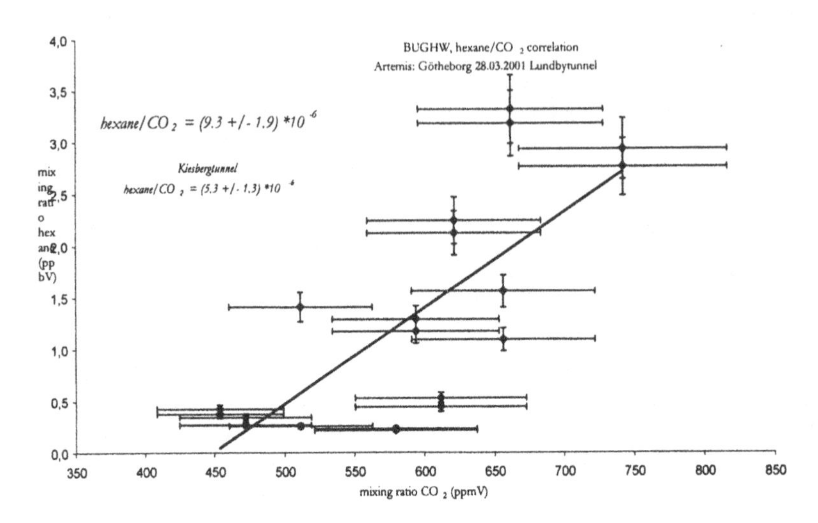

Figure 4. Correlation n-hexane/CO_2

The calculated ratios were compared to those of Kiesberg tunnel study. A good correlation of the results in the two tunnels was found.

Table 1. Correlation of the results in the Lundby and Kiesberg tunnels

	Lundby tunnel	Kiesberg tunnel
Hexane/CO_2	$(9.3\pm1.9)*10^{-6}$	$(5.3\pm1.3)*10^{-6}$
Toluene/CO2	$(5.4\pm1.5)*10^{-5}$	$(6.1\pm1.5)*10^{-5}$

Conclusions

More than 50 compounds were detected and 24 of them quantitatively analysed.in the Lundy tunnel, Sweden.

The NMVOC, especially toluene and iso-pentane, which were observed in high concentrations at the tunnel exit, are primary polutants emitted by road traffic. Among the aromatic hydrocarbons, toluene showed the largest mixing ratio, whereas iso-pentane and propene had the largest mixing ratios among the alkanes and alkenes, respectively.

The NMVOC profile observed in the Lundby tunnel is in a good agreement with the NMVOC profiles of other tunnel studies. The measured NMVOC/CO_2 ratios in the traffic tunnel are in a good agreement with those from other studies.

Acknowledgement

This work was supported by the European Union under Assessment and Reliability of Transport Emission Models and Inventory Systems (ARTEMIS) project.

References

Becker K.H., Lörzer J.C., Kurtenbach R., Wiesen P., Jensen T., Wallington T.J; Contribution of Vehicle Exhaust to the Global N_2O Budget, *Chemosphere -Global Change Science* **2**, (2000) 387-395

Gomes J. A. G., Impact of Road Traffic Emissions on the Ozone Formation in Germany, Ph.D. Dissertation, Bergische Universität-Gesamthochschule Wuppertal, (2001)

Hassel D., Jost P., Weber F. J., Dursbeck F., Sonnborn K. S., Plettau D; Abgas-Emissionsfaktoren von Pkw in der Bundesrepublik Deutschland; Abgasemissionen von Fahrzeugen der Baujahre 1986 bis 1990, *Abschlußbericht zum Vorhaben UFO-PLAN-Nr.: 104 05 152 und 104 05 509. Umweltbundesamt, Berlin* (1994)

John Ch., Friedrich R., Staehelin J., Schläpfer K., Stahel W.A; Comparison of emission factors for road traffic from a tunnel study (Gubrist tunnel, Switzerland) and from emission modelling, *Atmospheric Environment* **33** (1999) 3367-3376

Kurtenbach R, Ackermann R; Becker K.H., Geyer A, Gomez J.A.G, Lörzer J.C, Platt U, Wiesen P; Investigation of Emissions and Heterogenous Formation of HONO in a Road traffic Tunnel" *Atmospheric Environment* **35 (20)** (2001a) 3385-3394

Kurtenbach, R; Ackermann, R; Becker, K.H.; Geyer, A; Gomes, J.A.G; Löerzer, JC; Platt, U. and Wiesen, P: Verification of the contribution of vehicular traffic to the total NMVOC emissions in Germany and the importance of the NO_3 chemistry in the city air, J. Atmos. Chem, (2001) (in press)

Lies K.H.; Nicht limitierte Automobil-Abgaskomponenten, *Volkswagen AG, Wolfsburg* (1988)

Lonneman, W.A.; Sella, R.L.; Meeks, S.A; Non-methane Organic Composition in the Lincoln Tunnel *Environ. Sci. Technol.*, **20** (1986) 790-796

Patyk A., Höpfner U; Komponenten-Differenzierung der Kohlenwasserstoffemissionen von KFZ. Ermittlung von Faktoren zur Bestimmung der differenzierten Kohlenwasserstoffemissionen bei KFZ zur Erfüllung des Anforderungen des § 40.2 BImSchG. Forschungsbericht (1995) 10506 069.

Schmid, H.; Puxbaum, H.; Pucher, E.; Biebl, P; Willau, E.; Heimburger, G; Tauerntunnel Luftschadstoffuntersuchung 1997; Report, Land Salzburg/Technische Universität ien (1998)

Staehelin, J.; Schläpfer, K.; Bürgin, T.; Steinemann, U.; Schneider, S.; Brunner, D.; Bäumle, M.; Meier, M.; Zahner, C.; Keiser, S.; Stahel, W.A.; Keller, C; Emission factors from road traffic from a tunnel study *The Science of the Total Environment* **169** (1995) 141-147.

Sturm P.J., Todler J., Almbauer R.A; Validation of emission factors for road vehicles based on street tunnel measurements; in *R. Jourmard (ed) Proceedings of the 9th Intern. Conf. Transport and Air pollution, Avignon 2000,* ISBN 2-85782-533-1, 2000

Season Variation of Near Ground Impurities of O_3 and CO in East Siberia

V. Potemkin and V. Obolkin

Limnological Institute of Siberian Branch of Russian Academy of Sciences,

Irkutsk 664033, Box 4199, Russia.

e-mail: klimat@lin.irk.ru

Monitoring site

Mondy station (51° 39' N, 100° 55' E, 2010 m above sea level) is located in the mountain ranges near the borderline of Russia-Mongolia in Eastern Siberia. The station is part of the astrophysics laboratory operated and maintained by the Limnological Institute of Russia Academy of Sciences, Siberian Branch. There are no cities within the 300 km radius from Mondy except small villages which are spread thinly all over the region. The nearest village to the station is Mondy village which is located downhill at 720 m elevation in a distance of 12 km away and has a population of about 7000. This area can be considered as one of the more thinly populated areas of the world. No major man-made pollutants, neither from point sources nor mobile sources, has been found to strongly affect this area. Thus, the Mondy station can be regarded as one of the first true continental remote stations for monitoring background atmospheric trace components in an unpolluted region of East Asia. Measurement of O_3 at Mondy has been conducted continuously since October 1996 whereas CO measurements have been carried out from March 1997.

Measurements and Analysis

At Mondy, ozone was measured using an UV absorption analyzer (Dylec model 1007-AHJ) while CO measurements were made using a non-dispersive infrared spectrometers (Kimoto model 541). The trajectories used in this work are isentropic 10-day backward trajectories. The basic meteorological data set is from the meteorological station in Mondy.

Results

Apparent seasonal cycles of O_3 and CO

The seasonal cycles of O_3 and CO observed at Mondy from the end of October 1996 until December 1998 are shown in Figure 1. The daily averages of the O_3 and CO concentrations are presented. Only days when 24 hours data could be taken are displayed. It can be seen that the O_3 seasonal cycles at Mondy reveal a spring maximum while the CO seasonal cycles show a spring maximum - summer minimum. The O_3 concentration increased gradually from the initial observation period in winter to its highest concentration in spring, and then decreased to its lowest concentration during the summer /autumn before increasing again in early-winter. It is interesting to note that despite the low concentration, the summer O_3 exhibits the highest variability. The variability is lowest during winter and somewhat higher in spring and autumn.

I. Barnes (ed.), Global Atmospheric Change and its Impact on Regional Air Quality, 261–266.

This variability is due to the diurnal and temporal change of O_3 as will be discussed below. CO, similarly, decreased from a maximum value in spring to a summer minimum and increased successively during autumn and winter but did not show large variability in any season.

The maximum O_3 monthly averaged concentration, 51-54 ppb, is observed constantly in April 1997 and April-May 1998. The minimum of O_3 concentration, 34-38 ppb, also appears constantly in both July-August 1997 and 1998. It may be noticed that there is an ozone drop in November which is due to the low ozone concentration in 1997 (measurement started in late-October 1997). This low O_3 concentration in 1997 probably reflects the year-by-year variation of the O_3 concentration in this region of Siberia/Russia.

It is worth mentioning here that the number of measurement hours of O_3 and CO data in April and May of 1997 were lower than for other periods. The O_3 and CO data in April were indeed collected regularly but due to the interference from forest fires which occurred near the station in spring, O_3 and CO data during the period of forest fires plumes were disregarded. Statistically, the unused data had no affect on the overall data in the case of O_3 but influenced strongly the CO concentration. The April maximum of O_3, 50.7±3.6 ppb, was obtained from 486 measurement hours. If the total data, 701 hours, were used, the April averaged concentration would be 51.6±5.0 ppb. Data in May and June 1997 were partly absent due to intense fires that shut down the power supply for the O_3 and CO analyzers. Overall, in the first year of measurement, 8016 hours of usable data were obtained. This equals 91% of the total hours and ranks comparatively high among other stations in East Asia. More measurement hours were achieved in 1998.

The histogram of the O_3 concentration distribution appears to be a left-skewed unimodal distribution. More than 70% of the O_3 data ranged between 35-50 ppb. The annual average is 41.4±5.0 ppb. This observed value at Mondy may be relatively high comparing with the concentration at some sites in Europe which were suggested at about 30-35 ppb (see for example, [Derwent et al., 1994], [Solberg et al., 1997]). However, the elevation of Mondy site must be taken into account to explain this difference. With the elevation of Mondy station, 2010 m, free tropospheric O_3 could play a significant role at the site. When the Mondy data are compared to other high elevation sites in Europe and America [Kley et al., 1994; Staehelin et al.,1994], the difference becomes less. O_3 concentrations at Mondy in spring and summer are even lower than other elevated sites but because it is known that sites in Europe and America are affected significantly by anthropogenic activities, a direct comparison is not meaningful. O_3 concentrations at Mondy are also comparable with those at other sites in East Asia [Sunwoo et al., 1994; Akimoto et al., 1996; Pochanart et al., 1999].

The spring maximum and summer minimum of O_3 at Mondy resemble the seasonal cycles of O_3 observed at many sites in the Northern Hemisphere. Mondy O_3 data is supposed to represent the background O_3 characteristic in the clean continental atmosphere in Eurasia. In addition, the small extent of spring elevated O_3 contains a stratospheric input. As for Mondy which is affected more by the free tropospheric motion and is located at a rather high mid-latitude, following Penkett et al. [1993], a potential source of the elevated free tropospheric ozone during the spring might be photochemical production from hydrocarbons that build up in the winter in the free troposphere at high latitudes and become reactive as photochemical production increases after the vernal equinox. The extensive reservoir of material north of the polar front could significantly influence ozone formation in the troposphere.

The summer minimum for O_3 seasonal cycles in the "background continental" category at Oki can probably be ascribed to possible photochemical loss process and partial influence from marine air masses. For the Mondy station, the contribution of marine air mass would certainly be much less. Although it has been proposed that the photochemical loss process in the very clean atmosphere can be responsible for the summer minimum of O_3 observed at some high latitudes sites in Europe [Solberg et al.,1997], this explanation is unlikely for the summer characteristic of O_3 at Mondy. In the remote Pacific, where low O_3, usually less than 10 ppb, and extremely low NO_x concentration are observed, the photochemical loss may dominate but in the continental site as Mondy where O_3 is over 35 ppb, the same mechanism should not be the case. Another feasible suggestion for the O_3 minimum in summer may be due to the O_3 sink by surface deposition and vegetation related activities which is known to favor summer. It has been pointed out that the natural environment such as the huge forest can act as a net sink for O_3 [Kirchhoff, 1988]. At present, the summer minimum of O_3 concentration at Mondy is still unidentified. Further investigation including the measurement of NO_x is definitely needed for the explanation in this aspect.

For the CO seasonal cycle, the spring maximum shows about a factor of two higher CO concentration than the summer minimum. CO data in March-April 1997 and 1998, excluding the forest fires plumes in the case of 1997 (Figure 2), show the average concentration of 162-163, respectively. In summer the CO concentration averaged 85-90 ppb during July and August. CO concentrations at Mondy are close to the background values reported for remote sites in the Northern Hemisphere [Novelli et al., 1992]. From trace gas measurements using the trans-Siberian railroad during the TROICA 2 campaign in Siberia/Russia in summer 1996, [Crutzen et al., 1998] found the lowest surface CO level of near 110 ppb over the sparsely populated area in west Siberia. This value is about 20-25 ppb higher than the Mondy summer CO concentrations. The lower CO at Mondy can probably be explained by the elevation of the site which represents the free troposphere and is less affected by the CO emission within the boundary layer.

Diurnal variation of O_3 and CO

No diurnal variation of O_3 was observed in winter. Because Mondy is located in a very clean atmosphere, no diurnal variation could certainly be expected. However, it is interesting that the diurnal variation can still be observed at Mondy in other seasons. Diurnal variation is insignificant in spring and autumn, with the observed amplitude of 1-3 ppb while in summer, it is more pronounced with afternoon maximum of about 3-6 ppb. Generally, the afternoon maximum, though with small amplitudes, can be related to several processes, the photochemical production, chemical destruction, deposition on ground, and entrainment caused by changing stability conditions [Dubois et al., 1997]. As for CO no diurnal variation is observed all through the year. The rather constant CO concentrations throughout the day and night imply that there is no diurnal change of CO sources from both anthropogenic and natural emission.

Figure 2 shows the hourly O_3 and CO data for some periods in spring when the diurnal variations were distinct. The obvious diurnal cycles were occasionally involved with the forest fire plumes, as could be observed from the abrupt change of the CO concentration. The data associated with these periods were removed but it is still possible that the accumulation of O_3 precursors produced during forest fire could induce photochemical O_3 production on other days. This influence from forest fires may be the main reason that the diurnal pattern in May in Figure 2 appears somewhat peculiar. In summer, the diurnal variation of O_3 is not associated with any change of CO concentration. CO remains low and relatively constant throughout summer.

Summary

The seasonal variation of O_3 and CO observed at Mondy reveal a spring maximum and a summer minimum, which are suggested to be mainly controlled by natural chemistry and dynamics in this region. The O_3 and CO concentration obtained at Mondy may represent the background O_3 and CO in the Eurasian continental mid-latitude. Diurnal variations of O_3 were observed during spring, summer, and autumn, with an increase of a few ppb of O_3 in the late-afternoon. This diurnal variation is suggested to be partly explained by *in situ* photochemical O_3 production. More investigation is definitely necessary to clarify the O_3 and CO characteristics in this remote region of Asia.

References

Akimoto, H., H. Mukai, M. Nishikawa, K. Murano, S. Hatakeyama, C. M. Liu, M. Buhr, K. J. Hsu, D. A. Jaffe, L. Zhang, R. Honrath, J.T. Merril, and R. E. Newell, Long-range transport of ozone in the East Asian Pacific rim region, *J. Geophys. Res.* **101** (1996) 1999-2010.

Crutzen., P. J., N. F. Elansky, M. Hahn, G. S. Golitsyn, C. A. M. Brenninkmeijer, D. H. Scharffe, I. B. Belikov, M. Maiss, P. Bergamaschi, T. Röckmann, A. M. Grisenko, and V. M. Sevostyanov, Trace gas measurements between Moscow and Vladivostok using the trans-Siberian railroad, *J. Atmos. Chem.* **29** (1998) 179-194.

Derwent, R. G., P. G. Simmons, and W. J. Collins, Ozone and carbon monoxide measurements at a remote maritime location, Mace Head, Ireland, from 1990 to 1992, *Atmos. Environ.* **28** (1994) 2623-2637.

Dubois, R., H. Flentje, F. Heintz, H. J. Karbach, and U. Platt, Regionally representative ozone monitoring at Cape Arkona, *J. Atmos. Chem.* **28** (1997) 97-109.

Kirchhoff, V. W. J. H., Surface ozone measurements in Amazonia, *J. Geophys. Res.* **93** (1988) 1469-1476.

Kley, D., H. Geiss and V. A. Mohnen, Tropospheric ozone at the elevated sites and precursor emissions in the United States and Europe, *Atmos. Environ.* **28** (1994) 149-158.

Novelli, P. C., L. P. Steele, and P. P. Tans, Mixing ratios of carbon monoxide in the troposphere, *J. Geophys. Res.* **97** (1992) 20,731-20,750.

Penkett, S. A., N. J. Blake, P. Lightman, A. R. W. March, P. Anwyl, and G. Butcher, The seasonal variation of nonmethane hydrocarbons in the free troposphere over the north Atlantic Ocean: Possible evidence for extensive reaction of hydrocarbons with the nitrate radical, *J. Geophys. Res.* **98** (1993) 2865-2885.

Pochanart, P., J. Hirokawa, Y. Kajii, H. Akimoto, and M. Nakao, The influence of regional scale anthropogenic activity in Northeast Asia on seasonal variations of surface ozone and carbon monoxide observed at Oki , Japan, *J. Geophys. Res.* **102** (1998) 28637-28649.

Solberg, S., F. Stordal, and Ø. Hov, Tropospheric ozone at high latitudes in clean and polluted air masses, a climatology study, *J. Atmos. Chem.* **28** (1997) 111-123.

Staehelin, J., J. Thudium, R. Buehler, A. Volz-Thomas and W. Graber, Trends in surface ozone concentrations at Arosa (Switzerland), *Atmos. Environ.* **28** (1994) 75-87.

Stull, R. B., An introduction to boundary layer meteorology, Kluwer Academic Publ., Dordrecht, 1988.

Sunwoo, Y., G. R. Carmichael, and H. Ueda, Characteristics of background surface ozone in Japan, *Atmos. Environ.* **28** (1994) 25-37.

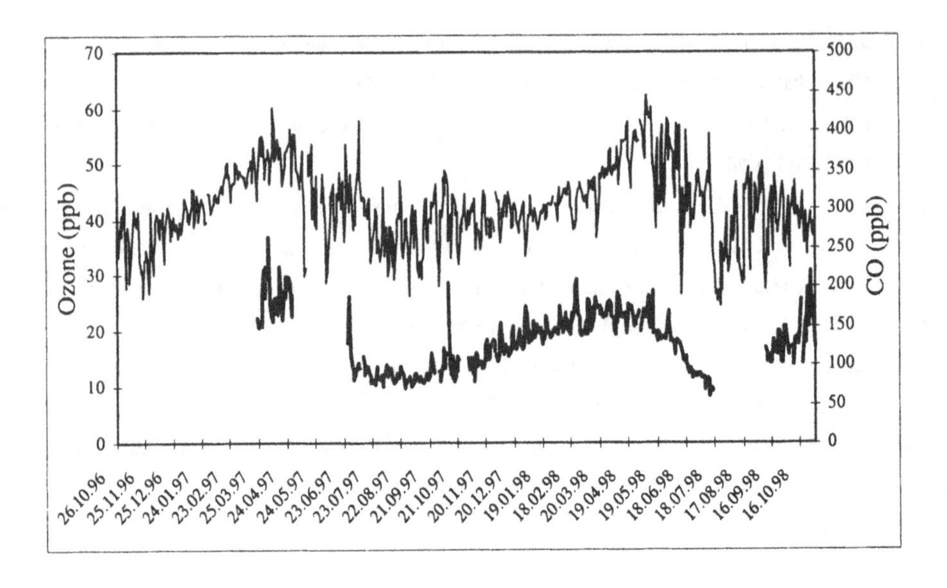

Figure 1. Seasonal variation of averaged O_3 and CO during October 1996 – October 1998.

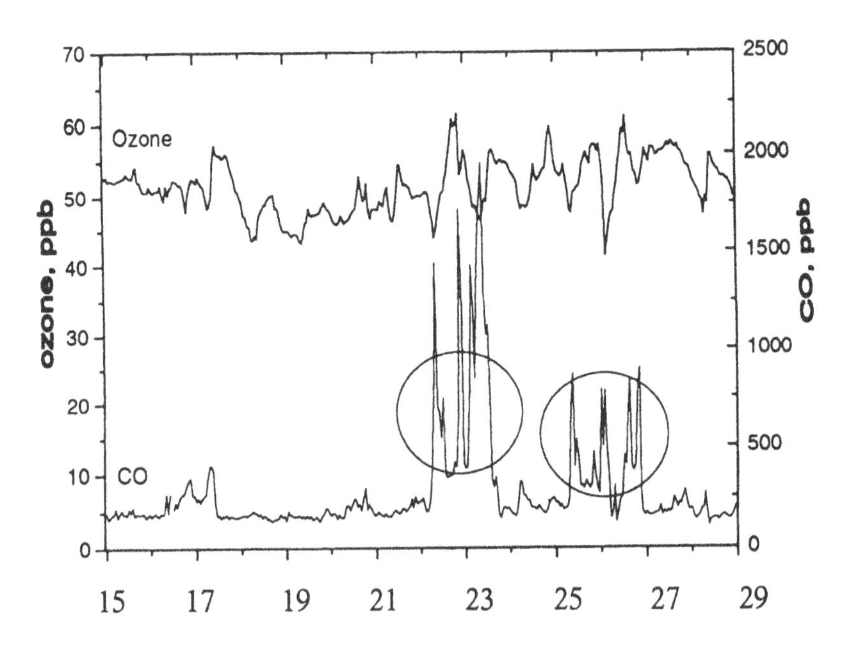

Figure 2. Measured 1-hr averaged concentrations of O_3 and CO during some periods in April 1997, when forest fires took place near the sites and could be observed by the abrupt increase of CO concentration.

Roadside Measurements of Traffic Emissions from Motorway BAB 656 in Heidelberg with Emphasis on Particles

Elisa Rosenbohm [1,2], Volker Scheer [1], Rainer Vogt [1] and Ole John Nielsen [2]

1: Ford Research Centre Aachen, Environmental Science, Suesterfeldstr. 200, D-52072 Aachen, Germany
2: University of Copenhagen, Department of Chemistry, Denmark

Introduction

Particles in the atmosphere originate either from direct emissions (primary aerosols) or from gas to particle conversion of gaseous compounds in the atmosphere (secondary aerosols). Atmospheric particles are involved in several global air pollution issues, e.g. reactions can take place on particle surfaces. Particles influence the Earth's radiation balance by directly scattering and absorbing radiation and indirectly by acting as cloud condensation nuclei. Furthermore, there is an intensive discussion on the health effects of ambient particulate matter. Also, new air quality standards differentiating between different sizes are foreseen.

Precise emission data are of essential importance for all kinds of atmospheric dispersion models as well as for climate change modelling on every scale. However, in many cases the estimates of emissions from traffic, industry and households on the one hand or biogenic emissions on the other hand are very uncertain. Traffic emissions are usually calculated using emission factors based on normalized driving cycles and statistical data on road traffic. Only a few investigations have compared calculated emission data for roadways with real world emissions measured at the roadside (Weingartner *et al.*, 1997; Vogel *et al.*, 2000). In particular, no comparisons are available for the emissions of particulate matter from motorways. In order to check the quality of simulated gaseous and particulate traffic emissions, the project BAB II (Second campaign at the Bundesautobahn 656 near the vicinity of Heidelberg) was organized by the Institute für Meteorologie und Klimaforschung (IMK) at the Forschungszentrum Karlsruhe, Germany. The aim of this campaign was to measure the variations in the NO, NO_2, CO, CO_2, VOC and particle concentrations at different distances from a busy motorway and the extension of the plume in both the vertical and horizontal directions. From the concentration profiles and the parameters describing the traffic situation (traffic density, driving speed, vehicle type, type of catalyst) the source strength of NO, NO_2, CO, CO_2, VOC and particles will be quantified in order to compare measured and calculated emission data.

Measurements

The Ford Mobile Laboratory (FML) was used in BAB II in Heidelberg from April 24th to May 29th 2001. BAB II was a large-scale campaign including 10 participating research groups. 50-meter high towers were situated on either side of the motorway in order to measure the vertical extension of the air plume before and after the wind crossed the motorway. In addition, measuring stations were situated directly at the motorway, at 50 meters distance and at 100 meters distance. The Motorway runs southeast to northwest (130° - 310°) and hence is located perpendicular to the main wind direction from southwest or northeast. The terrain near the motorway is flat and subject to agricultural use. The FML was located next to the tower 50 meters from the motorway on the downwind side. The instruments were operated 24 hours a day for 3 weeks continuously with 5 min time resolution.

I. Barnes (ed.), Global Atmospheric Change and its Impact on Regional Air Quality, 267–271.

The FML is equipped with a TEOM (Tapered Element Oscillating Monitor, R&P, Model 1400a) for total particulate mass measurements, an APS (Aerodynamic Particle Sizer, TSI Inc. Model 3320) for measuring particle number densities of the aerodynamic diameter in the range from 0.5 μm to 10 μm, an SMPS (Scanning Mobility Particle Sizer, TSI Inc.) for measuring particle number densities in the mobility diameter range of 10 nm – 400 nm and an Elemental/Organic carbon analyser (R&P, Model 5400). In addition to the particle measuring instruments the FML is also equipped with trace gas monitors (NO, NO_2, NO_x, CO and O_3) and meteorological instruments to measure wind speed, wind direction, temperature, and relative humidity. Furthermore, in this work, two SMPS set-ups were operated in the tower, one at 17 meters and one at 30 meters height, respectively.

Preliminary Results

The results presented in this abstract are preliminary. Full quality assurance has not been possible until now, since not all the data from the participating research groups was available at the time of writing.

The data obtained show a distinct difference in the particle number concentration at the 3 measuring heights: ground level using the inlet on the roof of the FML, and sampling points at the tower at 17 m and 30 m. A typical particle background concentration is approximately 10,000 particles/cm^3, which was observed at nighttime. The level increased during the day with peaks during rush hour.

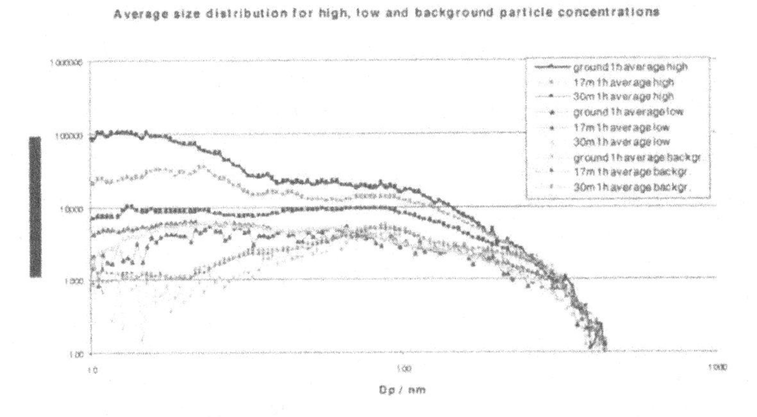

Figure 1: Size distributions showing 1-hour averages during high traffic (squares), low traffic triangles and background air (circles).

Figure 1 shows examples of one-hour average size distributions in all three heights during different traffic loads, high traffic load, low traffic load and no traffic load (background). The high traffic load was measured on Thursday May 17th from 06:45 – 07:45 during morning rush hour with wind coming from southwest (200°; downwind of the motorway). The low traffic load episode was Saturday May 19th from 18:45 – 19:45 and background was measured on Sunday May 20th 16:30-17:30, with wind coming from north east (50°) i.e. the FML was upwind. The particle concentration fall in a range from 4000 to 20,000 particles/cm^3 in the fine particle size range around 100nm. In the ultra-fine particle range, around 20 nm, number concentrations in the range from 1000 particles/cm^3 during no traffic, to about 100,000 particles/cm^3 during high traffic load were observed. The background measurements, obtained when the FML was upwind, showed no concentration gradient with height. For the low traffic load measurements the overall particle concentration increased compared to background values, however there was no obvious concentration gradient. High traffic load measurements show a clear concentration gradient with height, with the highest concentrations found at ground level. Maximum concentrations of 100,000 particles/cm^3 were found at ground level and low concentrations of 30,000 particles/cm^3 (17m height) and 10,000 particles/cm^3 (30m height) were measured.

a)

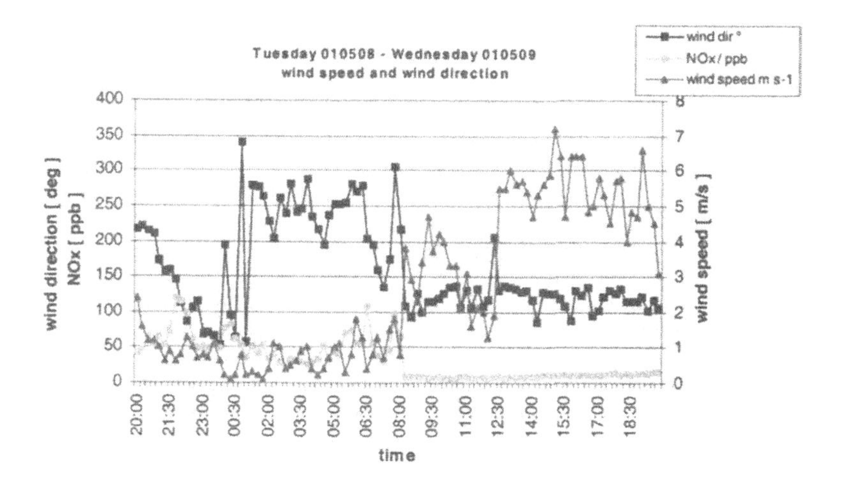

b)

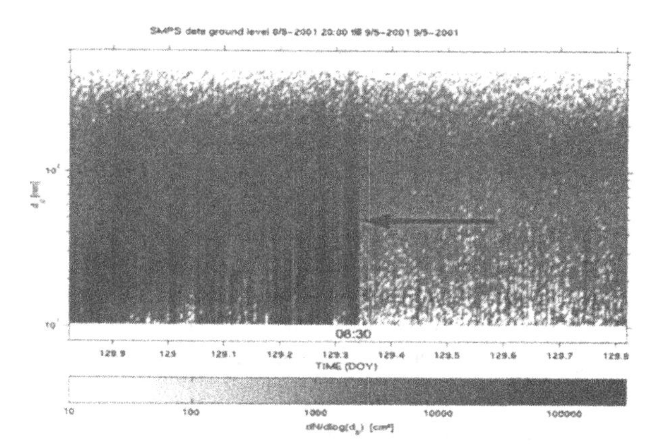

Figure 2: a) Wind direction [deg], wind speed [m/s] and NO_x [ppb] measured on 8/5-2001 20:00 until 9/5 2001 20:0; b) Contour plot of all SMPS size distributions for the same time period as shown in a).

In Figure 2a and 2b an episode is shown that illustrates how the air masses sampled at the motorway depended on the wind direction. At 8:30 on Tuesday May 8th there occurred a drastic wind change from southwest (250°; downwind of the motorway) to southeast (100°; upwind of the motorway) while at the same time the wind speed increased from 0.5 m/s to 5.5 m/s ,(Figure 2a). During southwest wind the particle concentration is in the order of 50000 particles/cm^3 (darkest areas in Figure 2b). At 8:30 the wind direction changed to southeast and the particle number concentration decreased to approximately 5000 particles/cm^3. The difference of a factor 10 was due to the change in wind direction and wind speed. The increased wind speed leads to a separation of the motorway plume and the background air mass. This causes the drastic change in number concentration.

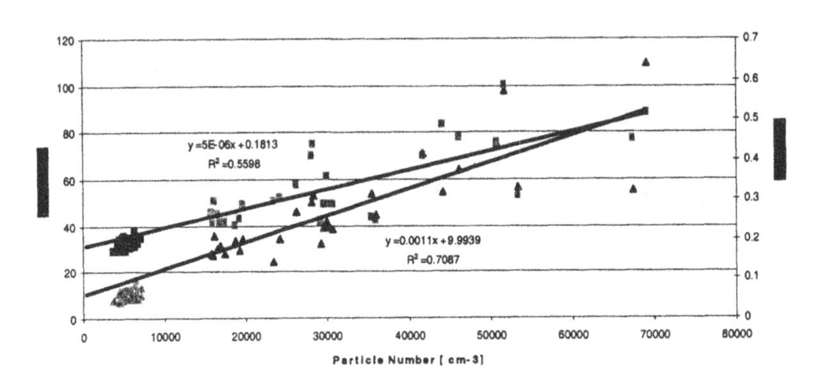

Figure 3: Tuesday May 8th. Particle number, NO_x (triangles, bottom graph) and CO (squares, top graph). The light triangles and dark squares were measured from 8:30 – 20:00 after the wind change had occurred air refers to background air

Figure 3 shows the correlations between NO_x and particle number (black triangles), and CO and particle number (grey squares) on May 8th in the time period from 01:30 to 8:30 when the wind was southwest. During southeast wind from 8:30 to 20:00 the gas concentrations are in the background range (NO_x grey triangles and CO black squares in Figure 3).

Summary

The measurements showed some general features. If the wind direction was perpendicular to the motorway coming from southwest and the wind speed was low, highest particle concentrations in all three heights were observed. Southwest wind and high wind speed led to a vertical gradient with decreasing particle concentration from ground to 30m level, whereas a vertical gradient was negligible during northeast wind and low particle concentrations, indicating a well mixed background air mass.

NO_x concentrations and to somewhat lesser degree CO showed a good correlation with particle number concentration. The best correlations between particle number / NO_x and particle number / CO where observed during southwest wind and wind speeds lower than 2 m/s. The background concentration varies with wind direction and wind speed from day to day.

Future evaluation

When all the data is available the real background concentration can be determined comparing the data from the upwind and downwind side at all times. Trace gas concentrations at all heights can be evaluated for possible correlations with particle number. The NO_x / particle correlation has to be investigated for possible differences corresponding to the measuring height. The concentration profiles and wind speed at different heights will serve as data input to a 3D dispersion model. Together with the actual traffic count and individual vehicle speed the determination of the vehicle emission factor will be the final outcome of the project.

Acknowledgements

We would like to acknowledge the great effort of the colleagues from the Forschungszentrum Karlsruhe for organising the BAB II campaign.

References

Vogel B., U. Corsmeier, H. Vogel, F. Fiedler, J. Kühlwein, R. Friedrich, A.Obermeier, J. Weppner, N. Kalthoff, D. Bäumer, A. Bitzer, K. Jay; Comparison of measured and calculated motorway emission data, *Atmos. Environ.* 34 (2000) 2437-2450.

Weingartner E., C. Keller, W.A. Stahel, H. Burtscher and U. Baltensperger; Aerosol Emission in a Road Tunnel, *Atmos. Environ.* **31** (1997) 451-462.

The Effect of Distant Sources of Ozone and PM_{10}s on Boundary Layer Air Quality in Cornwall – a rural area in the UK

Barbara Parsons and Leo Salter
Cornwall College, Cornwall TR15 3RD, UK. (l.salter@cornwall.ac.uk)

Introduction

Cornwall is an isolated rural area in the far south-west peninsula of the UK (EA, 2001) – see Figure One. Frequent ventilation of the peninsula with atlantic air suggests that air quality in Cornwall should be good. However, because of significant contributions from distant sources, boundary layer concentrations of ozone and airborne particulate matter (PM_{10}s) can exceed the objectives set by the UK National Air Quality Strategy (NAQS) viz. Ozone: 50 ppb as a daily maximum of running 8-hour means (1999) (not now included in the NAQS due to contributions by transboundary influences), PM_{10}: 50 $\mu g\ m^{-3}$ as a running 24-hour average, (DETR, 2000).

This paper discusses the data obtained from monitoring ozone and PM_{10}s in Cornwall and the possible effects of distant sources on the concentrations observed.

Method

Ozone was monitored with an Environnement S.A. 0341-LCD UV Photometry Ozone Analyser supplied by Siemens (Dorset, UK). The instrument performs a daily self-calibration check and has a six-monthly calibration recheck by the manufacturers.

Airborne particulate matter was monitored using a 1400a TEOM (Tapered Element Oscillating Microbalance) providing 15 minute PM_{10} masses from a 3 $dm^3\ min^{-1}$ airflow. Occasionally a co-located Partisol Model 2000 gravimetric unit (16.7 $dm^3\ min^{-1}$ airflow through a 47 mm filter) was used to support the real-time TEOM measurements and to collect material for examination by electron microscopy. Calibration is undertaken by the manufacturers every six months.

A NM950 Environmental Monitoring Weather Station (ex-ELE International Ltd.) was co-located with the TEOM.

Details of the equipment and methodology have been published elsewhere (Salter and Parsons, 1999; Salter *et al.*, 1999).

Results and Discussion

Ozone:

Much of the coastline of Cornwall supports nationally and internationally important vegetation with large numbers of rare plant and animal species. Its uniqueness as a repository of biodiversity is reflected by the definition of over 400 km of the coast by the UK Countryside Commission as Heritage Coast and by the fact that it contains over 160 SSSIs (Sites of Special Scientific Interest) and 13 SACs (Special Areas of Conservation).

The average diurnal concentration range of boundary layer ozone in Cornwall during the 1998 summer ranged from 30-35 ppb (Salter *et al.*, 1999). These ozone concentrations are therefore close to EU Council Directive 92/72/EEC 24 h average values for the vegetation protection threshold of 33 ppb (NEGTAP, 2001) and suggest that significant habitat damage has been and is occurring. Much of the ozone over Cornwall is from transnational sources (see below).

Figure 2 compares ozone concentrations (12-30 June, 1998) measured at a rural (R) site (S1; SW 604 293) in the west of Cornwall with those from the urban (U) centres Bristol (B), Plymouth (P) and Nottingham (N) – see Figure One. (Data for the non-Cornwall sites was

I. Barnes (ed.), Global Atmospheric Change and its Impact on Regional Air Quality, 273–278.

obtained from the National Environmental Technology Centre, (NETCEN)). The ambient boundary layer ozone concentrations decrease as the air mass moves north-westerly P→ B→ N as ozone deposition occurs. This is typical of the movement from the south coast of the UK of air masses originating in continental Europe in which the photochemically driven reactions of the primary pollutants generate ozone. One of the consequences of these incursions is that concentrations of ozone over UK south coast rural areas are high. For instance, in Cornwall the average ozone concentration measured during the summer of 1998 (12 June – 22 October) at the monitoring site in west Cornwall (32 ppb) is higher than elsewhere in the UK (e.g. Bristol (U) 20 ppb, London Brent (U) 20 ppb, Yarner Wood (R) 28 ppb – Figure One), (NETCEN).

Given that control over local air quality resides within UK local government structures there is an urgent need for such bodies to develop mechanisms to address the abatement at source of transnational ozone pollution impacting on their areas of responsibility.

PM_{10}s: Figure 3 shows a comparison (March, 1998) between PM_{10} concentrations monitored at a rural site in Cornwall (S2; SW 953:581) and those from Bristol (U), Plymouth (U) and Nottingham (U) – Figure One, (NETCEN). Two features are worthy of attention.

First, as reported by other workers (King and Bennett, 1997), there is some agreement in the temporal variation between the concentrations monitored at all four sites irrespective of their type (R (no traffic) or U) and location in the UK. This suggests non-local sources of pollution.

Second, the figure shows a 'particulate matter event' during the period 14-18 March 1998. Concentrations recorded in Cornwall were of the same order as those at the other UK sites and, as has been reported elsewhere (King and Bennett, 19987), although such events in the UK are typically associated with easterly and south-easterly winds the March 1998 event was during a period of westerly anticyclonic winds of low speed in Cornwall (inset, Figure 3).

The annual (1/04/97-31/03/98) average PM_{10} concentration measured in Cornwall was 21 $\mu g\ m^{-3}$ and is similar to annual averages measured at urban centres such as Bristol (24 $\mu g\ m^{-3}$), Nottingham (23 $\mu g\ m^{-3}$) and London Brent (21 $\mu g\ m^{-3}$), (NETCEN). This similarity between rural and urban areas suggests that non-local sources of PM_{10}s make a significant contribution to locally monitored concentrations of airborne particulate matter – though locally high traffic flows will have a great impact at certain times.

The impact of airborne PM_{10} on health is an issue of some importance (Maynard and Howard, 1999; DoH, 1995). Samples taken in Cornwall had demonstrable cytotoxicity and genotoxicity (Curnow *et al.*, 2001). Because source apportionment for PM_{10}s is difficult the precise nature of the contribution of distant sources to their health effects in Cornwall cannot be evaluated. However, evidence suggests that health effects are independent of the nature of the particulate matter and hence any increase in PM_{10} concentrations will have an impact on the health of the local population. Once again attempts to abate these local effects will founder because of the lack of effective mechanisms to control transnational air pollution.

Cornwall has the most important china clay extraction operations in the UK and, as well as being the recipient of transnational airborne particulate matter it may also contribute to such pollutants. The industry produces some 5 million m^3 of waste annually and waste material (silica, quartz, mica and possibly cristobolite) is stored in open 'dams'. Under appropriate weather conditions (dry, high pressure) dust storms occur in the area and because 40% of resuspended material is PM_{10} long range transport will be possible (Parsons *et al.*, 2001).

Conclusion

Cornwall is one example of an isolated rural area that is exposed to air pollution from distant sources. There are many similar examples worldwide. Statutory mechanisms need to be developed to require abatement of such air pollution at source so these areas and their habitats are protected.

Acknowledgements The authors would like to thank Cornwall Air Quality Forum and the Duchy Health Charity Ltd. for financial support.

References

Curnow, A., B. Parsons, L. Salter, N. Morley and D. Gould; The mutagenicity of airborne PM_{10}s, *Proceedings of a Joint ERIC/RSC/ISBE Conference, Monitoring Indoor Air Pollution*, Manchester Metropolitan University, 18-19 April (2001) 22.

DoH (Department of Health); Committee on the Medical Effects of Air Pollutants; Non-Biological Particles and Health, London:HMSO, (1995). http://www.doh.gov.uk/hef/airpol/airpolh.htm

Environment Agency; *Environment South West 2001*, (2001). http://www.swenvo.org.uk/index.htm

King, A. and S. Bennett; Comparisons of Non-urban Levels of Particulate Matter, *Urban Network and Meteorological Data ETSU Report Number ETSU N/01/00033/REP DTI* (1997).

Maynard, R. L. and C. V. Howard (eds); Particulate Matter: Properties and effects upon health, BIOS Scientific Publishers Ltd., Oxford (1999).

DETR; The National Air Quality Strategy (NAQS) Stationery Office, Norwich (2000).

NETCEN: www.aeat.co.uk/netcen/airqual/welcome.html

NEGTAP (National Expert Group on Transboundary Air Pollution); Transboundary Air Pollution – Acidification, Eutrophication and Ground-level Ozone in the UK (2001). http://www.nbu.ac.uk/netgap/

Salter, L.F. and B. Parsons: Field Trials of the TEOM® and Partisol for PM_{10} Monitoring in the St Austell China Clay Area, Cornwall, UK, *Atmospheric Environment* **33** (1999) 2111-2114.

Salter, L.F., B.Parsons and F.Fisher; Ozone and PM_{10} monitoring in Cornwall, UK, *Clean Air and Environmental Protection* **29** (1999) 107-110.

Parsons, B., T.P.Jones, T.Coe, S. Mansfield, R. Mathias and R.J.Richards; Airborne minerals in the vicinity of china clay 'mica dams': St Austell, Cornwall, *Proceedings Mineralogical Society Conference, Mineral Particles and the Environment*, Kingston University, London, April 19, (2001).

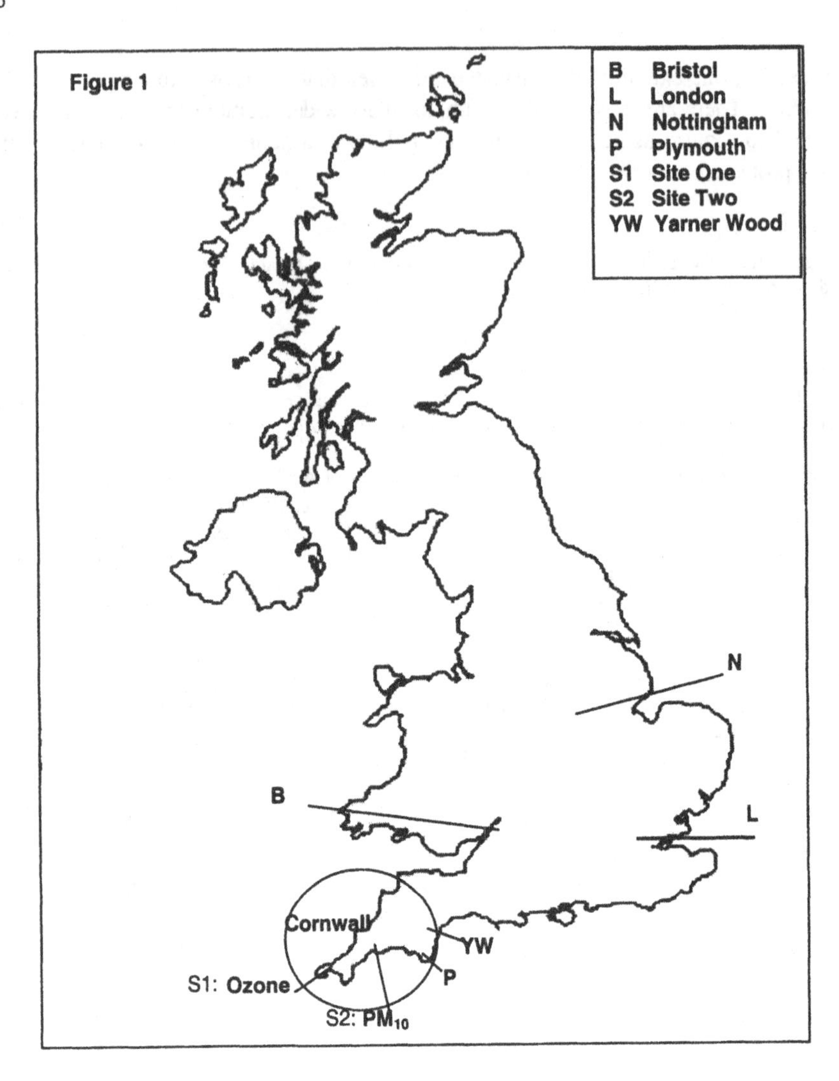

Figure 1: UK Site Location Map.

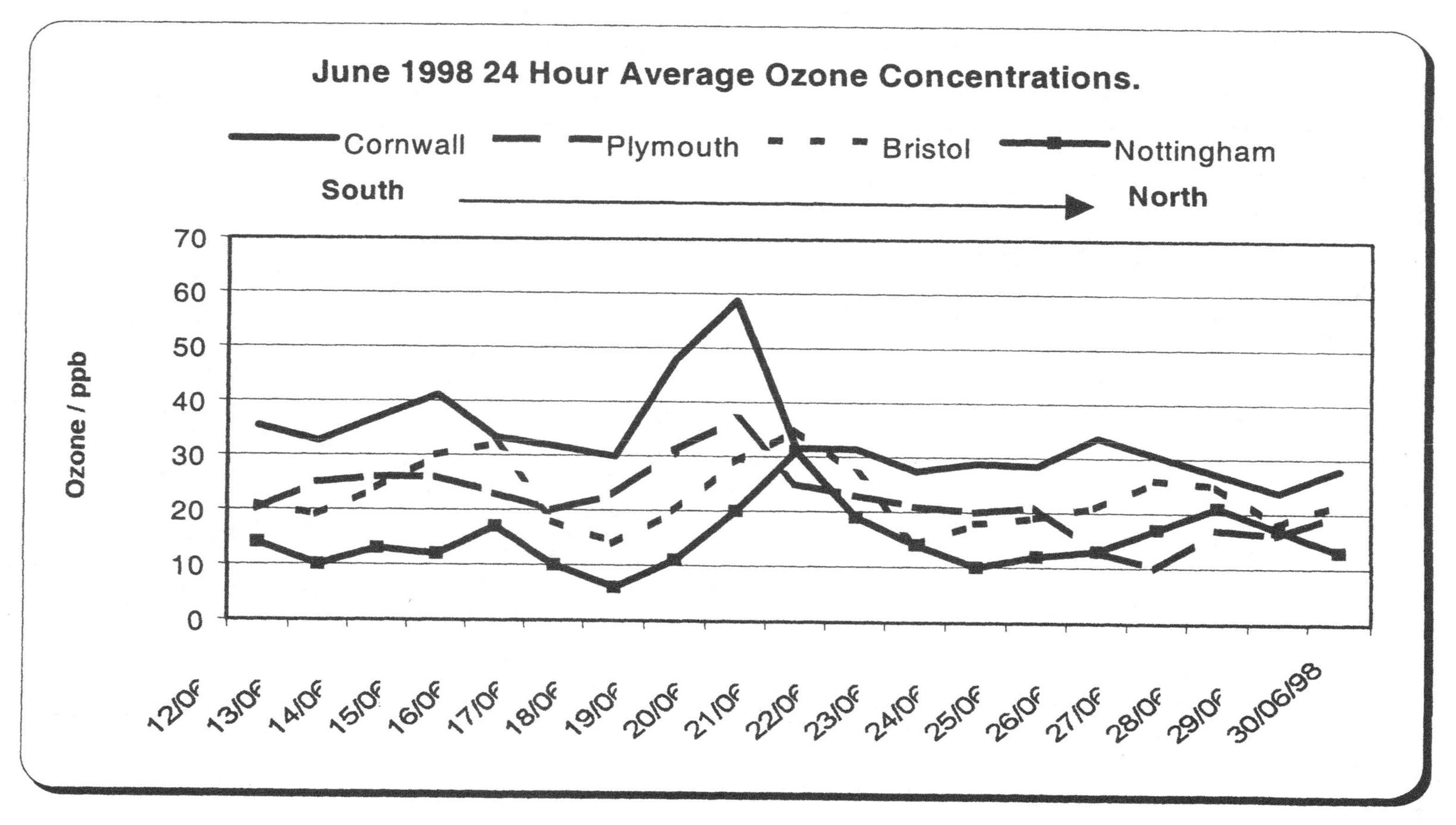

Figure 2. Ozone concentrations (June, 1998)

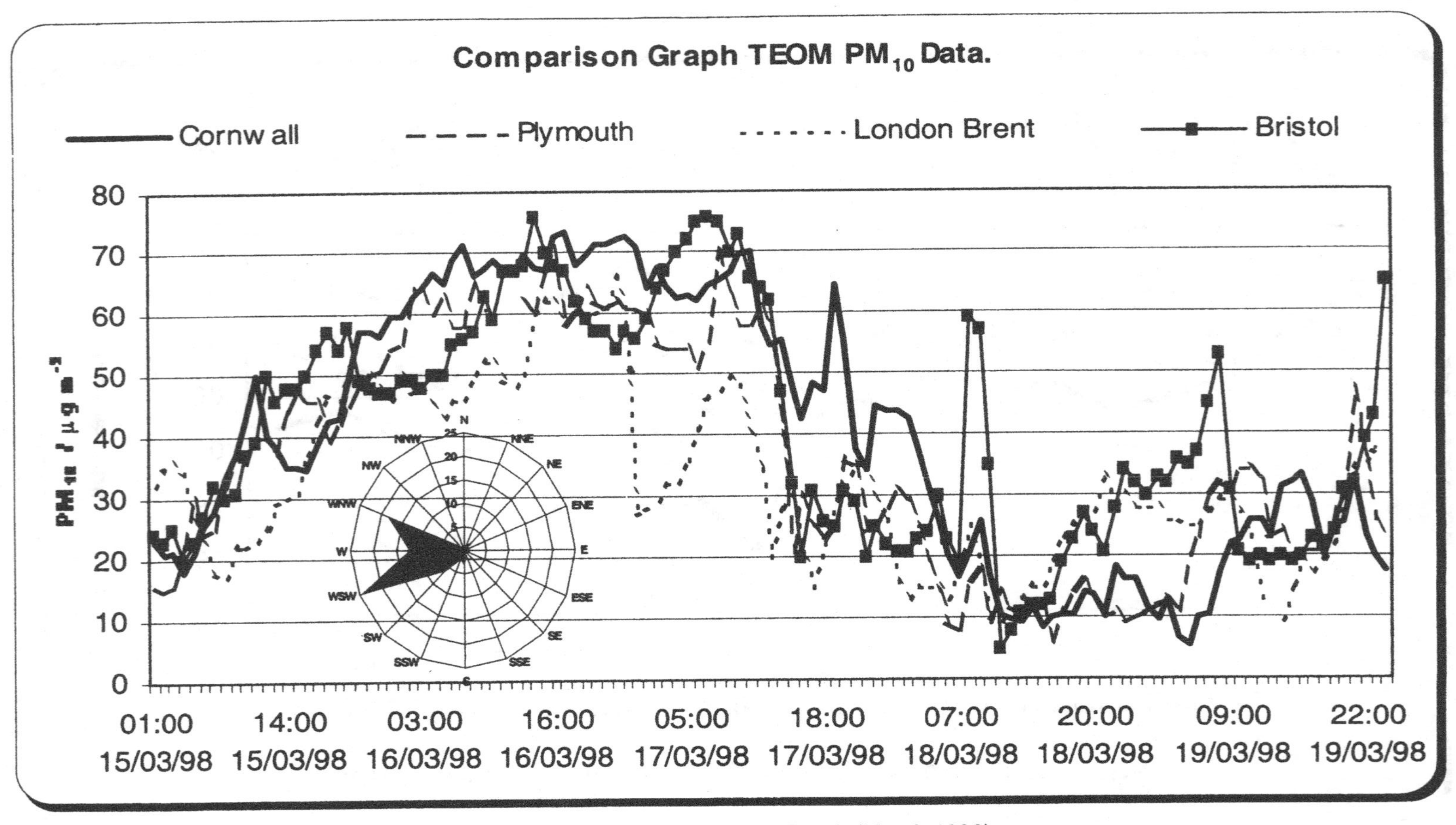

Figure 3. PM_{10} Comparison Graph (March 1998)

Chemical Composition of Atmospheric Fall-Out as an Indicator of the Impact of Local Aerosol Emission Sources on Global Air Quality Change

Y.I.Kovalevskaya, G.A.Tolkacheva, T.Y.Smirnova
Central Asian Research Hydrometeorological Institute

Introduction

Historical monitoring of the physical-chemical characteristics of atmospheric air in the Central Asian region has revealed that aerosols of natural origin (soil and marine ones) are its natural component.

Comparative characteristics of the chemical atmospheric composition in background and regional scales are given in the table 1.

Table 1: Chemical atmospheric composition (background, regional)

Atmospheric composition	Background atmospheric composition, %	Atmospheric composition in the region, %
Oxygen	20.99	20.99
Nitrogen	78.03	78.03
Argon	0.94	0.94
Carbonic gas	0.3	0.35
Hydrogen	0.01	0.01
Water vapor	От 0,1 до 4	До 4
Aerosols traces	0.02	0.04
Traces of other gases: helium, neon, krypton, xenon		
Polluting substances: (SO_2, NO_x, CO, dust, organic components, heavy metals etc)	traces	0.2

From the data of table 1 one can see that for the atmospheric composition within the region the portion of pollutants including aerosols exceeds the background level.

The role of aerosols is known in atmospheric processes such as cloud- and precipitation formation. Aerosols impact significantly on the Earth's radiation balance and consequently on climate.

Within the last years interest in the aerosol problem from the viewpoint of their influence on human health and environmental objects has increased. On a global scale the assessment of aerosol impact on atmospheric air quality is a complicated problem since there is not enough actual data about the contribution of local aerosol emission sources to the total regional aerosol background.

The Central Asian region is a region with a unique combination of natural climate zones with sharp-continental climate, loose soil cover, and plenty of solar days. These factors set conditions for intense natural emissions of dust aerosol into the atmosphere (table 1). In this connection an assessment of the impact of local aerosol emission sources on global and regional background is of great importance.

The objective of our long-term investigations in the Central Asian region has been to assess the impact of local aerosol emission sources on air quality and air global characteristics.

The investigations were performed at various sites characterizing different natural-climate zones:

I. Barnes (ed.), Global Atmospheric Change and its Impact on Regional Air Quality, 279–285.

- soil-desert aerosol emission sources (meteorological station Muynak);
- industrial zone with predominant contribution of anthropogenic aerosols (Tashkent province);
- background zone (Chatkal Biospheric Reserve, Abramov glacier).

The general physic-geographical characteristics of observational sites in the region are presented in the table 2.

Table 2: Physical-geographical characteristics of observational sites

Characteristics	Tashkent		Muynak		Chatkal		Abramov glacier	
	winter	summer	winter	summer	winter	summer	winter	summer
Mean seasonal temperatures, ^{0}C	- 1	+ 28	- 8	+ 26	- 6	+ 22	- 14,3	6,2
Mean monthly relative air humidity, %	60 – 70	20 – 30	More 80	40 – 50	50 and less	20 – 30	59	55
Characteristic of radiation, a number of hours of solar shine per year	2 600 – 2 800		2 800 – 3 000		2 700		2131 – 1741	
Precipitation, мм	400 - 550		100 -120		800		447 – 1029	
Soil cover	Grey-oasis loamy on loess		Meadow and marsh, flood-land –alluvial, brackish and saline		Brown slightly leached, rough-skeleton on alluvium and talus(Fertsiger,1996)		Parent material (Atlas, 1982)	

High and extremely high values of the atmospheric pollution potential (APP) are registered for territories with a large urban agglomeration (for example, the Tashkent province, which is characterized by a high population density and where ecologically "dirty" enterprises are located). The high complex Index of Atmospheric Pollution (IAP) is observed for the territory of the Tashkent province. IAP includes pollutant concentration and their toxicity. Figure 1 shows the IAP changes according to years for Tashkent and Almalyk cities.

A reduction of the total level of atmosphere pollution by main admixtures has been observed. The location of large industrial cities along comparatively narrow mountainous valleys (cities: Chirchik – Tashkent - Yangiuyl) with specific local (mountain-valley) air circulation promotes interchange by polluting substances emissions into the atmosphere.

Imprinted on the background of the main polluting substances entering into the atmosphere, there are local zones of emissions of specifically hazardous substances from the enterprises of chemical industry, non-ferrous and ferrous metallurgy and machine building. Emissions of these enterprises contain lead, vapors of sulphuric and hydrochloric acids, vanadium oxide, benzapirene and other substances of the first class of danger (more than 150). (Review, 2000)

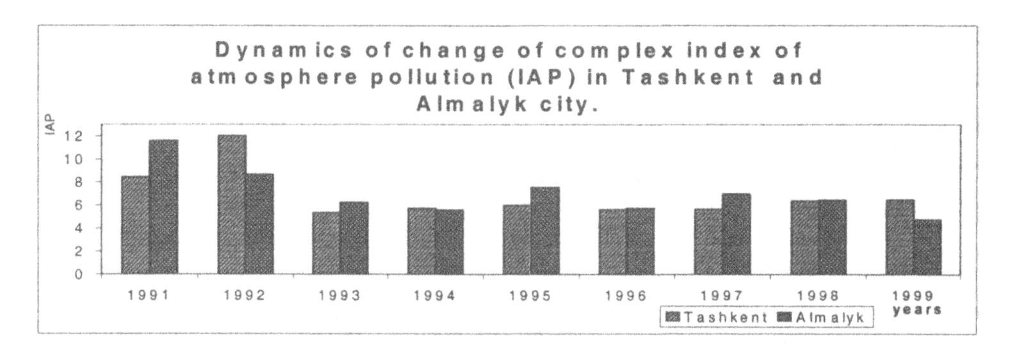

Figure 1. IAP change by years for Tashkent and Almalyk cities

Chemical and disperse composition of aerosol and atmospheric fall-out (precipitation, dry fall-out) were detected using the methods of ion chromatography, atom absorption, photocalorimetry, microscopy and some others. According to the data from regular observations of atmospheric air quality, high pollution levels by sulphur dioxide are registered in the regions the highest emissions (Almalyk mining enterprise). Precipitation with acidity 4.15 – 4.50 are periodically observed even in the Chatkal Biospheric Reserve, which is situated at a distance of 60 km from Almalyk city.

The highest nitrogen dioxide content in air is registered in the cities with intensive traffic and also with large heat plants and the enterprises producing chemical fertilizes (Tashkent, Chirchik and Almalyk cities).

High pollution levels by carbon oxide are registered sporadically near machine works and at the crossroads with intensive traffic in Tashkent city.

Heightened pollution levels by phenol are observed in the cities where there are enterprises using hydrocarbon raw material for production.

A high content of fluorine compounds is characteristic for the cities with metallurgical enterprises and phosphate manure production. Maximum and mean yearly concentrations of fluorine compounds evaluated as hydrogen fluoride are registered in the Almalyk city.

Spatial distribution of heavy metals in the atmospheric air corresponds to the location and intensity of the main emission sources. The highest concentrations of lead, cadmium, copper, zinc are observed annually on the territory of the Tashkent province, especially near the Almalyk mining enterprise. The amplitude of the mean yearly concentrations of lead is rather high.

All the above mentioned testified to a definite influence of anthropogenic emission sources on the chemical composition of the atmosphere on local and regional scales. It should be mentioned that pollutants listed above can also be sources of secondary atmosphere pollution. On the other hand, land cover and the marine surface can be a source of natural aerosol emissions into the atmosphere.

In Central Asia the role of soil aerosols in total atmospheric pollution is rather great. Powerful sources of soil aerosols are underlying surface of deserts and semi-deserts, irrigated lands, and the old dried coastal area of the Caspian and Aral Seas. (Tolkacheva, 2000)

The basic mechanisms of suspended particles and humid drops removal from of the atmosphere are physical ones: dry and wet deposition and fall-out as wet and dry precipitation.

Wet atmospheric precipitation is rainfall, snow, and hail. Dry atmospheric fall-out is dust particles, and depositing at environmental objects by gravitational settling. Dry fall-out, which is the main mechanism of particle removal out of troposphere below 100 m, consists of two processes: gravitational settling and particle accumulation on surfaces of objects, for example, buildings, plants, or soil.

Particle dimensions and properties of their surface (sorption and desorption, roughness, humidity, shape) impact considerably on the velocity of particle removal from the atmosphere.

The efficiency of particle removal out of the atmosphere is connected closely with physical changes of aerosol particles: their upsizing with time, and coagulation.

The quickness of the processes occurring with aerosol particles in the atmosphere calls forth difficulties when studying their properties. (Tolkacheva, 2000)

Systematic investigations of the chemical composition of wet precipitation and dry fall-out on the regional territory have shown that to some degree their chemical composition can be an indicator of the influence of local aerosol emission sources and reflects some parameters of global air quality change.

The multi - annual of the total precipitation mineralization change and seasonal change of total mineralization at listed stations are presented in Figures 2 and 3.

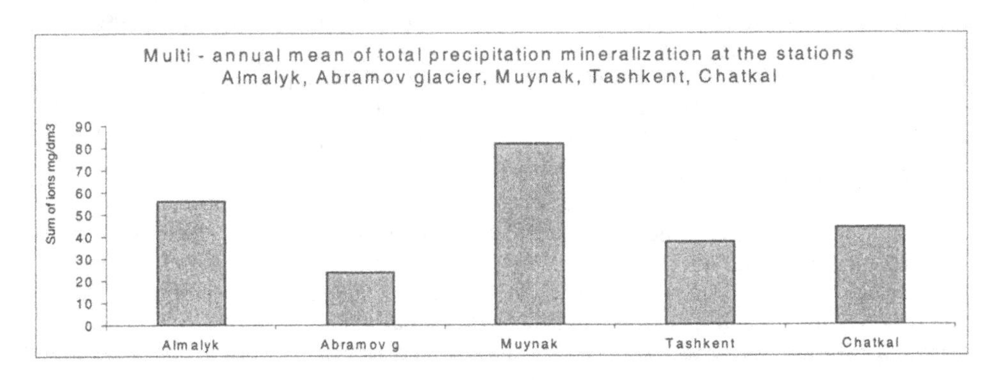

Figure 2

It has been observed that the greatest mineralized precipitation is wet fall-out at the Muynak station where natural aerosol sources prevail, followed by Almalyk town where there is a considerable contribution of anthropogenic pollutant emissions into the atmosphere.

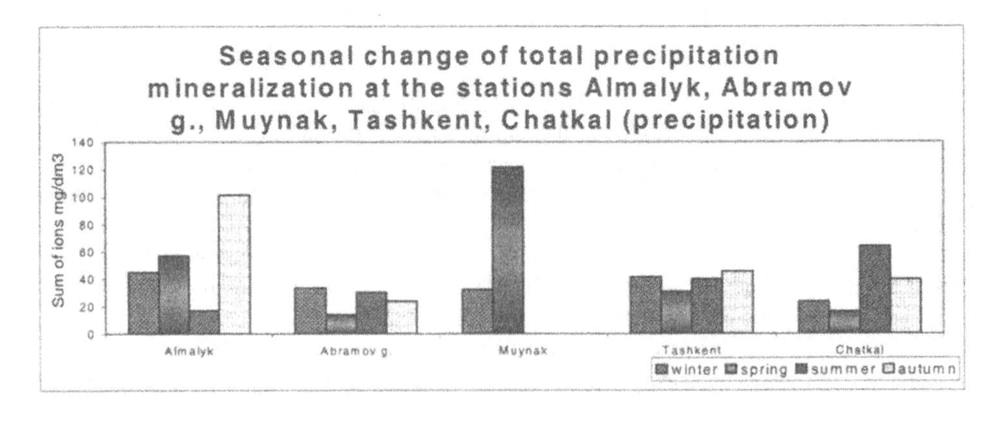

Figure 3

The most mineralized precipitation is observed in Almalyk city in autumn and at Muynak station in spring (as shown in Figure 3).

The change in concentration of sulfates, nitrates, hydrocarbons ions, and total precipitation mineralization using Tashkent city as an example is given in Figure 4.

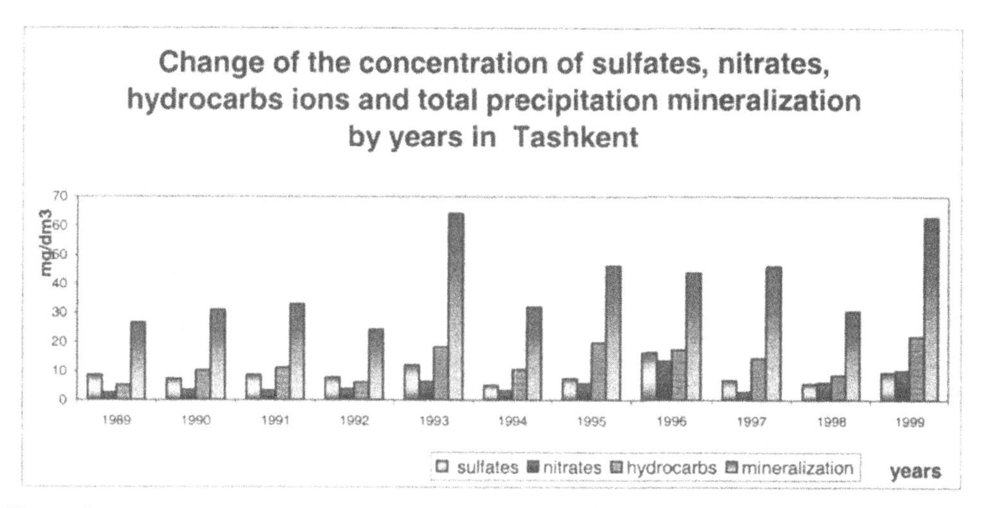

Figure 4

The change of total mineralization of water extracts of dry atmospheric fall-out (DAF) by years at several stations is shown in Figure 5.

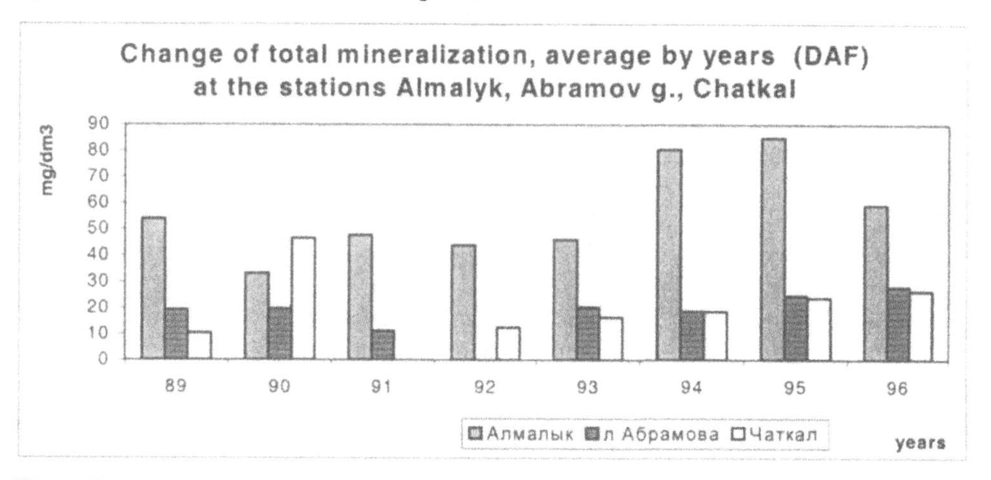

Figure 5

One can see from Figure 5 that the highest values of total mineralization are observed in Almalyk sity and at Myinak station.

The aerosol composition of the region's atmosphere is characterized by heightened background as compared with global one by 2-3 times. (Report, 1993)

The chemical composition of the aerosol is close to an arid one, which is of soil origin, their sub-micron and dispersed fractions are similar by components.

In the industrial zones the fine-dispersed fraction, organic components and heavy metals content, increase in the aerosol composition. The fine-dispersed fraction of the aerosol is formed due to interaction of reaction gases (SO_2, NO_x , NH_3 and others) with subsequent transfer and deposition on the underlying surface as salts and also owing to emissions from hot sources under the elevated temperature of the background.

The change of precipitation pH at the listed stations is presented in Figure 6. The highest values of precipitation acidity is observed in autumn in Almalyk sity, in spring at the Abramov glacier and in Tashkent city. It probably results from wash-out of aerosol accumulated in atmospheric air during the summer period.

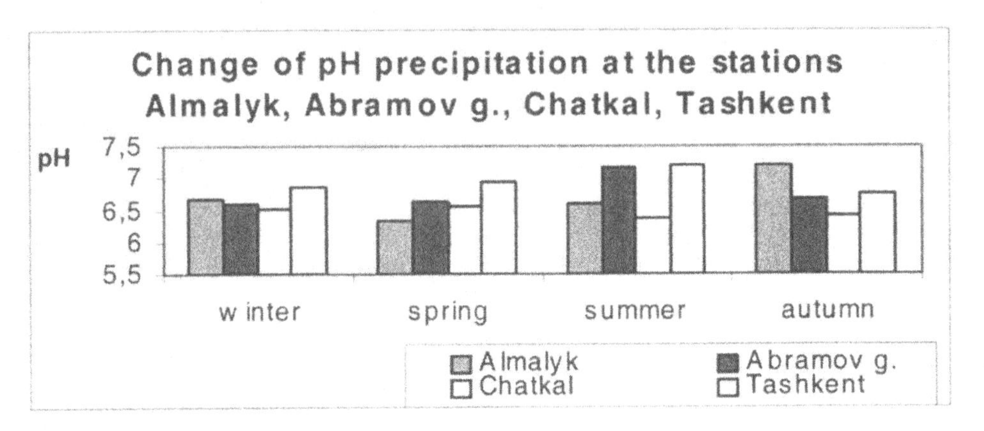

Figure 6

The changes of dry precipitation pH at the listed stations are shown in Figure 7.

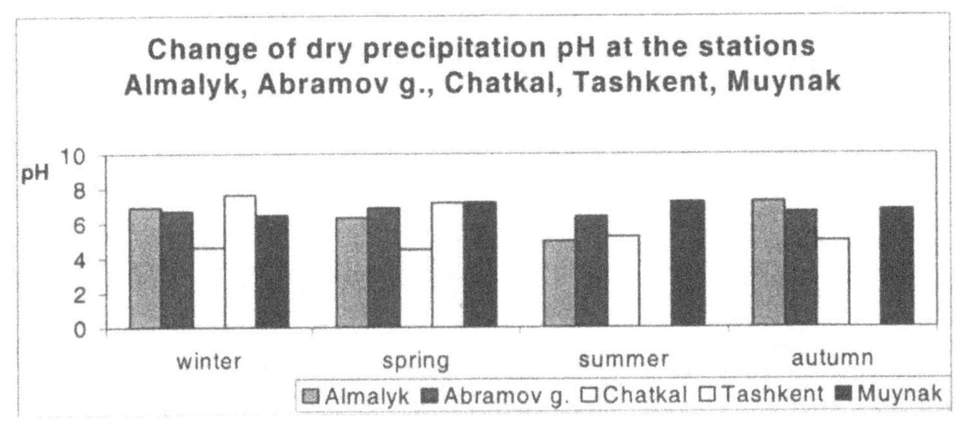

Figure 7

According to the data presented one can see that there is a relationship between qualitative and quantitative emissions composition and their content in the atmospheric air. Dry and also wet fall-out can be indicators of the impact of local aerosol and gas pollutant sources on global air quality change.

Figure 8 shows for the purpose of comparison, the long-term change of precipitation and dry fall-out pH at the listed stations.

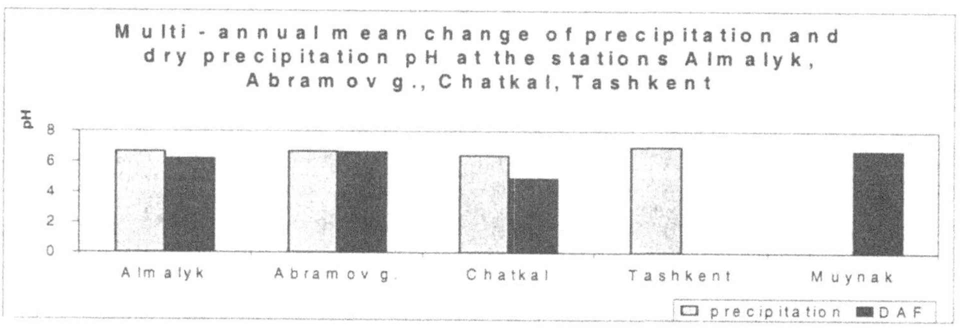

Figure 8

As was mentioned above main mechanisms of aerosol and gas deposition out of the atmosphere are fall-out with wet precipitation and gravitation settling of hard particles. It should be emphasized that the chemical composition of precipitation reflects the impact of anthropogenic local aerosol emission sources and can be treated as an indicator of regional and global atmosphere pollution. Trends of precipitation mineralization have shown a common tendency of increase and as a consequence, rising atmosphere pollution in the region.

The data of dry fall-out is of special interest. Their chemical composition enables to reveal local and global aerosol emission sources into the atmosphere.

Analyzing long-term data on the chemical composition of atmospheric fall-out in the region enables it to be treated as an indicator for the assessment of the impact of local aerosol emission sources on some characteristics of global air quality change.

References

Atlas of Uzbek Republic Moscow – Tashkent (1982) 64-68

F.I.Fertsiger Regime – Reference yuidance "Abramov Glacier – climate, runoff, mass balance" T.-(1996) 22-28

Review of atmospheric air pollution state and hazardous substances emissions on the territory of activity of Glavgidromet of the Republic of Uzbekistan during 1999 – Tashkent (2000) 13-26

Tolkacheva G.A. Scientific – methodical basis of atmospheric falls-out in Central Asion region (2000) 23,25

Report. Environment Protection and Use of Natural Resources of the Republic of Uzbekistan (1993) 55-59

Influence of the Anthropogenic Emissions and Atmospheric Chemical Processes on Climate in the XXI Century

Igor K. Larin and Anton A. Ugarov

Leninskii avenue 38, bdlg 2, Institute of Energy Problems of Chemical Physics of RAS, 117829, Moscow, Russia

Abstract

Using four scenarios of the emissions of greenhouse and other antropogenic gases for the period 2000 – 2100 developed by the Intergovernmental Panel on Climate Change (IPCC), the tropospheric content of these gases for the period has been calculated with the help of a one-dimensional mathematical model of the middle atmosphere and other approaches. Data for carbon dioxide (CO_2), methane (CH_4), nitrous oxide (N_2O), ozone (O_3), chlorofluorocarbon (CFC), hydrochlorofluorocarbon (HCFC) and hydrofluorocarbon (HFC) have been obtained. With help of the data obtained and literature data on the greenhouse gas concentrations and associated radiative forcing and the Earth's surface temperature, total radiative forcing and temperature changes for 2000 – 21000 have been calculated for all the IPCC scenarios. In addition, the contributions of tropospheric and stratospheric chemical processes to the results obtained have been estimated.

Introduction

In the XXth century indications for a climate change have appeared. So, the last century has appeared to be the warmest in the millenium (Mann *et al.*, 1998; Pollack *et al.*, 1998; Mann *et al.*, 1999) and, if for the previous 900 years temperature on the average has fallen by 0.2 degrees, then for the last 100 years it has increased by 0.6 – 0.2 degrees (Jones *et al.*, 1999). The analysis of these changes by various mathematical models allows it to be stated with a fair degree of confidence, that the observed temperature change is due to the action of the anthropogenic factors, i.e. growth in emissions of carbon dioxide and other greenhouse gases in the atmosphere.

An important part of the global warming problem is the influence of atmospheric chemical processes in climate change. This influence is explained both by future changes in stratospheric ozone and also changes in the chemical composition of the troposphere, especially its chemical active components. In this connection the prognosis's of the future climate change taking into account both direct, and indirect (chemical) effects is taking on special significance.

On the basis of recent data on the emissions of anthropogenic greenhouse and other gases to the atmosphere an attempt has been made in the present work to evaluate the effect of anthropogenic emissions and chemical processes on climate over the next hundred years.

Estimation of the Earth's surface temperature change in the XXI century

Modern climatic prognosis's are made with the help of complicated mathematical models (Bell *et al.*, 2000; Brenner *et al.*, 2001; Eatheral, 1997; IPCC, 1995) taking into account an atmospheric transformation of the greenhouse gases, a transfer of short-wave and long-wave radiation and numerous feed-backs in a system ground - atmosphere, which (note!) is not yet completely understood and consequently can not be properly taken into account in models. A simpler task is the evaluation of the radiative forcing of some

I. Barnes (ed.), Global Atmospheric Change and its Impact on Regional Air Quality, 287–293.

greenhouse components, which is a change of a radiative flux of IR-radiation at a level of the tropopause, originating as a result of a change of the tropospheric concentration of the given component. Such evaluations are fulfilled with the help of radiative - convective models without feed-backs, that allows analytical expressions to be derived connecting radiative forcing and a change of a greenhouse gas concentration (IPCC, 1995). This connection has a different character for different groups of greenhouse gases and is determined mainly by optical density, reached to the present time by a given greenhouse components in the Earth's surface IR- radiation of wavelengths range. Calculation of the Earth's surface temperature change ΔT_S through a change of radiative forcing ΔF_R is carried out using

$$\Delta T_S = \lambda_C \cdot \Delta F_R, \tag{1}$$

where the factor λ_C is selected as from common theoretical reasons on the radiative balance, and also by a comparison of the obtained results with those from more complicated models with feed-backs (WMO, 1998).

In our calculations of greenhouse gas concentrations we have used four anthropogenic gas emission scenarios (A1, A2, B1, B2) developed by the Intergovernmental Panel on Climate Change (IPCC, 2000), which are based on different variants of the development of the world in 2000-2100. Using these data we have calculated the changes of the atmospheric concentrations of CO_2, CH_4, N_2O, O_3, CFCs, HCFCs and HFCs in the XXI century with the help of one-dimensional photochemical model and other approaches developed in the Institute of Energy Problems of Chemical Physics RAS. Further, with the help of (IPCC, 1995) we have defined ΔF_R and, using equation (1) have found ΔT_S for all four scenarios of IPCC (IPCC, 2000), using for λ_C interval of values from 0,3 K/(W·м$^{-2}$) up to 1,1 K/(W·м$^{-2}$) (WMO, 1998).

The results on the change in radiative forcing and the Earth's surface temperature over 2000 - 2100 are shown in Figure 1 and Figure 2, respectively. In Figure 2 the thick curves show average magnitudes, and the thin ones a possible range of temperature change. Vertical segments to the right of Figure 2 show the maximum range of possible temperature change in 2100 for all the IPCC scenarios.

Estimation of the possible contribution of chemical processes in climate change

In an evaluation of the contribution of atmospheric chemical processes to radiative forcing and a change of temperature, the direct and indirect effects caused by a depletion and recovery of stratospheric ozone, by the growth of tropospheric ozone in the XXI century, and also by additional future increases in the concentrations of methane, HCFCs and HFCs because of a drop in concentration of OH radicals have been taken into account.

As to the role of stratospheric ozone, it is now known that a depletion of the ozone layer in the latter half of the XXth century has reduced the Earth's surface temperature by approximately 0.1 K due to the direct influence of ozone on the greenhouse effect (WMO, 1998). The indirect influence of stratospheric ozone on the greenhouse effect is explained by reasons given in Table 1 (Krol, 1997).

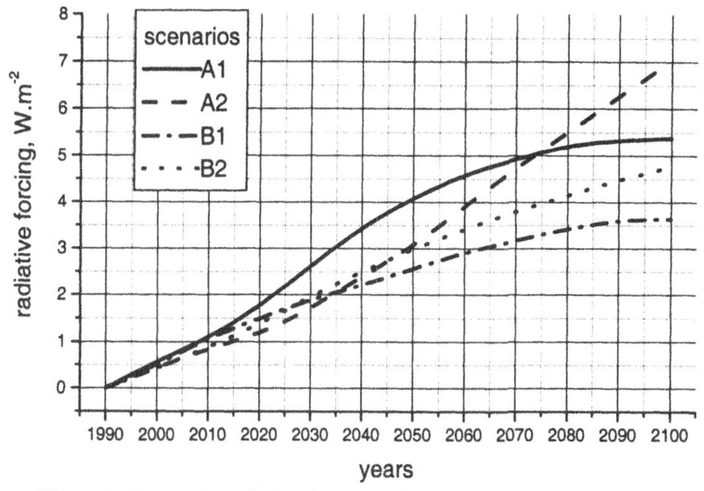

Figure 1. Change in radiative forcing over 2000-2100 according to A1, A2, B1 and B2 emission scenarios of IPCC.

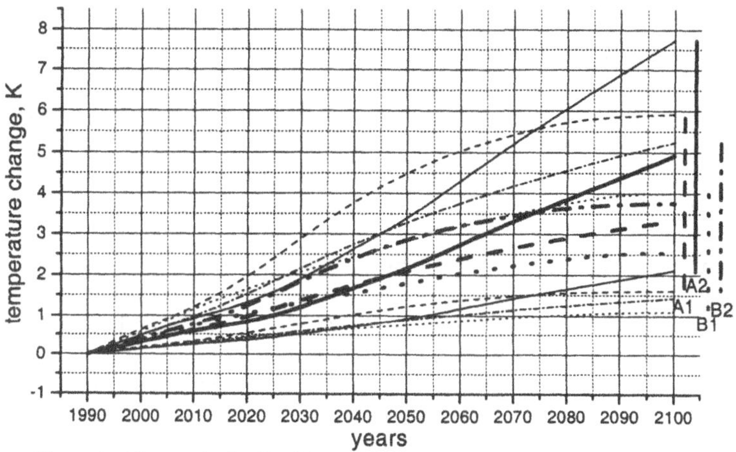

Figure 2. Change in the Earth's surface temperature by 4 scenarios of IPCC (A1, A2, B1, B2).

Table. 1. Influence of the stratospheric ozone depletion on tropospheric chemical composition and the greenhouse effect.

1. Growth of the level of the UV-radiation in the range of 290-320 nm in the troposphere;
2. Increase in the rate of a photodissociation and related processes in the troposphere:
   $O_3 + h\nu \rightarrow O(^1D) + O_2$;
   $NO_2 + h\nu \rightarrow O \rightarrow O_3$;
   $H_2O_2 + h\nu \rightarrow OH + OH$;
   $HNO_3 + h\nu \rightarrow OH + NO_2$;
   $CH_2O + h\nu \rightarrow H + HCO$ or $H2 + CO$;
   $H + O_3 \rightarrow OH + O_2$;
   $O(^1D) + H_2O \rightarrow OH + OH$.
3. Growth of tropospheric concentration of OH radicals and as a result a diminution of tropospheric concentrations of hydrochlorofluorocarbon (HCFCs) and hydrofluorocarbon (HFCs):
   CH_4, HCFCs, HFCs + OH $\rightarrow$ loss of CH_4, HCFCs, HFCs.
4. Additional loss of tropospheric ozone in a HO_x cycle:
   $O_3 + OH \rightarrow HO_2 + O_2$;
   $HO_2 + O_3 \rightarrow OH + 2O_2$.
   Net: $O_3 + O_3 \rightarrow 3O_2$.

A final result of the perturbation is an increase in the tropospheric concentration of OH radicals, that decreases the concentrations of methane, HCFCs and HFCs, reducing their contribution to the greenhouse effect (Bekki, 1994). Our calculations have shown, that by a depletion of the ozone layer by 6,8 % (typical situation for middle latitudes at the end of the XX century) the content of tropospheric ozone decreases by 2,6 %, and methane by 6,2 %. Simultaneously, the concentration of OH radicals in the troposphere is increased by 5 %, and the concentration of $O(^1D)$ by 12%. On recovery of the ozone layer the effect will interchange its sign and the greenhouse effect will increase appropriately in magnitude.

Troposphere chemical processes can to a larger degree affect the greenhouse effect in the XXI century. This influence is related to the following reasons: 1) growth of tropospheric ozone because of the growth of emissions (and concentrations) of ozone creating compounds – CH_4, CO and NO_x (see Figure 3); 2) diminution of the concentration of OH radicals because of their loss in reactions with CH_4 and CO (see Table); 3) additional increase in concentrations of CH_4, HCFCs and HFCs because of a diminution of the concentration of OH radicals, determining the atmospheric lifetime of these components and their tropospheric concentration.

The results of our calculations for a column of tropospheric ozone in the XXI century according to the A2 IPCC scenario (maximum growth of the emissions of all gases) are shown in Figure 4. One can see that on the average the column growth is nearly 100 %, and that the growth is higher for the lower temperature (see also Figure 1).

Figures 5 and 6 show data on the change of radiative forcing of methane and HFC-134a, respectively, over 2000 - 2100 with and without taking into account tropospheric chemical processes according to IPCC scenarios A1, A2 and B2. It can be seen that the chemical effects for these different components are in the range of 20 % - 55 %, and for scenarios A2 and B2 they are continuously increasing over 2000 - 2100.

The total relative contribution of chemical processes to global warming until the end of the XXI century can contribute 6,6 % (scenario A1), 16,0 % (A2), 4,6 % (B1) and 12,9 % (B2).

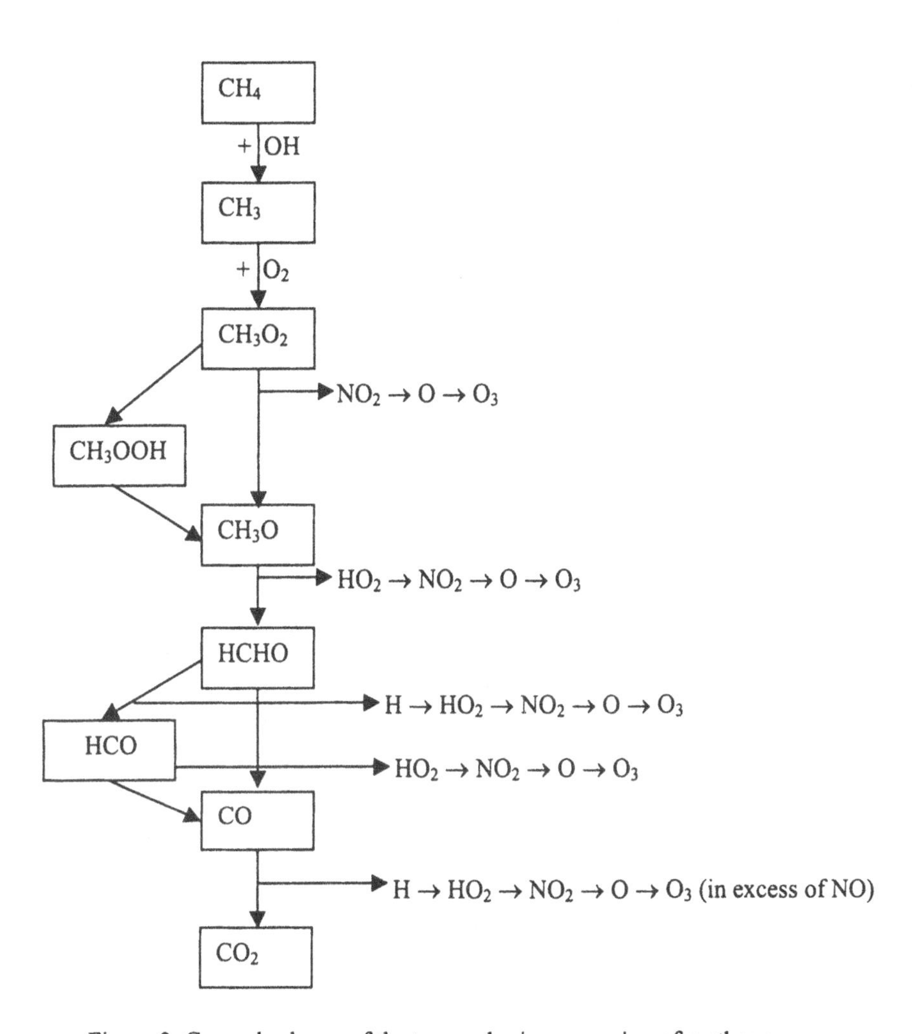

Figure 3. General scheme of the tropospheric conversion of methane.

In conclusion we should like to emphasise that the results presented and also the various data of other research into climate change unequivocally indicate that at further growth in the emissions of antropogenic greenhouse gases to the atmosphere will inevitably result in a considerable global warming, first indications of which are already evident.

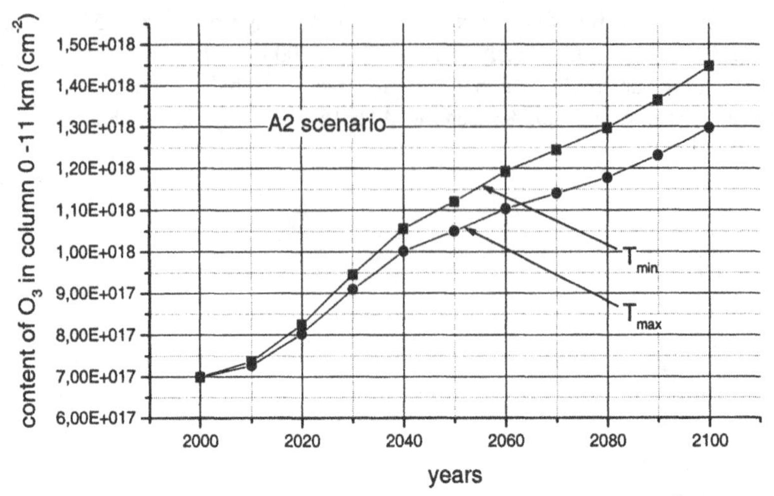

Figure 4. Change in tropospheric column content of ozone (cm^{-2}) over 2000 - 2100 according to A2 IPCC scenario.

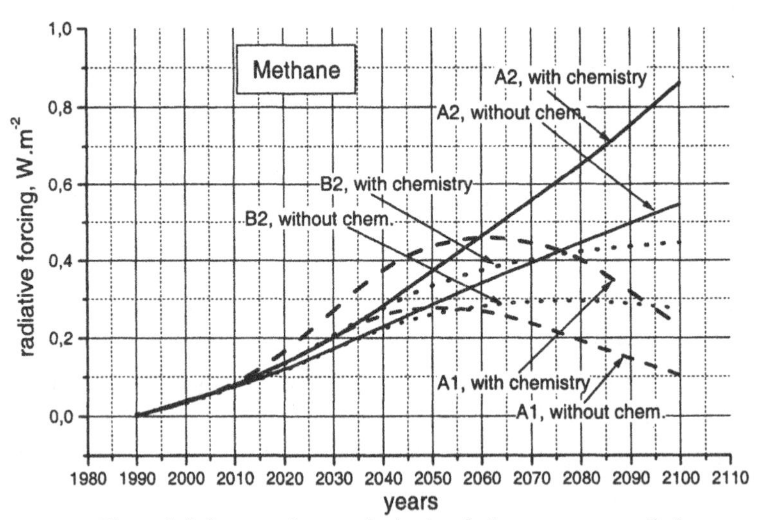

Figure 5. Influence of tropospheric chemical processes on radiative forcing of metane according to IPCC emission scenarios in XXI century.

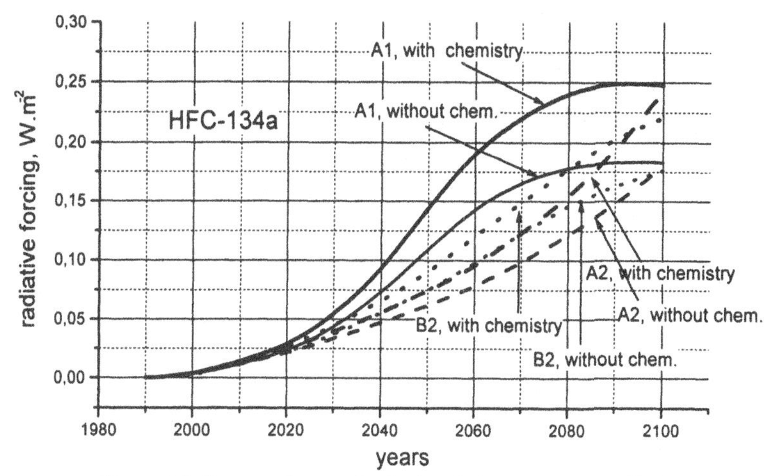

Figure 6. Influence of tropospheric chemical processes on radiative forcing of HFC-134a in XXI century according to IPCC emission scenarios.

References

Bekki S., K.S.Law, J.A.Pyle; Effect of ozone depletion on atmospheric CH_4 and CO concentrations, *Nature* **371** (1994) 595-603. Bell J., P.Duffy, C.Covey, L.Sloan *et al.*; Comparison of temperature variability in observations and sixteen climate model simulations, *Geophys.Res.Lett.* **27** (2000) 261-264.

Brenner S., S.Havlin, A.Bunde, H-J.Schellnhuber, R.B.Govindan, D.Vjushin; Long-range correlations and trends in globalclimate models: Comparison with real data, *Phys.A: Stat. Mech.Appl.* **294** (2001) 239-248.

Eatheral A.; Modeling ckimate change impacts on ecosystems using linked models and a GIS, *Clim.Change* **35** (1997) 17-34.

Intergovernmental Panel on Climate Change (IPCC, 1995), *Climate Change 1995: The Science of Climate Change.* Ed. by J. T. Houghton *et al.*, Contribution of Working Group I, Cambridge University Press, 1996.

Intergovernmental Panel on Climate Change (IPCC, 2000), *Emissions Scenarios.* Special Report of the Intergovernmental Panel on Climate Change. Ed. by Nebojsa Nakicenovic and Rob Swart, Cambridge University Press, UK, 2000.

Jones P. D., M. New, D. E. Parker, S. Martin, I. G. Rigor; Surface air temperature and its changes over the past 150 years, *Rev.Geophys.* **37** (1999) 173-199.

Krol M.C., M van Weele; Implications of variations in photodissociation rates for global tropospheric chemistry, *Atmos.Environ.* **31** (1997) 1257-1273.

Mann M.E., R.S.Bradley, M.K.Hughes; Global-scale temperature patterns and climate forcing over the past six centuries, *Nature* **392** (1998) 779-787.

Mann M.E., R.S.Bradley, M.K.Hughes; Northern Hemisphere temperatures during the past millennium: Inferences, uncertainties, and limitations, *Geophys.Res.Lett.* **26** (1999) 759-762.

Pollack H.N., Shaopeng Huang, and Po-Yu Shen; Climate Change Record in Susurface Temperatures: A Global Perspective, *Science* **282** (1998) 279-281.

World Meteorological Organization (WMO, 1998), Global Ozone Research and Monitoring Project – Report No44, *Scientific Assessment of Ozone Depletion:1998.* UNEP, U.S. Department of Commerce, NOAA, NASA, Geneva, Switzerland, 1998.

Methods and Techniques

Digital Photogrammetric Method for Ecological Monitoring

L.Trubina[1] , K.Koutsenogii [2], A.Guk[1]
[1]-Siberian State Geodetic Academy, 630108, Novosibirsk, Plakhotnogo, 10
[2]Institute of Chemical Kinetics and Combustion SB RAS, Novosibirsk
koutsen@ns.kinetics.nsc.ru

Introduction

Promotion of the digital treatment of images enlarges the range of possible applications of photogrammetric methods. The use of digital cameras and photogrammetric stations has qualitatively changed the production and treatment of images, and increased the operative potentialities and degree of automatization of photogrammetric technologies. This has made them available not only for specialists in image treatment but also for those specialized in concrete application areas in studies of the environment. To practically realize the application of digital Photogrammetry for solving nonstandard problems it is necessary to develop additional techniques taking into account the peculiarities of different types of study objects, which are rather simple for a user without a special photogrammetric background. In this respect, our main concern is the adaptation of digital photogrammetric technologies to collection and treatment of spatial information during complex studies of the different objects composing a biocenoses. The advantage of photogrammetric methods is the possibility of forming volumetric models of objects by their stereoscopic digital images. In photogrammetric treatment, a stereoscopic model is used to measure the spatial coordinates of the points of photographed objects their morphometric characteristics (length, width, configuration, form) and to visualize objects almost adequate to studied ones, in the form of three-dimensional models preserving their metric characteristics.

Mathematical methods and model parameters

Mathematical models of different objects can be obtained by measuring and mathematically treating a single stereo pair or several stereo pairs of the images of photographed objects for the objects of great extension. A classical mathematical model of treatment of a stereopair of images resulting from central projecting of object points is based on conditions of co linearity and co planarity of projecting rays (Dubinovsky, 1982). Let us write down a functional dependence of model parameters in the general form,

$$F(C_n, A_\Gamma, x_n, y_n, X, Y, Z)=0,$$

where n=1,2 which corresponds to the left and right images of the stereo pair. C_n -are the coordinates and angles of the orientation of a surveying system and the parameters characterizing its geometry are the support data, x_n, y_n -are the coordinates of the corresponding points of the left and right images, X,Y, Z are the spatial coordinates of the determined object points.

In a particular case, parameters determining position of the surveying system can be measured by shooting to within sufficient accuracy. However, for the general case, they should be determined, i.e., one must solve the problem of calibration. This requires a set of image points with the known coordinates in the object space or the linear dimensions of objects, which can be known with the necessary degree of accuracy. These values form a basis for the support A_r data.

Coordinates of the corresponding points x_n, y_n of the stereo pair are measured at a digital photogrammetric station in the stereoscopic regime and the spatial coordinates of the points of XYZ object are calculated using the basic modules of the digital photogrammetric station. Thus, treating a stereo pair at the digital photogrammetric station provides primary

I. Barnes (ed.), Global Atmospheric Change and its Impact on Regional Air Quality, 297–301.

data for a subsequent formation of mathematical models of the photographed objects, which can be solved using the programmed graphic products. Thus, the general problem of the modeling of different objects by image stereo pair includes a series of sub-problems each having its own peculiarities depending on the shooting scale and characteristics of the object studied.

1. Determination of shooting parameters (calibration): focus distance of pictures, coordinates of a shooting point and angles of inclination of pictures.
2. Determination primary data: spatial coordinates of characteristic points, boundaries of objects, digital height model.
3. Determination secondary data: geometric characteristics of object form and configuration.
4. Reflection of the results obtained: tree-dimensional visualization of object models.

Different algorithms for realizing each problem have been developed. The choice of optimum algorithm with respect to the available technical and programmed means providing the necessary accuracy of reflection of a concrete object is a very important problem which should be solved for each type of studied object. In addition to the mathematical aspects of the choice of the best solution, there is a series of technological problems. In particular, to provide good stereoscopic plastics of a stereo model, it is necessary to carefully choose shooting parameters and conditions for object lighting to work up the surface of studied objects. The necessary accuracy can be reached with a correct choice of either the method for determining support data or the preparation of special t test-objects.

Main technological processes

The following operations should be realized to create the mathematical models of separate biocenoses objects by their stereo images:

1. The choice of optimum shooting parameters on a basis of calculation experiments on model pictures and their correction during experimental shooting of real objects.

2. The preparation of test-objects corresponding, in their size, to each group of studied objects.

3. The shooting of studied objects, i.e., the production of qualitative digital stereo pairs. In this case, the shooting should be performed with the help of both digital and photographic cameras with subsequent scanning.

4. The mathematical treatment of digital images with the help of a digital photographic station for collecting primary data. It includes two stages: calibration and measurement of a stereoscopic model.

5. The reflection of results in the form of three-dimensional mathematical models.

In developing this technique, we used a Siberian Digital Stereoplotter (SDS) station created at the Department of Photogrammetry and Remote Sensing of the Siberian State Geodesic Academy (Guk, 1996)

The files of digital and graphic data obtained at the digital photogrammetric station can be exported to GIS and other graphic programmed products for further treatment. The three-dimensional visualization of results was realized with the help of such packets as Mat Lab and Surfer. Initial data were collected by a stereo pair using SDS means either automatically or interactively. In this case, the different forms of their representation were used according to the peculiarities of the photographed object: for studying relief - Grid, and for micro-objects - a matrix of three-dimensional coordinates of object points. The point wise description of surface is necessary because of the complex and unusual form of biological objects photographed by electron microscopes. A detailed representation of their numerous geometric characteristics is possible only with the help of this approach. Requirements on the density of point distribution on the surface depend on the peculiarities of the form of photographed objects.

Results

The digital photogrammetric method for collecting spatial data for modeling the different objects of biocenoses and other objects of the environment has been realized in the form of technologies developed in detail for solving the following problems :

- The study of anthropogenic landscape disturbance using aerophotographs;
- The determination of morphological relief characteristics;
- The study of morphometry of pollen grains;
- The recovery of the spatial form of seeds (using pink family as an example);

The estimated maps modifications of a landscape of separate areas of Novosibirsk area are created (Trubina and Bykova, 1999). For show ground allotment of a various type in view of anthropogenous influence the aerial photographs of scale 1:20000 were used, and for geobinding aerial photographs and the formations of a rarefied topographical basis were used topographical maps of scale 1:25000. The layers of the thematic information were formed by means MapInfo by the digital photomap. Thus, the metric information was received with accuracy by way of the order 2m on district, and at a height 1m that corresponds to an accuracy of maps of scale 1: 25000.

As a result of the analysis of the thematic information the derivative layers are generated. In particular, estimated map, with the generalized information on the certain blocks of territory (percentage ratio) those or others ground allotment and map modifications, on which the parameters modifications, calculated are represented on the basis of a ratio of the areas of grounds not broken and undergone to various influences. For example, modification of vegetable cover is given on Figure 1.

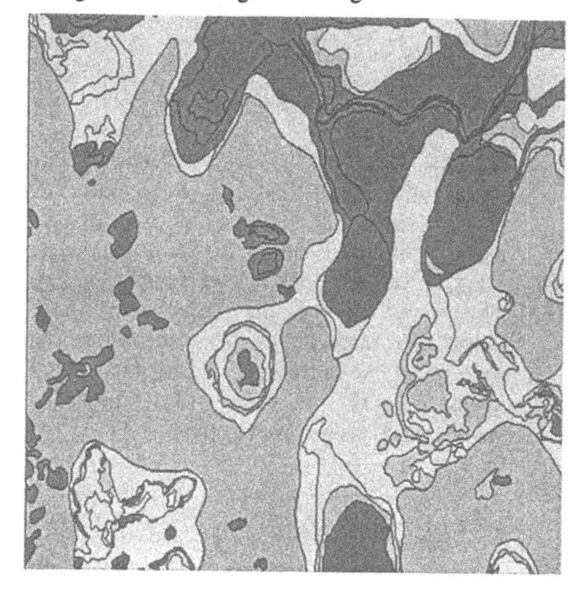

Ladge modifications
Middle modifications
Temparate modifications
Faint modifications

Figure 1. Vegetable cover appraisal of modification.

The digital photogrammetric method is allowing the determination of morphological relief characteristics used in the detail digital relief model (DEM). Some DEM was created by SDS for the complex problem of the estimation of the ecological state of the terrain.

The ecological maps for some regions of Novosibirsk were created.

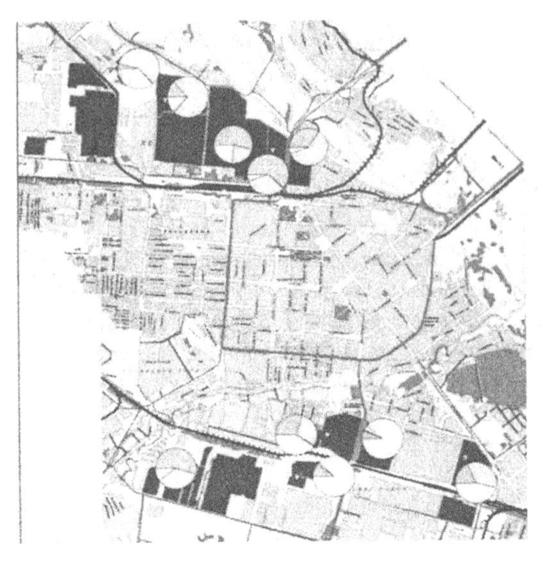

For example, the quota of substances for class of danger is shown in Figure 2.

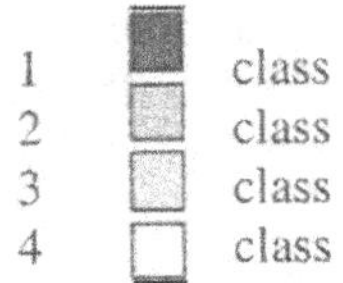

Figure 2. Thematic map of the content of danger substances in the atmosphere.

A series of experiments was carried out to study the shape of seeds and pollen grains by their microscopic images. The method was developed to get stereo images using both optical and electron microscopes (Golovko, 2000). The images were treated at a digital photogrammetric station, which allows one to determine for the first time, the parameters characterizing the spatial form of pollen grains and seeds of some plants. Figure 3 shows the larch pollen placed on the test-object (the distance between dashes being 100 mkm). During photogrammetric treatment, in the process of stereoscopic observation, the isolines of the same height were traced each 5 mkm. The plane of the test-object was taken as the initial (zero) plane. These data made it possible to determine the volume of the surface of the visible part of pollen grains necessary for modeling aerodynamic properties of pollen.

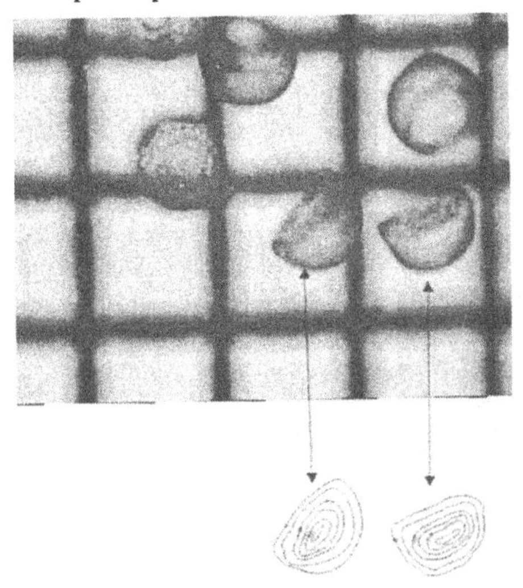

Figure 3. The larch pollen placed on the test-object.

A similar technique was used to obtain spatial characteristics for studying the seeds of complex structure (Vlasova, 2000). This technique was worked out using the seeds of pink family characterized by particular elements of surface structure (hills or papillae of different height and form (Figure 4)). The number of hills, their form and height on dorsal and lateral surfaces are of importance for morphofunctional analysis and are readily identified by their stereoscopic model.

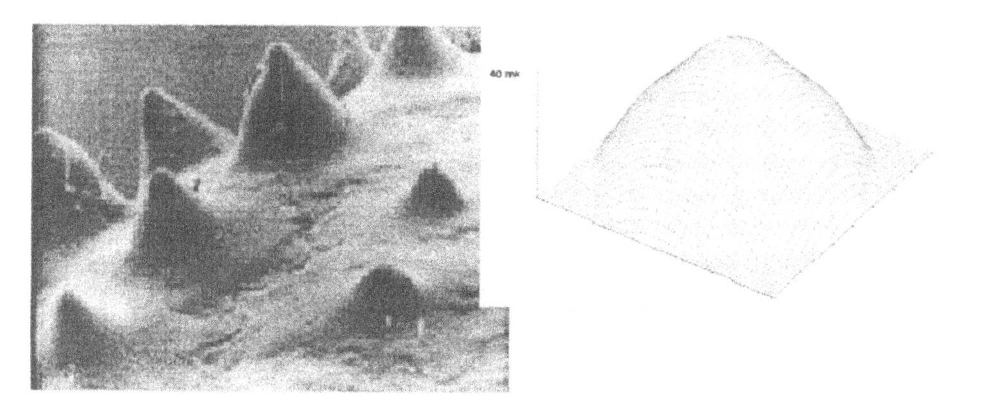

Figure 4. The electronics photography of papillae of different height and form on the seed surface.

Conclusions

Digital photogrammetric technologies are one of the means of informational supply of various problems on the study of biocenoses and biodiversity. Stereoscopic images provide additional information on various biological objects necessary for determining phenological features and systematics as the objects of plant and animal world and allow one to study the spatial structure of biocenoses. The application of digital photogrammetric technologies for treating images of different scale provides a three-dimensional conjugate modeling of the objects studied and the space-time continuity of the analysis of different objects composing biocenoses.

References

Golovko V., Koutzenogii K., Kirov E., Trubina L., Guk A.; The use of photogrammetry for determining pollen characteristics, *Optika armosfery i okeana* **13** (2000) 882-885.

Dubinovsky V. Calibration of pictures. Moscow, "Nedra", 1982.

Guk A., Korkin V., Samushkin V. et al.; Digital photogrammetric complex for creating and restoring maps, *Geodeziya i kartografiya* **12** (1996) 39-48.

Trubina L., Bykova O.; Potentialities of usage of photogtammetric technologies on ecologo-geographic cartography, *Proc. of Intern. scientific conf. dedicated to 65th anniversary of SDGGA-NIIGAiK, November 23-24,1998*, Novosibirsk: SGGA (1999) 73-77.

Vlasova N., Koutzenogii K., Trubina L., Guk A. Use of photogrammetry for studying the morphology of seeds of pink family (Caryophyllacea-Alsinoideae), in: Problems of studying the plant cover of Siberia. *Abstracts of reports at the II Russian scient. conf. dedicated to the 150th anniversary of P.N.Krylov* , Tomsk: izd.TGU (2000) 23.

Hydrogen Technology: H_2 Storage and Release

O.S. Morozova

N. Semenov Institute of Chemical Physics of Russian Academy of Sciences, Kosygin st. 4, 117334 Moscow, Russia, FAX: 007/095/938-2156, e-mail: om@polymer.chph.ras.ru

Introduction

Owing to the negative impact of fossil fuels on the environment, it is imperative to find renewable and clean energy sources. Of all the alternative fuels, hydrogen is now known to be a potential renewable substitute for petroleum and gasoline. Hydrogen is attractive as a fuel because it has the highest density of energy per unit of weight of any chemical fuel, is essentially nonpolluting (the main byproduct of combustion is water) and can serve in variety of energy converters ranging from internal-combustion engines to fuel cells. The present storage methods for hydrogen are suitable and save for the present industrial uses. However, appropriate storage facilities both for stationary and for mobile applications, are complicated because of the very low boiling point of hydrogen (20.4 K at 1 atm) and its low density in the gaseous state.

Metal hydrides with a large hydrogen storage capacity (both in mass and in volume) can be successfully used as rechargeable hydrogen storage media (Reilly et al., 1980). This interest stems from the fact that this is safest means of storing hydrogen as compared to hydrogen liquefaction and hydrogen storage at high pressure. However, hydrogen storage materials suffer from the fact that they must be first activated to adsorb hydrogen. This results from natural surface contamination of the material, usually in the form of oxides and hydroxides. The activation of hydrogen absorbing materials may take a long time depending on the material and the extent of its surface contamination (Schulz et al., 1995).

A breakthrough in the hydrogen storage technology was achieved by preparing nano-crystalline hydrides using high-energy ball milling (Liang et al., 1998). These new materials are stable enough and show very fast adsorption and desorption kinetics that qualify metal hydrides for storage application. It was found that temperature and fullness of hydrogen recovery depends on the particle size, microstructure, and morphology of the material. Due to this, nano-crystalline compounds demonstrate the best properties. Nowadays, mechano-chemical treatment is widely used for modification and preparation of metal hydrides (Guether and Otto, 1999). There are two ways of preparation of materials for hydrogen storage: mechanical destabilization and compositional modification of metal hydrides to reduce their stability. Ball milling is known to change various properties of metal hydrides, as a result of the formation of special microstructures, metastable phases or modified surfaces. Hydrogenation – dehydrogenation properties and kinetics are very sensitive to this modification. For example, formation of nano-cristalline structures produced by ball milling results in dramatic changes in the stability of ZrH_2 (Morozova et al., 2000). Another way of affecting hydrogenation properties is by preparation of mixtures of composites of hydrogen-absorbing materials during ball milling. Unfortunately, the hydrogen sorption capacities of Zr- Ni and Zr-Fe hydrides prepared by this technique drops down after one cycle of sorption/desorption due to Ni precipitation (Butyagin et al., 1998). In another approach, a hydrogen-absorbing material can be hydrogenated in the ball-milling process with simultaneous modification of hydrogen-absorbing metal by graphite. In this case, ball milling is performed under hydrogen pressure, which causes simultaneous hydrogen uptake and modification of metal surface (Bouaricha et al., 2001; Morozova et al., 2001).

In this paper it will be shown that ball milling of zirconium hydride changes the sorption-desorption properties of the material due to re-distribution of hydrogen atoms in the

I. Barnes (ed.), Global Atmospheric Change and its Impact on Regional Air Quality, 303–308.

ZrH_2 lattice and that ball milling of zirconium in a hydrogen flow with a small amount of graphite results in the formation of promising nano-composite material for hydrogen storage.

Experimental

All the experiments were carried in a flow mechano-chemical reactor fixed to the vibrator. The following parameters were used for the milling process: a vibration frequency of 50 Hz; an amplitude of milling of 7.25 mm; the average energy intensity was 1.0 kW/kg.

A stainless steel container was loaded with 1.8 g of ZrH_2 (99%, Aldrich), Zr (99%) or Zr and graphite (spectral pure) reaction mixture (1.5 g Zr + 0.3 g graphite) together with 19.8 g of hardened steel balls (diameter 3 – 5 mm). The input of the reactor was connected to a setup for preparing gas mixtures; the outlet was combined on-line with a gas chromatograph (or mass-spectrometer in H – D exchange experiments) to analyze the effluent gases. On mass-spectral analysis, the intensity of lines corresponding to molecular peaks of H_2 (m/e = 2), HD (m/e = 3), and D_2 (m/e = 4), respectively, were measured every minute. H_2 or D_2 (99.5 %) were used as received, without purification. A gas flow rate (~ 0.01 l/min) was measured before and after the reactor. The milling was started when the hydrogen concentration at both the input and the outlet of the reactor became equal. All the experiments were carried out at room temperature and atmospheric pressure. The specific surface area of the powders (S) was measured by the BET method using low - temperature Ar adsorption. XRD patterns of the solids were recorded before and after the treatments using a Dron-3 diffractometer with a Cu (K_α) anode. Temperature-programmed desorption (TPD) measurements were carried out at a heating rate of 12^0/min from 20 to 550^0C under flow conditions (flow rate 100 ml/min.) using a H_2/Ar mixture with 7 vol. % of H_2 or Ar. SEM measurements were made on a Cameca MBX-1 microprobe in a regime of scanning microscopy. Transmission electron microscopy (TEM) measurements were carried out on a Philips EM 420 ST electron microscope with a resolution limit of 0.3 nm, and an accelerating potential of 120 kV. For high resolution TEM (HRTEM) measurements, a JEM 2010 electron microscope with a resolution of 0.14 nm and accelerating potential of 200 kV was used. The samples for TEM and HREM were prepared from an ethanol suspension and placed on copper grids covered by amorphous carbon.

Results and Discussion

Mechanical Destabilization of ZrH_2.

Reaction of H – D exchange between ZrH_2 and D_2 during ball milling of ZrH_2 in a D_2 flow, as well as TPD from the sample obtained in this process and HREM were applied for studying the mechanism of ZrH_2 destabilization under ball milling. Figure 1 shows H – D exchange kinetics measured under the flow milling conditions. Brittle crushing of ZrH_2 particles in the first 400 – 500 sec of milling was accompanied by a drastic increase in the specific surface area of the sample from 0.34 to 1.8 m^2/g and by the formation of strong low-coordinated centers of adsorption on the newly formed surface. Owing to this, a significant sorption of D_2 on ZrH_2 was observed. Simultaneously, H_2 and DH were detected in the reaction products. The kinetic curves of H_2 and HD formation match almost exactly in shape, but differ in intensity. The concentration of HD is ~ tenfold higher than that of H_2. This indicates the single-type

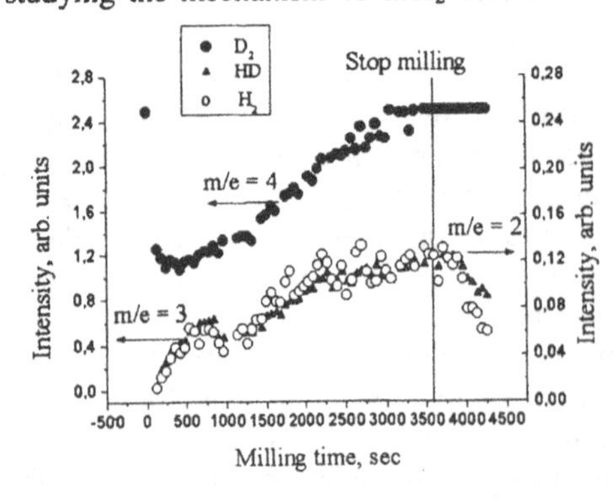

Figure 1. Time dependence of peak intensities for m/e = 2 (H_2), m/e = 3 (HD), and m/e = 4 (D_2).

processes of formation for both products. From experimental data (see Figure 1 and Table 1) we estimated that the mechanism of $ZrH_2 - D_2$ exchange should be as follows: deuterium was firstly adsorbed on particles created by crushing and in the second step, the D_2 adsorbed diffused into the bulk due to a well-developed micro-block structure generated under the ball milling. As a result, the total content of H and D atoms in the Zr-hydride lattice was found to be ~ 20% higher than stoihiometric. Plastic deformation and lattice stress caused by high-energy ball impacts are well known to accelerate the H diffusion from the bulk of ZrH_2 to the surface (Hemplemann and Rush, 1985; Morozova et al.,. 2000). In the first step (~1300 sec of milling), both H_2 and HD were formed by the surface recombination of H atoms diffusing from the bulk with H or D atoms. It is very likely that the latter occupied ~ 90% of the surface. Fracturing and particle size reduction result in an increase of the specific surface area for H – D exchange reaction. Some portion of HD may be produced on the micro-grain or micro-block boundaries. Figure 1 shows that the concentrations H_2 and HD in reaction products increased even though the D_2 sorption decreased after ~ 1300 sec of milling (concentration of D_2 in a gas phase increased). This indicates that deuterium absorbed also took part in H – D exchange. Steady –state conditions were reached after ~ 3500 sec of milling.

During this time, about 47% of hydrogen from Zr-hydride ran out as HD and H_2. It means, that the reacting solid was only partially destabilized.

Table 1. D_2 consumption and H_2 and HD formation: balance estimations.

Milling time, s	D_2 consumption, molec	HD in a gas phase, molec	D_2 sorbed on ZrH_2, molec/g	H_2 in a gas phase, molec	H_2 total consumption (H_2 + 0.5 HD), molec/g ZrH_2	Specific surface area, m^2/g
420	7.8×10^{20}	1.3×10^{20}	7.2×10^{20}	1.6×10^{19}	8×10^{19}	1.8
900	1.78×10^{21}	5.3×10^{20}	1.50×10^{21}	6.1×10^{19}	3.25×10^{20}	3.6
1300	2.56×10^{21}	8.3×10^{20}	2.11×10^{21}	1.0×10^{20}	5.2×10^{20}	

The TPD experiments further support this conclusion. The TPD technique is very useful for obtaining information about the kind and type of structural defects generated during ball milling of ZrH_2 in a D_2 flow and about the distribution of H and D atoms in the lattice of Zr-hydride. As can be seen, each curve in the TPD spectrum (Figure 2) has three well-defined peaks with T_{MAX} 160 – 168°C, ~ 380°C, and ~ 610°C, respectively. The area of an individual peak is strongly related with a number of molecules (atoms) bonded to specific defects.

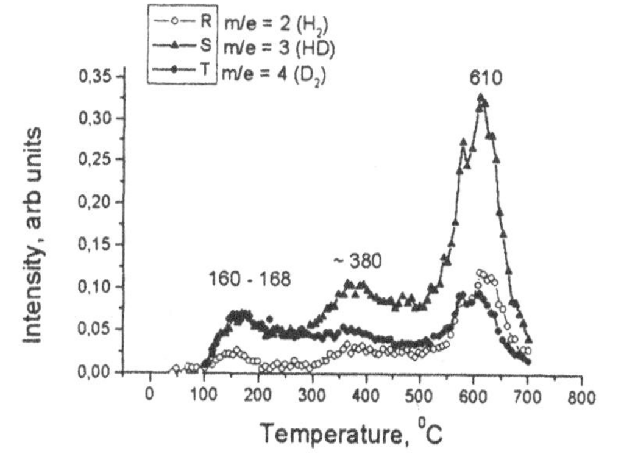

Figure 2. TPD spectra of H_2, HD and D_2 for ZrH_2 activated in D_2 flow (see Figure 1).

According to HREM data, a rather perfect microstructure of original ZrH_2 was radically modified to a well-developed micro-block structure due to mechanical treatment. "Polycrystalline" type of micro-diffraction with diffuse rings was indicative of distorted lattice that is in good agreement with XRD data. The Zr – hydride micro-crystals consisted of fairly small (~ 10 – 20 nm) patches well-ordered inside but disordered against each other. Besides these fragments, other types were observed: large enough areas (~ 50 – 60 nm) perfectly crystallized and characterized by a particularly high concentration of micro-strains being manifested through the typical striped "tiger" contrast. Both types of defects are responsible for TPD peaks with T_{MAX} 160 – 168^{0}C, ~ 380^{0}C, respectively. The peak with T_{MAX} ~ 610^{0}C belongs to atoms distributed in the ordered hydride lattice (mostly, in tetrahedral positions). Some characteristics estimated from TPD spectra are summarized in Table 2. Calculations were made as in (Kislyuk and Rosanov, 1995). It is clear that desorption temperature for a portion of hydrogen decreased from ~ 600 to ~ 230^{0}C due to mechanical destabilization of the ZrH_2 lattice. However, the other atoms (~ 45%) still occupied the regular positions. It was estimated that between them ~ 6% belong to ZrH_4, ~ 63% to ZrH_3D and ZrH_2D_2 configurations, respectively. The relaxation of structural defects was observed during heating in TPD experiments. Thus, mechanical destabilization by ball milling is a not promising way for preparing Zr- containing hydrogen storage materials.

Table 2. Effect of the ZrH_2 treatment on parameters of TPD of H_2, HD and D_2.

T_{MAX} ^{0}C	Binding energy, kJ/mol	$H_2 + HD + D_2$ molec/g Zr-hydride
Original ZrH_2		
≅ 600	165	6.5 x 10^{21}
ZrH_2 after treatment in D_2 flow		
160 - 168	~ 88	1.1 x 10^{21}
380	132	~ 2.5 x 10^{21}
610	178	~ 2.9 x 10^{21}

Mechanical synthesis of Zr – graphite composite material.

Ball milling of zirconium in a H_2 flow during 2 hours resulted in the formation of highly dispersed tetragonal phase ε-$ZrH_{1.8-2}$, according to XRD. The behavior is radically changed, when metal is milled with graphite (20 wt. %): total transformation to the same phase of Zr – hydride occurred in 80 min of milling (see Figure 3). No phases of ZrC or ZrH_XC_{1-X} with a low concentration of carbon were detected in the reaction products. The specific surface area of the sample increased from 0.6 to 58.6 m^2/g. According to the TEM data, the powder milled consisted of large loose aggregates (200nm - 250 nm) formed from small composite nanoparticles of more or less oblong shape about 20 nm - 50 nm in projection. Each nanoparticle contained small ε-$ZrH_{1.8-2}$ "fragments"(2 – 30 nm in size) randomly distributed in amorphous carbon. However, prolonged ball milling in H_2 flow stimulated decomposition of ε-$ZrH_{1.8-2}$ phase to Zr metal and Zr-hydride phases with lower hydrogen content (see Figure 3). This incredible transformation seems to be induced by the drastic morphological modification: the loose particles of reactive powder stick together to form the compact and large particles with a high concentration of strains. Specific surface area of the powder falls down from 58.4 m^2/g to 1.8 and even to 0.3 m^2/g. Simultaneously, the intermixing of the Zr-containing fragment with carbon on a microscopic

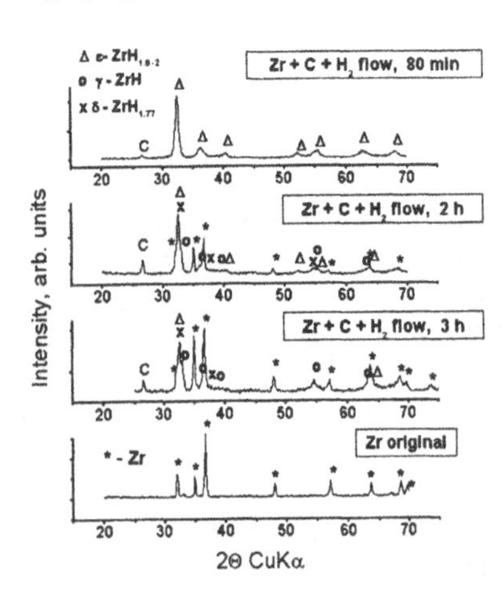

Figure 3. XRD patterns of original Zr and Zr – graphite powders after different time of ball milling.

level was observed by TEM. The particles consist of thin carbon and Zr-containing nano-layers (2 – 8 nm in width) chaotically intermixed. These particles are believed to be the something like a transition state of material. The interaction between carbon and ε-$ZrH_{1.8-2}$ is realized due to close contact of solid surfaces and a significant extend of interfaces. As a result, carbon atoms incorporate through the surface into the Zr-hydride lattice to make the surface inaccessible for H_2. The decrease in hydrogen content initiates the ε-$ZrH_{1.8-2}$ decomposition. Once the phase of Zr metal was re-formed, its content in the sample increased up to ~ 80 at %. Prolonged ball milling (up to 3 hours) resulted in the further increase in the particle size and partial crystallization of Zr-containing fragments, as well. According to TEM, ribbon-like composite particles, several micrometers in size, were formed. Carbon whiskers and nanotubes (20 – 80 nm in diameter) were also observed on the particle surface. The Zr-containing fragments from ~ 400 nm till several nm in size were randomly distributed in a carbon matrix (Figure 4).

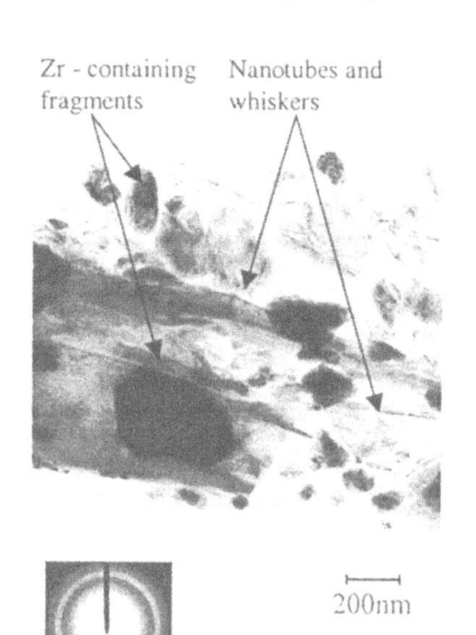

Figure 4. TEM micrograph and corresponding diffraction pattern of Zr-graphite powder after 3 h of milling

This composite material was found to be stable and promising for hydrogen storage. According to TPR – TPD data, the as-hydrogenated state (full transformation from Zr to ZrH_2) can be reached by heating in 7% H_2/Ar flow during several minutes. Very fast hydrogen sorption occurred in the range between 300 and ~ 420^{0}C. Desorption temperature for the ZrH_2 obtained was ~ 430^{0}C (as compared to ~ 600^{0}C for both original and destabilized ZrH_2). Total and fast hydrogen release was observed at this temperature. XRD confirm the absence of Zr – hydride phases after desorption. Sorption – desorption cycles can be repeated several time. TEM investigation demonstrated that morphology of the composite material is rather stable. Hydrogen uptake and release by Zr-containing fragments were carried out with no modification of their microstructure. Usually, volume changes occur during hydrogenation with associated volume expansion, in contrast with hydrogen desorption with volume contraction that leads to material crushing. In our case, encapsulation of active Zr – containing fragments in the compact ribbon-like carbon matrix seems to be of special importance for the material stability.

Conclusions

A promising hydrogen-storage material was synthesized by the ball-milling technique on the basis of complex investigation of Zr – H and Zr – H –graphite systems. Combination of micro- and nano-sized Zr-containing fragments embedded in a carbon ribbon-like matrix affords the stability of the material. Zr-containing fragments can be transformed in ZrH_2-containing fragments of the same morphology. The bulk hydrogen can be completely removed from and introduced into these fragments at 420 – 430^{0}C.

Acknowledgments

The author is thankful to Prof. P.Yu. Butyagin , Dr. A.N. Streletskii (Institute of Chemical Physics RAS), and Dr. A. Pundt (Institute of Material Physics of the University of Goettingen) for valuable comments and discussion, to Dr. T.I. Khomenko (Institute of Chemical Physics RAS) for performing TPD experiments, Dr H. Jander (Institute of Physical Chemistry of the University of Goettingen) for the preparation of TEM samples, Dr. Ch. Borchers (Institute of Material Physics of the University of Goettingen) for performing TEM measurements, Dr. G.N. Kryukova (Boreskov Institute of Catalysis SO RAS) for performing HREM measurements and Dr. A.V. Leonov (Chemical Department of Lomonosov Moscow State University) for XRD measurements. This work is supported in part by Ministry of Education, Science and Culture of Japan (C) No. 12650692 and by Russian Foundation for Basic Research, project No 01-03-32803.

References

Bouaricha S., J.P. Dodelet, D. Guay, J. Huot, and R. Schulz; Activation Characteristics of Graphite Modified Hydrogen Absorbing Materials, J. Alloys Comp. 325 (2001) 245 - 251.

Butyagin P.Yu., A.N. Streletskii, O.S. Morozova, A.V. Leonov, I.V.Berestetskaya, and A.B. Borunova; Mechanically Induced Chemical Conversion in α-Zr- CO – H_2 System. Basic Reactions, Chem. Phys. Rep. 17 (1998) 521 - 534.

Hemplemann R. and J.J. Rush, in: G. Bambakidis and R.C. Bowman, Jr. (eds). Hydrogen in Disordered and Amorphous Solids: Plenum, N.Y. (1985) 283 - 290.

Kislyuk M.Y.and V.V. Rozanov; Theoretical description of Temperature-Programmed Desorpton and Temperature-Programmed Reaction Spectra, Kinetika i Kataliz 36 (1995) 89 - 96.

Liang G., S. Boily, J. Hout, A. Van Neste, R. Schulz; Mechanical Alloying and Hydrogen Absorption Properties of Mg – Ni system, J. Alloys and Comp. 267 (1998) 302 - 335.

Morozova O.S., A.N. Streletskii, I.V. Berestetskaya, A.V. Leonov, and G.N. Kryukova; Effect of Reaction Medium on the Mechanochemical Decomposition of ZrH_2 and on the Reactivity of Hydride Hydrogen, J. Metastable and Nanocryst. Mat. 8 (2000) 429 - 435.

Morozova O.S., A.V. Leonov, T.I. Khomenko, and V.N. Korchak; Mechanism of H_2 Activation on Metals under Mechanical Treatment, J. Metastable and Nanocryst. Mat. 9 (2001) 415 - 420.

Reilly J.J. and C.D. Sandrock; Hydrogen Storage in Metal Hydrides, Scientific American 242 no 2 (1980) 98.

Schulz R., S. Boily, L. Zaluski, A. Zaluska, P. Tessier, and J.O. Streom-Olsen; Innovation in Metallic materials (1995) 529 - 539.

Impact on Ecosystems

Effects from Industrial Emissions on the Lake Baikal Region Forests

Tatiana Mikhailova and Nadezhda Berezhnaya

664033, P.O.Box 1243, Irkutsk, Russia
e-mail:mikh@sifibr.irk.ru

Introduction

The Baikal region is directly connected with the world-famous Lake Baikal, and lies in the middle of Asia on the territory of two states, Russia and Mongolia. Its total area is about 800 000 km^2. An up-to-date zoning of this large region is reported in a monograph "The Present and Future State of the Lake Baikal Region" (Koptyug and Belov, 1996). The region's territory is subdivided into the central zone, the buffer zone (lying on the territory of Russia and Mongolia), and the zone of atmospheric impact outside the catchment area, westward and north-westward of the lake. The zone of atmospheric impact is about 200 km wide; it includes the territory from which the lake's catchment area receives atmospheric emissions from industrial enterprises of the Irkutsk region, as well as from adjacent territories, which has an adverse effect on the lake's ecosystem. The region's industrial centers that are located on the territory of the zone of atmospheric impact, are emitting a large amount of acidic pollutants (sulfur dioxide and carbon oxides), and solid aerosols to the atmosphere; a smaller share goes to nitrogen oxides and hydrocarbons.

An overview of publications devoted to the impact of atmospheric industrial emissions on forests in the three zones shows that the most attention has been given to the conditions of fir *(Abies sibirica Ledeb.)* and Siberian stone pine *(Pinus sibirica Mayr)* which stand in the southern part of the central zone and partly in the buffer zone affected by atmospheric emissions from the Baikal pulp-and-paper plant. The scarcity of information about the forest conditions in a very vast zone of atmospheric impact has been substantially remedied; specifically, surveys were made of the forests adjacent to the Irkutsk-Cheremkhovo industrial area, and to a number of major cities (Mikhailova *et al.*, 1994; Mikhailova and Berezhnaya, 2000). Results of these efforts showed that most of the surveyed coniferous forests are, to a considerable extent, weakened by industrial emissions. The question arose as to how far from the industrial territories and cities the decline of the coniferous forests spreads, which are predominant in the zone of atmospheric impact. To answer this question it is necessary to have an understanding of the extent to which the environment-forming and environment-protective functions of these forests are lowered. Therefore, the objective of this study was to assess, using a set of parameters, the conditions of the coniferous forests growing in the zone of atmospheric impact, within different distances from the industrial area.

Materials and Methods

Testing surveys covered the southern and middle parts of the zone of the atmospheric impact of the Baikal region, making up about 50% of the territory of this zone. The main forest-forming species here is Scots pine (*Pinus sylvestris L.*). During 1996-1999, naturally growing pine stands were surveyed at different distances from the industrial territory, according to the following transects: I. south-eastern (from Cheremkhovo to the source of the Angara river), II. north-eastern (from Usolye-Sibirskoye to the upper Kuda river), III. north-eastern (from Angarsk to the upper Murin river), IV. southern (from Shelekhov to the source of the Angara river), V. south-western (from Angarsk to the piedmonts of the East Sayan mountains), and VI. south-western (from Usolye-Sibirskoye to the upper Kitoi river) (Figure 1).

I. Barnes (ed.), Global Atmospheric Change and its Impact on Regional Air Quality, 311–315.

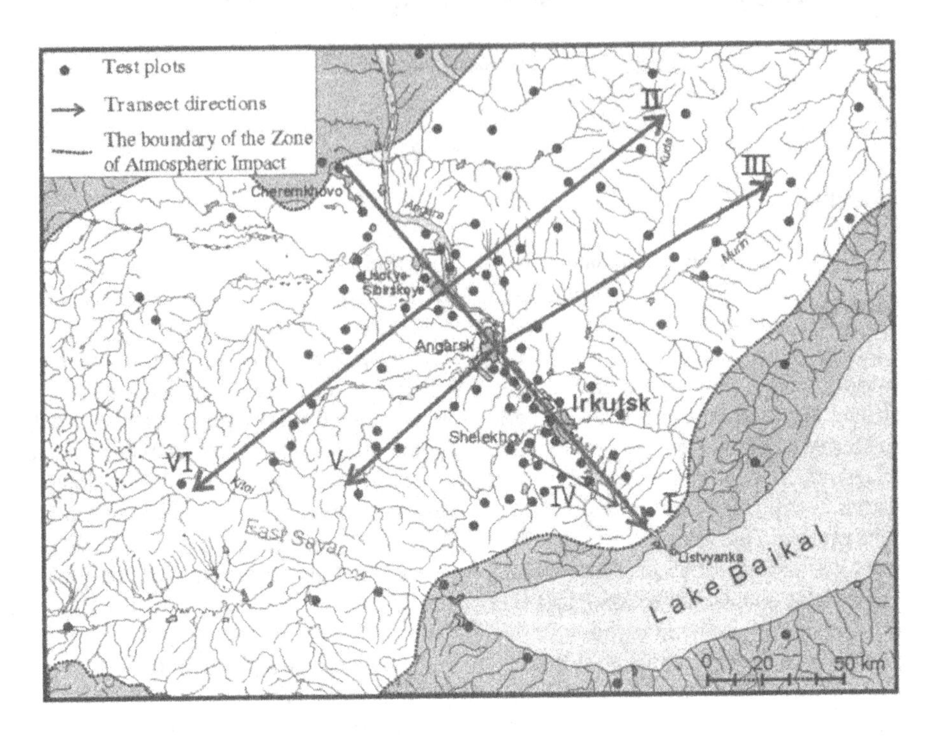

Figure 1. A schematic map of the transect directions surveyed and location of sample plots in the Zone of Atmospheric impact in the Baikal region

The surveys were carried out by setting up sample plots with an area of 400 m^2, following commonly accepted procedures developed within an International Cooperative Program (ICP Forests). The transects were laid with due regard for the prevailing directions of winds conducive to the transfer of emissions from the industrial area, the peculiar features of terrain, specifically exposures of leeward slopes of the heights; allowance was also made for the extensive occurrence of emissions along river valleys. A total of 48 sample plots were set up in transects. The crown defoliation level and the degree of discoloration were determined for all trees on each of the sample plots; to determine the element composition and nitrogen fractions, samples of the 2-year-old needles were taken from 40-year-old trees.

In the pine needles, the following elements were determined: fluorine, sulphur, heavy metals: iron, zinc, lead, cadmium, mercury, copper, manganese, as well as aluminum, silicon, potassium, calcium, magnesium, sodium, and nitrogen-containing compounds. Element composition was measured by conventional methods: flame photometry, and by the atomic absorption method. Protein and non-protein nitrogen was measured by the Nessler reagent (Ermakov, 1987). The composition of elements and nitrogen fractions in the needles was calculated in percentage of dry weight. Calculations were made of the mean value of concentration of a particular element or substance, and of the standard error.

Results

The sample plots for all survey transects exhibited a clearly pronounced tendency for a decrease of the defoliation level (Figure 2a). The mean level of defoliation makes up 20-25% within distances of 80-100 km, 55-60% near the industrial area, and as much as 70% on some of the sample plots.

Transect I (especially near the cities of Usolye-Sibirskoye) and IV (near Shelekhov) showed a discoloration of the pine needles within 10-25% and 20-40%, respectively.

When determining the sulphur content, it was found that almost all transects reveal a similar tendency: a decrease in sulphur concentration with the distance from the industrial area and large cities to 70-100 km (Figure 2b). For transect I lying along the main transfer of emissions, the level of sulphur is high for all sample plots, and somewhat decreases only on the sample plot at the source of the Angara, i.e. virtually on the shore of Lake Baikal. The amount of fluorine accumulated in the pine needles on the surveyed sample plots is also high near the industrial area, and decreases drastically at a distance of 40-60 km (Figure 2c). Particularly noteworthy is transect IV starting from Shelekhov where a large aluminum smelter is located. On the sample plots of this transect, the tree needles show a high level of fluorides, even within 30 km of the smelter.

A high content of heavy metals in the needles is also found near the industrial area and the cities. Thus, the concentration of lead, cadmium and copper in this case makes up, respectively, $10x10^{-5}$ %, $3\text{-}7x10^{-6}$ % and $5\text{-}12x10^{-4}$ % of dry weight. A sharply pronounced peak in mercury content was observed near Usolye-Sibirskoye (as much as $15x10^{-6}$ % of dry weight). With the distance from the industrial area, the concentration of heavy metals in the needles declines significantly, and their background levels are detected at the distance of 70-100 km: $1.5x10^{-5}$ % of lead, $1.6x10^{-6}$ % of cadmium, $1.1x10^{-4}$ % of copper, and $0.9x10^{-6}$ % of mercury. A different pattern is revealed for manganese: a rather abrupt decrease in its concentration near the industrial area when compared to background concentration. Thus, the pine needles on the sample plots near the cities contains $0.1\text{-}0.9x10^{-2}$ % of dry weight of manganese, while this figure is $3\text{-}6x10^{-2}$ % at a distance of 70-100 km. A slight increase in iron concentration in the tree needles is recorded near the industrial area. The zinc content in the pine needles remained virtually unchanging within different distances from the industrial area. Calcium and sodium showed an explicit increase in concentration in the needles as the industrial area is approached: by factors of 3 and 5-6 for calcium and sodium, respectively. A decrease in potassium level was observed only in the pine needles for transect IV as the aluminium smelter is approached. None of the surveyed sample plots revealed any pronounced changes in magnesium content in the pine needles.

Results derived from determining the content of protein and non-protein nitrogen have shown that the dynamics of these fractions in the pine needles differs according to the direction: as the industrial area is approached, the level of protein nitrogen decreases, and that of non-protein nitrogen, on the contrary, increases. The N protein/N non-protein ratio in the needles has also a tendency to decrease near the industrial area and large cities (Figure 2d).

Discussion

The results obtained from the above survey bear witness to a decay of coniferous forests both near the industrial area and within relatively long distances from it. This is suggested by the high defoliation level of pine trees, the frequently occurring discoloration of the needles, the exceedance of background concentrations of sulphur, fluorine and heavy metals, and the low N protein/N non-protein ratio caused by a decrease in protein nitrogen content in the pine needles.

If the stand condition is assessed from visual parameters (percentage of defoliation and discoloration), then most of the stands should be categorized as the second class of injury (Procedures, 1987). For more accurate diagnostics of the physiological condition of trees and tree stands, we have suggested that, in addition to using the index of defoliation level, such

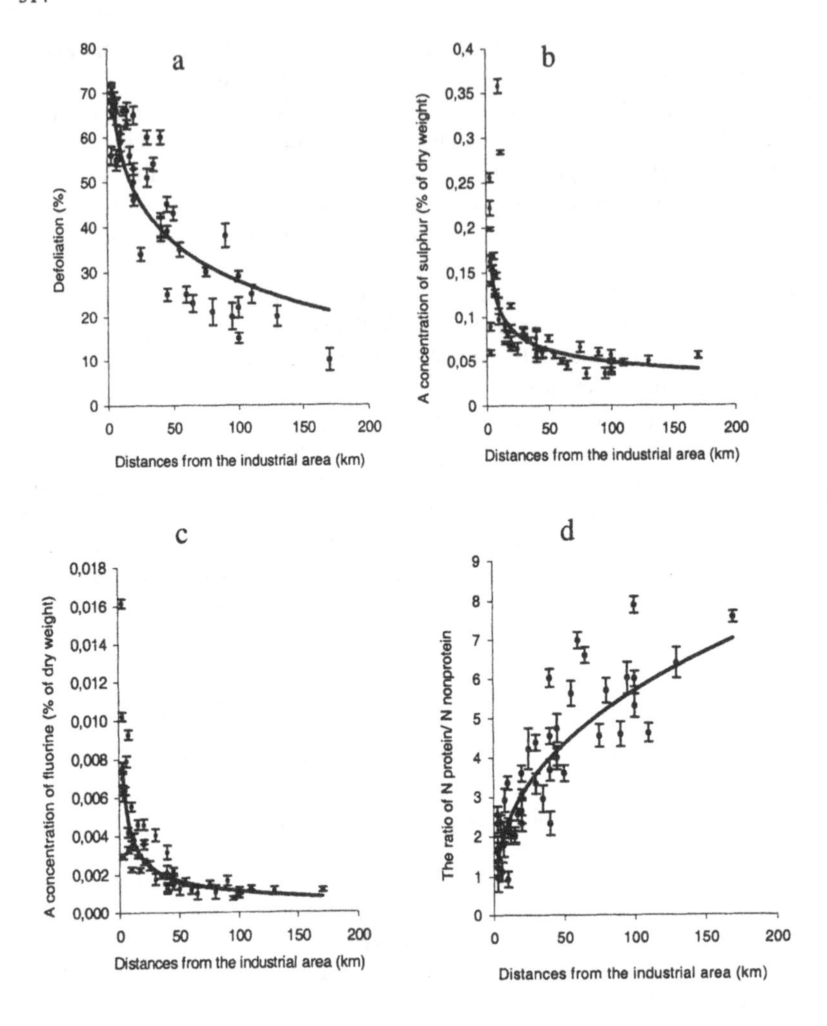

Figure 2. Trends of changes in tree crown defoliation levels (a), sulphur concentration (b), fluorine concentration (c) and the ratio N protein/ N non-protein (d) in Scots pine needles within different distances from the industrial area. Error bars indicate one standard deviation of the mean

a parameter should be employed as the N protein/N non-protein ratio in the needles. The ratio reflects the intensity of growth processes (N protein) and a degree of the adverse effect of pollutants (N non-protein). These two indices (% of defoliation and N protein/N non-protein) were used to develop a diagnostic scaling, and six classes of stand condition identified: background, low-weakening, moderate, moderate-heavy, heavy, and dying (Mikhailova, 2000). According to this scaling, the surveyed stands are grouped together as follows: heavily weakened stands occur at a distance of up to 10 km from Angarsk, Usolye-Sibirskoye and Shelekhov; moderately-heavily weakened stands are found along almost entire transect I and within distances of up to 20-25 km from the industrial area for the other transects; moderately weakened stands are observed as far as 40-50 km away from the industrial area; and low weakening is found within up to 60-80 km. Background stands are of relatively limited occurrence, and are normally found within 80-110 km of the industrial area, although local areas of comparatively healthy stands may be found at shorter distances, provided that they are protected from emissions by natural barriers such as mountains or watershed divide heights.

Conclusion

Pine forests of the most polluted part of the zone of atmospheric impact in the Baikal region have been surveyed, based on six transects on 48 sample plots. Acidic pollutants make up a significant proportion of the region's emissions. It was found that the injurious effect of industrial emissions on the forests spreads as far as 80 km from the industrial territory. The stand conditions were assessed from a set of parameters: the percentage of crown defoliation and discoloration, the change in element composition of the needles (14 elements were determined), and the protein N/non-protein N ratio in the needles. Stands of heavy, moderate-heavy, moderate and low weakening were identified. Results derived from carrying out the survey suggest that there exist disturbances of the environment-forming and environment-protective role of this zone in the Baikal region.

Acknowledgement

We are profoundly grateful to engineer S.Yu. Toshchakov for technical assistance in carrying out the survey.

References

Ermakov A.I. (ed); Methods of Biochemical Investigation of Plants, Agropromizdat, Leningrad (1987) 430pp.

Mikhailova T.A.; The physiological condition of pine trees in the Prebaikalia (East Siberia), *Forest Pathology* **30** (2000) 345-359.

Mikhailova T.A. and N.S. Berezhnaya; Assessment of forest condition under prolonged air pollution from an aluminium plant, *Geografiya i prirodnye resursy* **1** (2000) 43-50.

Mikhailova, T.A., N.S. Kochmarskaya, L.V. Antsiferova, and A.S. Pleshanov; The impact of industrial emissions on coniferous forests of the Angara region, in: O.M. Kozhova (ed), The Assessment of the Condition of Water and Terrestrial Ecosystems. Institute of Biology, Novosibirsk (1994) 127-131.

Koptyug, V.A. and A.V. Belov (eds),The Present and Future State of the Lake Baikal Region. The Russian Academy of Sciences Siberian Branch Irkutsk and Buryat Scientific Centers, Novosibirsk (1996) 111pp.

Assessment of Siberian Forest Ecosystem Stability Against the Acidification

Mikhail Semenov
Limnological Institute of Siberian Branch of Russian Academy of Sciences, Irkutsk 664033, Box 4199, Russia
semenov@irigs.irk.ru

Introduction

In the perspective of worldwide agreement on atmospheric pollutants emission abatement it is very important to assess the forest ecosystem sensitivity to acid deposition in such a vast and forested area as the Asian part of Russia - Siberia. This region is situated between the Ural Mountains and the Pacific Ocean. At the present significant inputs of acid forming pollutants are related mostly to trans-regional and trans-boundary pollution from Europe because of circumpolar wind directions, which are predominant in the Northern hemisphere. In the future, ecosystem damage due to the atmospheric deposition in southern Siberian regions caused by trans-boundary pollutants transport from China is also possible. During last years, some research has been conducted to give quantitative estimates of Asian ecosystems sensitivity to acidity loading (Bashkin *et al.*, 1995, 1996). These estimates have been based on semi quantitative approaches to the assessment of soil/ecosystem buffering mechanisms like chemical weathering, base cation and nitrogen uptake by vegetation, nitrogen leaching and denitrification. As usual, the input information for the calculation of the given parameters was based predominantly on expert conclusions. The uncertainty of both input parameters and output results was very great.

Our latest publications have been devoted to the adaptation of quantitative methods for the integrated assessment of forest ecosystem stability in Siberia (Semenov and Bashkin, 2000; Semenov *et al.*, 2001). As the measure of stability the critical load of acidity was chosen. The biogeochemical model PROFILE (Warfvinge and Sverdrup, 1995) and the mineralogy reconstructing model UPPSALA (Manual, 1996) were used for calculations.

Initial information and its adaptation

The values of nutrient cycling in Siberian ecosystems (coniferous, deciduous and mixed forests on different soils) were calculated using experimental data from the literature (Arganova, Elpatievsky, 1990; Bazilevich, 1993; Chernyaeva *et al*, 1978; Gadjiev, 1982; *etc.*). The net nutrient uptake i.e. the nutrients in the biomass compartments that are expected to be removed from the system at harvest were determined to be zero, as all the ecosystems were regarded as unmanaged (virgin). The information on the chemical composition of the solid and liquid phases of various soils and their physical parameters was also extracted from the literature.

One of the most important climatic features of Siberia is the long period of stable snow cover, 6-7 months on average. It is well known that the snow is the most informative, integral index of element input to the ecosystem. All calculations of atmospheric deposition were made using the data on the element content in snow.

The total potential acidity input to the system ($Ac(pot)_{dep}$) was calculated according to the existing formula:

$Ac(pot)_{dep.} = SO_{xdep}+NO_{xdep}+NH_{xdep}-BC_{dep}+Cl_{dep}$ (Manual ..., 1996).

Critical loads calculation

Critical loads of acidity were calculated according to the formula: $CL(Ac) = BC_w - ANCle_{(crit)}$, where BC_w is the base cation weathering and $ANCle_{(crit)}$ is the critical Acid Neutralizing Capacity leaching calculated using the value of 1 of Bc/Al ratio.

I. Barnes (ed.), Global Atmospheric Change and its Impact on Regional Air Quality, 317–320.

Mapping procedure

The maps of deposition were drawn using isolines. A critical load map was prepared in two steps. In the first step the base map of territorial complexes was drawn by combining soil and vegetation maps. In the second step CL values were marked on the base map and the contours of equal values were united. The underlying principle of the mapping method was to characterize each contour by minimum 2 points (sites) taking into account the square and ecological conditions diversity.

Results and discussion

During the mapping procedure in order to show the deposition of base cations as an intrinsic ecosystem property, a sea-salt correction with Na as a tracer was applied. However, in some areas like Yamal, Central Yakutia, Far East with an excess Na deposition due to continental sources, a sea-salt correction led to underestimated non-sea-salt deposition and to a distorted acid-base balance of total deposition (Table 1).

Table 1: Real and sea-salt corrected values of atmospheric deposition and acid neutralizing capacity (keq ha^{-1} yr^{-1}). BC - non sea-salt corrected basic compounds (Ca, Mg, K, Na), Na - deposition of sodium, Bc - sea-salt corrected basic compounds (Ca, Mg, K), Ac - acidic compounds excluding nitrates, $\underline{\text{ANC}}$ = Bc - Ac excluding total nitrogen, ANC - total acid neutralizing capacity

Location	Measured values				Sea-salt corrected values			
	BC-Na	Ac	$\underline{\text{ANC}}$	ANC	Bc	Ac	$\underline{\text{ANC}}$	ANC
Sites with marine originated sodium only								
Yakutsk	0.14-0.03	0.09	0.05	0.04	0.10	0.05	0.05	0.04
Mondy	0.24-0.02	0.78	- 0.54	- 0.56	0.21	0.75	- 0.54	- 0.56
Ilchir	0.50-0.02	0.20	0.30	0.24	0.47	0.17	0.30	0.24
Baikal	0.47-0.03	0.24	0.23	0.15	0.43	0.20	0.23	0.15
Sites with both marine and continental sodium								
Mirny	0.36-0.31	0.30	0.06	- 0.03	0.017	0.15	- 0.13	- 0.22
Neryungri	0.49-0.26	0.10	0.39	0.35	0.16	0.00	0.16	0.12
Yamal	0.98-0.84	0.73	0.25	0.21	0.03	0.00	0.03	-0.01
Far East	1.31-0.55	0.49	0.82	0.62	0.60	0.25	0.35	0.15

The spatial distribution of atmospheric deposition compounds are in agreement with the existing data on atmospheric mass transfer. Due to it's predominant direction - from the Athlantic to the Pacific and Ural Mountaines location both acid and base deposition values are highest in the South of West Siberia decreasing with the longitude increase (East direction) up to the Yakutia (100°Lo) and latitude increase up to the ocean. After the 100°Lo Athlantic transfer is strongly affected by the Pacific ocean. This result in reallocation of deposition isolines in South - North (meridional) direction.

The South of the Far East region (near the pacific ocean) is charactetrized by a high intensity of atmospheric deposition.

It must be noted that the mountain systems affect the spatial distribution of deposition all over Siberia because of air masses bypassing them, moreover bypass zones are characterized by high deposition values (Kuznetsova, 1978). Despite the common transport direction the spatial distribution of base compounds (base cations) is different from that of acid ones. The main difference is the compact localization of high deposition values, whrereas the acid deposition is more dispersed. This is due to the origin and weight of the compound matter - aeolian particles in the case of base and gases in case of acid ones. The spatial distribution of Ac(pot) values shows the picture of total acidity input as the result of deposition compound interaction (Figure 1).

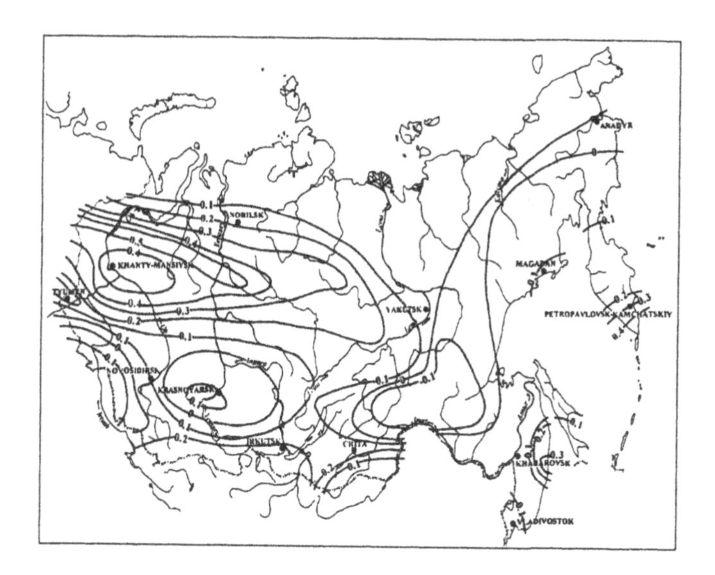

Figure 1. Ac(pot) deposition (S + N + Cl - Bc), keq ha^{-1} yr^{-1}.

The maximum weathering rates (up to 2 keq ha^{-1} yr^{-1}) are obtained for carbonate and low-weathered soils in southern plain regions, minimum rates (down to 0.01 keq ha^{-1} yr^{-1}) are typical for the tundra zone and highland areas.

In total, information on the data necessary for CL calculations using the PROFILE model for characterizing more than 200 sites in Siberia was collected. Due to the variety of ecological conditions the critical load values vary in the range of 0-7 keq ha^{-1} yr^{-1} (Figure 2). The most sensitive ecosystems within the investigated area are the ecosystems of the Tundra zone (Longitude 60°-160°, Latitude 80°-70°) and ecosystems of the East Sayan mountains, coniferous forests (Lo 100° -120°, La 60°-50°) with dystric cambisols and gleysoils - critical loads of actual acidity (CL(Ac)) = 0-0.3 keq ha^{-1} yr^{-1}. The most tolerant ecosystems are ecosystems of deciduous forests with podsoluvisols, luvisols and humic luvisols of South Taiga zone in West Siberia (Lo 60°-90°, La 60°-50°) - CL(Ac) = 3.5-7.0 keq ha^{-1} yr^{-1}. Generally the values of critical loads are increasing from the North to the South and from the East to the West following the bioproductivity, annual soil temperature and deposition alkalization increase. Thus, the picture of the spatial CL distribution is similar to the intensity of biological turnover complicated by the spatial geology distribution.

Preliminary evaluation of exeedances followed according to the simplest formula: Ex = Ac_{dep}-Bc_{dep}-CL(Ac), showed negative values all over Siberia. Both (Ac_{dep}-Bc_{dep}) and CL(Ac) values are close to each other for the territory of the East Sayan mountain highland zone (1000 m and higher).

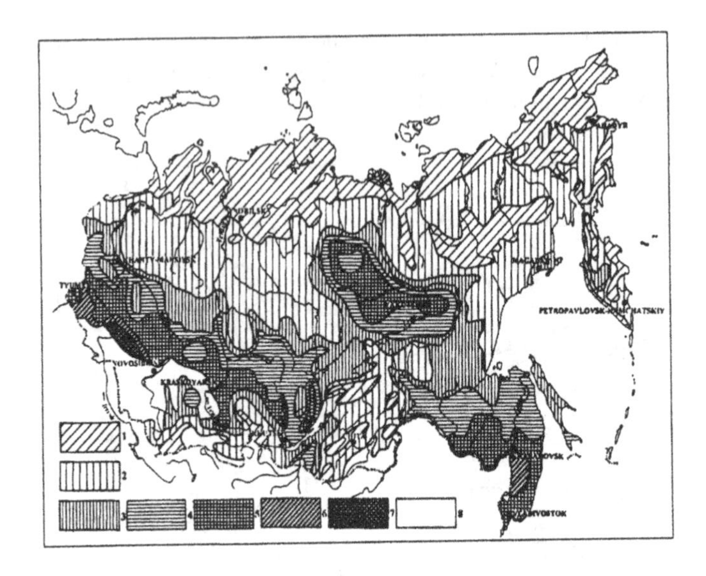

Figure 2. Critical loads of acidity, keq ha^{-1} yr^{-1}: 1 – 0-1; 2 – 1-2; 3 – 2-3; 4 – 3-4; 5 – 4-5; 6 – 5-6; 7 - >6; 8 – steppe and forest-steppe ecosystems.

References

Arganova, V.S., P.V.Elpatievsky; Geochemistry of landscape and technogenesis. "Nauka" publishers, Moscow (1990) (in Russian).

Bashkin, V.N., M.Ya.Kozlov, I.V.Priputina, A.Yu.Abramychev and Dedlova; Calculation and Mapping of Critical Loads of S, N and Acidity on Ecosystems of the Northern Asia. *Water, Air, and Soil Pollution* **85** (1995) 2395-2400.

Bashkin, V.N., M.Ya.Kozlov, and A.Yu.Abramychev; The Application of EM GIS to Quantitative Assessment and Mapping of Acidification Loading in Ecosystems of the Asian Part of the Russian Federation. *Asian-Pacific Remote Sensing and GIS Journal* **8(2)** (1996) 73-80.

Bazilevich, N.I.; Biological productivity of Northern Eurasian ecosystems. "Nauka" publishers, Moscow (1993) 350 (in Russian).

Chernyaeva, L.E., A.M.Chernyaev, A.K.Mogilenskih; Chemical composition of atmospheric deposition (Ural Mountines and around-Ural region). Gidrometeoizdat publishing house, Leningrad (1978) 177 (in Russian).

Gadjiev, I.M.; Soil evolution in South Taiga of West Siberia. "Nauka" publishers, Novosibirsk (1982) 278 (in Russian).

Kuznetsova, L.P.; Moisture transport in the atmosphere over the territory of USSR. "Nauka" publishers, Moscow (1978) 115 (in Russian).

Manual on Methodologies and Criteria for Mapping Critical Levels/Loads and Geographical Areas They Are Exceeded. UN ECE Convention on Long-range Transboundary Air Pollution: Texte 71/96. Berlin: Federal Environmental Agency, (1996) 142.

Semenov, M.Yu., V.N.Bashkin, H.Sverdrup; Application of Biogeochemical Model "PROFILE" for Assessment of North Asian Ecosystem Sensitivity to Acid Deposition, *Asian Journal of Energy and Environment* **2** (2000) 143-162.

Semenov, M.Yu., O.G.Netsvetaeva, N.A.Kobeleva, T.V.Khodzher; The Actual and the Permissible Acidity Loads on the Asian Part of Russia, *Atmospheric and Oceanic Optics* **6-7** (2001) 499-504 (in Russian).

Warfvinge P. and H.Sverdrup; Critical Loads of Acidity to Swedish Forest Soils: Methods, Data and Results. Lund, Sweden (1995) 104.

Air Pollution in the Adirondack Park; Public-Private Efforts to Reduce Acidity in Soils and Water

Barbara Jancar-Webster

State University of New York, 282 Atateka Drive, Chestertown, NY 12817

The Adirondack Park

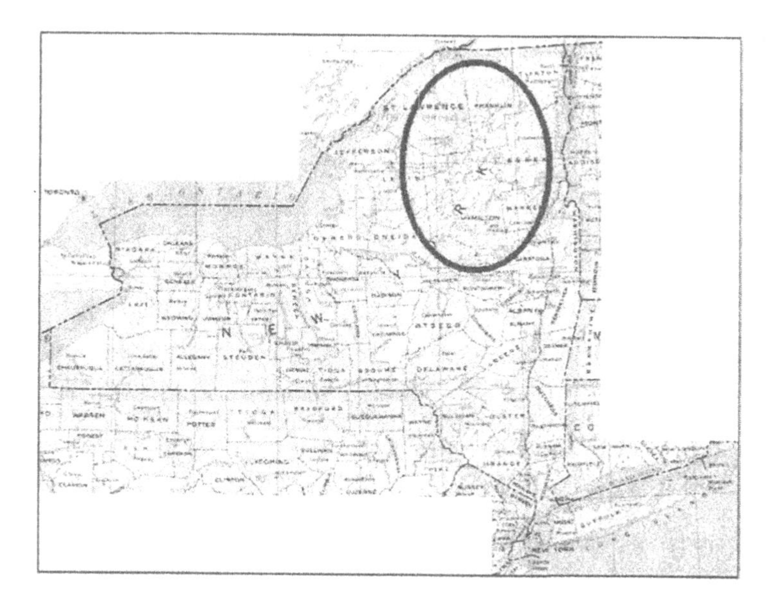

Figure 1: Map of New York State showing location of Adirondack Park

The Adirondack State Park occupies most of the northern third of New York State and comprises 6.1 million acres. It is the largest U.S. park outside Alaska. It has more acreage than any three national parks in the lower 48 combined. The Adirondack Park differs in several important ways from most public lands. The park comprises all land, public and private, within a circular boundary called "the Blue Line." Only about 48% of the park, known as the Forest Preserve, is public property. The rest is owned by individuals, corporations, clubs, or municipalities. There are no user fees and no "entrances," other than occasional road signs marking boundaries. Park services are relatively limited. There are just two official visitor centers, and they're both in the middle of the park.

Sidestepped by the westward migration, the Adirondack region was not carefully explored until the mid-1800s. Pike's Peak, Colorado, was tackled before climbers reached the Adirondacks' highest peak, Mt. Marcy. And the source of the Nile was discovered before the source of the Hudson River--again, near the peak of Mt. Marcy. Once people settled the region, intense logging transformed New York into the nation's leading lumber state. White pines were turned into ship masts, spruce into paper pulp, and hardwoods into either furniture or charcoal necessary to the success of the region's secondary enterprise, iron mining.

It was predicted that the resulting deforestation would wreak havoc upon the Erie Canal, New York's commercial lifeline, whose water level depended heavily upon the Adirondack watershed. In 1885, state business leaders and politicians, on the advice of America's pioneer

I. Barnes (ed.), Global Atmospheric Change and its Impact on Regional Air Quality, 321–326.

conservationists, instituted one of the country's first measures of enlightened conservation. The legislature established the Adirondack Forest Preserve, declared it state-owned, and disallowed the removal of trees. The purpose was clear: Protect the state's business interests. Resentful timber cutters scoffed at the preserve law, inspiring the legislature to create a second layer of protection in 1892, called the Adirondack Park.

The original Forest Preserve and much of the private land around it fell within this new designation. With loggers still disregarding these measures, however, an amendment was added to the state constitution in 1894, declaring the state-owned Forest Preserve portions of the park "forever wild" and rendering it unconstitutional to take down a tree on preserve lands. Subsequent alteration, such as construction of the highway to the top of Whiteface Mountain, the site of the 1980 Olympics alpine ski competition, has required state voter approval. No other wilderness area in the world enjoys such protection. (NYS Department of Environmental Conservation website 2001)

The Problem

The Adirondack State Park contains some 2800 lakes and ponds. Of these almost 25 percent are now so acidified that they no longer support plant or fish life. The acidity is of rather recent venue. Native brown trout have declined dramatically over the past 50 years in streams all through the park. (Adirondack Lake Survey Corporation, 1999) Today, trout streams are stocked with fish resistant to acidity but which do not reproduce. While forest die-back is not as severe as in parts of Europe, native conifers, such as pine and spruce, as well as the sugar maple are experiencing severe stress. Acidic deposition leaches cellular Ca from red spruce foliage, which makes trees susceptible to freezing injury, leading to over 50% mortality of canopy trees in the Park. Extensive mortality of sugar maple has resulted from deficiencies of Ca^{2+} and Mg^{2+}. Acidic deposition has contributed to the depletion of these cations from the soil.

Why? Regional Impacts of Acid Rain

In the United States the prevailing winds are from west to east, and from south to north. In the long-range transport of acid deposition to the Adirondack Mountains of New York State, the principle sources of SO_2 and NO_x emissions are the coal-fired power plants of the Ohio River valley, and Virginia and West Virginia. In comparison to total SO_2 and NO_x emissions nationwide, the emissions from these two regions are among the highest and thus not only impact on the Adirondack region, but their impact is perhaps more severe than that of any other emission sources on other regions of the United States. Indeed, while the patterns of sulfate and nitrate wet depositions have improved markedly in other parts of the country since the enactment of the 1990 Clean Air Act Amendments, they have remained essentially the same for northeastern New York. (EPA, *Airtrends 2000*: 6,10,16-17)

Actions of Federal and State Governments Urged on by Scientists, National and Regional Environmental and Non-Governmental Organizations and Public Opinion

1885 Forest Preserve Formed to protect headwaters of Hudson River and watershed for New York City.
1892 Adirondack Park established.
1894 Forever Wild clause attached to the Forest Preserve by the New York State legislature.
1970 Federal Clean Air Act passed. Act identifies six „criteria“ pollutants as especially dangerous. Among these are SO_2, NO_x, and O_3.

1980 Scientists identify 200 lakes in the Adirondacks where all fish had disappeared as a result of acid rain.

1984 New York State passes nation's first acid rain control law. The legislation requires a 245,000 ton reduction in state SO_2 emissions by the early 1990's; 40% of this must be achieved by January 1, 1988.

1990 Federal Clean Air Act Amendments passed. Congress establishes Acid Rain Program under Title IV of the Act with provisions to cut SO_2 and NO_x through a system of trading pollution permits, known as „cap and trade". Then President George Bush, Sr. claims new legislation will eliminate acidity in Adirondack streams and lakes.

1992 Energy Policy Act passed ending price controls on energy. Between 1992 and 1997, production of energy at power plants goes up 28%.

1993 State of New York, The Adirondack Council and the Natural Resources Defense Council sue the federal Environmental Protection Agency (EPA) over the new Acid Rain Program. The Federal Government agrees to produce a report in 1996 showing whether the Acid Rain Program is working.

1992 Because of 1984 legislation, New York State power plants overfulfill their target reduction in emissions. New York State and environmental groups move to prevent power plants from selling their permits to Midwestern and Southern plants whose trans-boundary pollution will pollute New York State. Figure 6 illustrates the direction of exports and imports of air pollution permits between New York and Midwestern and Southern power plants. Clearly, the New York State power plants' interest was to export their permits to the highest bidder. The problem was that that those bidders were the very plants that were polluting New York State and the Adirondack Park.

1997 Governor George Pataki and State Attorney General Dennis C. Vacco notify the EPA that New York State intends to sue the Agency for its failure to seek reductions in emissions of NO_x from Midwest and Southeast power plants.

1998 April State of New York and the Long Island Lighting Company reach an unprecedented agreement under which the utility will not sell sulfur dioxide allowances for use by power plants in 15 states whose pollution causes acid rain in New York State. Environmental groups who were active in promoting this agreement were National Resources Defense Council, Environmental Defense Fund, the Adirondack Council, Association for the Adirondacks, Adirondack Mountain Club

1999 *September*: New York State announces it will sue 17 power plants in 5 states.
November EPA sues American Electric Power, Ohio Edison, Cinergy and a number of other power plants in violation of the Clean Air Act.
October: Governor George Pataki orders power plants in New York State to cut SO_2 and NO emissions by 50 percent. New, all year round, not just in peak season
November: New York, New Jersey and Connecticut move to intervene in federal case against American Electric Power.

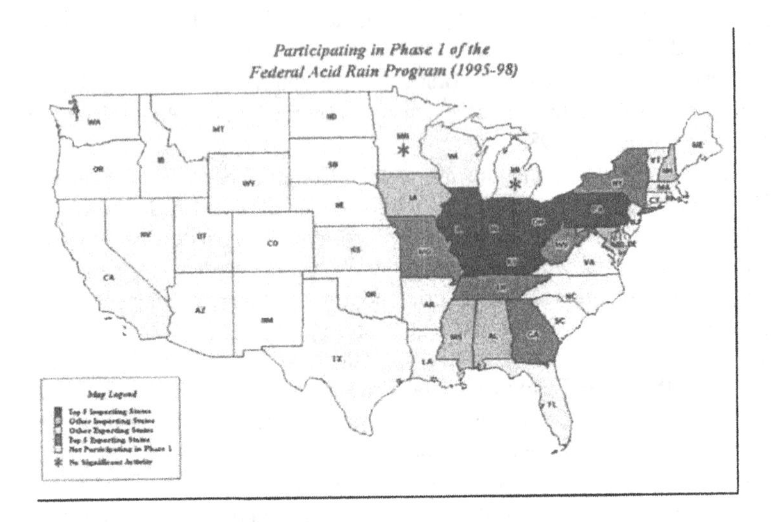

Figure 2: Map of US showing imports and exports of air pollution

2000, *January*: New York State cites Ohio Edison for non-compliance

April: Vermont, Massachusetts, New Hampshire, Rhode Island and Maryland join New York's legal complaint against American Electric Power

April: New York and Connecticut sue Ohio Edison, later joined by New Jersey

April: New York State Department of Environmental Conservation sends notices of violations to New York State power plants.

May: New York State Assembly and Senate pass legislation barring trading of pollution permits to out of state power plants that pollute New York State.

July: New York State sues Virginia Electric Power Company in federal court in New York City.

November: New York state and federal government announce agreement with Virginia Electric power under which company will cut sulfur dioxide emissions by 70 percent and NO_x emissions by 71 percent from its eight coal-fired power plants in Virginia and West Virginia. Figure 7 shows the number and location of the power plants compelled to cut emissions by this amount.

2001, *March*: New York State congressional delegation bill introduced in both houses of the US Congress entitled the Acid Rain Control Act. The bill requires 50 percent cuts in SO_2 and 70 percent reduction in NO_x emissions from electric power plants. (Adirondack Council, 2001)

July: New York State settles with Cinergy Company of Ohio to cut its emissions by 50% and 70% respectively.

Results

Over the 30 years since the Clean Air Act was passed, constant pressure from public and private agencies has contributed to a dramatic reduction of 17 percent in SO_2 emissions on the part of power plants upwind of the Park. There has been reduction per unit in NO_x but NO_x emissions actually increased 2 percent since 1992 due to the higher production of electricity and an increase in SUVs and light trucks on the road. (Environmental News Network, 2000) These vehicles were not regulated by the 1970 Clean Air Act.

Despite these achievements in emission reduction and atmospheric deposition since the Clean Air Act Amendments of 1990, a report on long-term research by the New Hampshire based Hubbard Brook Research Foundation (HBRF) in cooperation with Syracuse University, utilizing sites within the Park and across the northeastern US, found that there had been no significant improvement in the ANC of surface waters in the Adirondack region. (Driscoll et al, 2001)

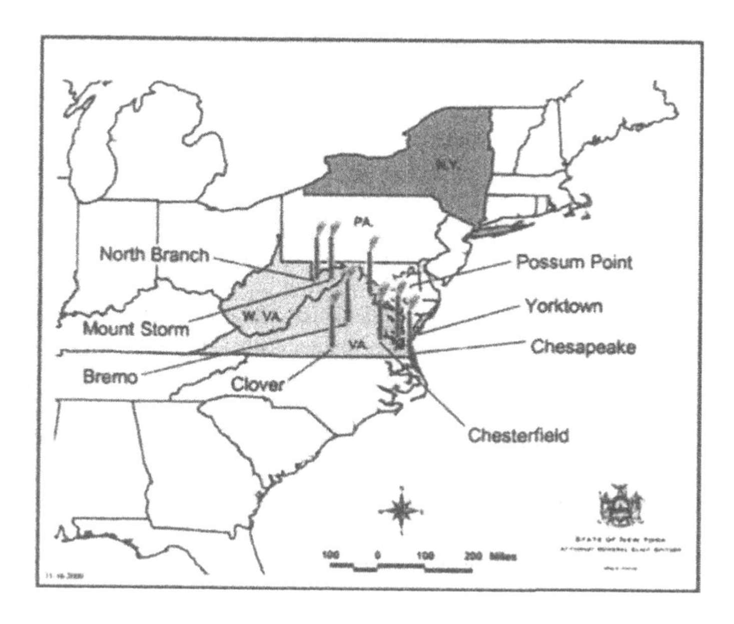

Figure 3: Location of power plants iv Virginia and West Virginia that signed the agreement to reduce emissions 71%

According to the report, acidic deposition has accelerated the leaching of base cations from soils, delaying the recovery of ANC in lakes and streams from decreased emissions of SO_2. At the HBRF, the available soil Ca pool appears to have declined 50% over the past 50 years. Sulfur and N from atmospheric deposition have accumulated in forest soils across the region. The slow release of these stored elements from soil has delayed the recovery of lakes and streams from emissions reductions. Acidic deposition has increased the concentration of toxic forms of Al in soil waters, lakes and streams.

The report found that the acid neutralizing capacity of the Adirondacks was not as strong as first believed, and that the recovery time would be much longer. The report described recovery from acidic deposition as a complex, two-phase process in which chemical recovery precedes biological recovery. HBRF model calculations indicated that the magnitude and rate of recovery from acidic deposition in the northeastern US was directly proportional to the magnitude of emission reductions. Model evaluations of policy proposals calling for additional reductions in utility SO_2 and NO_x emissions, year-round emission controls, and early implementation (2005) indicated greater success in facilitating the recovery of sensitive ecosystems than the targets set by the Clean Air Act Amendments of 1990. The report gave three scenarios based on the level of additional reductions in utility SO_2 and NO_x emissions. If the most stringent reduction scenario were put in effect, the report estimates that the recovery of lakes would take 10-15 years and the recovery of forests, 50-70 years.

Even before these findings were made public, New York State congressmen spurred on by environmental organizations, like the Adirondack Council, were pushing in Congress for newer and stricter amendments to the Clean Air Act, particularly with regard to NO_x emissions. (U.S. Senate, Committee on Environment and Public Works, Testimony,1999)

Summary

The paper demonstrates that the combined effort of lobbying by environmental and non-governmental organizations, diligent public prosecutors, attentive state executives, and national and state legislatures reacting to a highly organized public opinion, can effectively push through the implementation of an environmental remediation program. This is the positive result of public-private action on environmental issues. The paper further demonstrates that environmental legislation occurs in the presence of a high degree of uncertainty with regard to the expected environmental outcome. In the case under review, state and federal laws were successfully implemented. The environmental results however were less than policy proposal models had anticipated. This is the unanticipated consequence of public-private efforts and could be viewed as a negative result. The HBRF report calls for more scientific research to synthesize data on the effects of acidic deposition and to assess ecosystem responses to reductions in emissions. However, the conclusion is not that public-private efforts are futile. The renewed push by New York State congressmen for stricter emission controls at the federal level indicates that combined public-private action is the most persuasive vehicle for the development and implementation of the environmental agenda.

References

Adirondack Council; Statement of the Adirondack Council at Air Pollution/Acid Rain Stakeholder Meeting, United States Senate, Committee on Environment and Public Works, October 14, 1999, http://www. Senate.gov/~epw/Adirondack_Council.htm.

Adirondack Lakes Survey Corporation; Big Moose Lake description, 1999, http://www.adirondacklakessurvey.org/monthly.html

Adirondack Lakes Survey Corporation: Big Moose Lake graphhic, 1999, http://www.adirondacklakessurvey.org/ grapphics/bmph.html

Driscoll, Charles T., et al; Acidic Deposition in the Northeastern United States: Sources and Inputs, Ecosystem Effects, and Management Strategies, *Bioscience* **58,3** (March 2001)

Environmental News Service; Acid Rain Eats Away at Northeast, April 1, 2000, http://www.enn.com/enn-news-archive /2000/04/04012000/acideast_11606.asp.

New York State, Department of Environmental Conservation, Bureau of Public Lands; The Adirondack Forest Preserve 2000 http://www.dec.state.ny.us/website/dif/publiclands/adk/

U.S. Environmental Protection Agency; *Latest Findings on National Air Quality: 2000 Status and Trends,* 2001, PDF format downloaded from http://www.epa.gov/airtrends

U. S. Senate, Committee on Environment and Public Works; Testimony of Adirondack Council, Bernard Melewski, before the Senate Subcommittee on Clean Air, Wetlands, Private Property and Nuclear Safety, October 14, 1999, http://www.senate.gov/~epw/mel_1014.htm.

Subject Index